MIND
ECOLOGIES

MIND ECOLOGIES

BODY, BRAIN, AND WORLD

MATTHEW CRIPPEN
AND JAY SCHULKIN

Columbia University Press
New York

Columbia University Press
Publishers Since 1893
New York   Chichester, West Sussex
cup.columbia.edu
Copyright © 2020 Columbia University Press
All rights reserved

Library of Congress Cataloging-in-Publication Data
Names: Crippen, Matthew, author. | Schulkin, Jay, author.
Title: Mind ecologies : body, brain, and world / Matthew Crippen and Jay Schulkin.
Description: New York : Columbia University Press, 2020. | Includes bibliographical
references and index.
Identifiers: LCCN 2020006020 (print) | LCCN 2020006021 (ebook) | ISBN 9780231190244
(cloth) | ISBN 9780231190251 (paperback) | ISBN 9780231548809 (ebook)
Subjects: LCSH: Philosophy of mind.
Classification: LCC BD418.3 .C745 2020 (print) | LCC BD418.3 (ebook) | DDC 128/.2—dc23
LC record available at https://lccn.loc.gov/2020006020
LC ebook record available at https://lccn.loc.gov/2020006021

Cover design: Julia Kushnirsky
Cover illustration: Adobe Stock

CONTENTS

PREFACE AND ACKNOWLEDGMENTS

For one of us, the origins of this book go back to the 1970s. For the other, the immediate impetus was more recent: a 2012 conference paper fusing pragmatism, phenomenology, and work on artificial intelligence (AI). This was combined with the suggestion from Lana Kühle (then a doctoral candidate under the supervision of Evan Thompson) that embodied, embedded, enactive, and extended perspectives—otherwise known as 4E cognitive science—be added to the mix.

The last decade has brought us both into repeated contact with the 4E idea. Numerous conferences have also brought reoccurring intercourse with rising and established scholars weaving comparable threads, and who have in various ways shaped the trajectory of this book. Among those we wish to acknowledge are Tony Chemero, Ewa Chudaba, Joerg Fingerhut, Mark Johnson, Matthias Jung, Oliver Kauffmann, Roman Madzia, Richard Menary, Donata Schoeller, and Tibor Solymosi. Roman Madzia and Matthias Jung in fact organized a conference on pragmatism and cognitive science at the University of Koblenz-Landau, which most of the preceding scholars attended. Less than a half year later, another conference on the same topic took place at the American University in Cairo, with some of the above contingent presenting and contributing to a special issue of *Contemporary Pragmatism*. During this same period, Roberto Frega and Pierre Steiner organized still another conference on pragmatism and 4E cognitive science in Paris, where many of us met once more. In the spring of 2016, a few of our paths crossed yet again at a conference on pragmatism and the brain, organized to some extent around the work of John Shook and Tibor Solymosi, and held at the University of North Carolina at Ashville. This is where the two of us began talking in earnest.

The two of us found that we got along personally and stayed in contact. It helped that we thought alike on most matters pertaining to mind. Over the next months we increasingly discovered that our interests intersected and mutually complemented one another's. By Christmas of 2016, we were toying with the idea of writing some papers together. A few months later, this evolved into plans for this book.

Any major academic endeavor owes debts to others. In addition to those already mentioned, we would like to acknowledge teachers and scholars whom we cite too little or not at all, but who have nonetheless had a lasting influence on our intellectual development and this book. These include Catarina Belo, Kent Berridge, Jeanette Bicknell, Evan Cameron, Matthieu de Wit, Chris Green, Henry Jackman, David Jopling, Scott Jordon, Alexander Kremer, David Moffat, Ian McGregor, Phillip McReynolds, Bob Neville, Diego Nigro, Hal Pashler, Bryony Pierce, Elaine Powney, Jeff Rosen, Paul Rozin, Stuart Shanker, Rebekah Smick, Mog Stapleton, Susan Stuart, and Bob Sweetman. Among others who have left an imprint on us are Patrick Heelan, Sam Mallin, Curt Richter, George Scanlon, Eliot Stellar, and George Wolf, all now passed on. We offer especial thanks to Rob Switzer for his friendship and encouragement, and for suggesting alternative approaches when funding for a preliminary project that was to lay the seeds for this book collapsed. We are grateful to John Shook for permission to use amended fragments of a 2017 article published in *Contemporary Pragmatism*. We likewise acknowledge the largely invisible work of those in support roles who make scholarship possible, including administrative and student assistants, along with individuals at Columbia University Press. Among these are Sherifa Amin, Cristal Del Biondo, Reem Deif, Christine DeMichieli, Salma El-Galley, Zack Friedman, Lowell Frye, Ahmed El-Kosheiry, Salma El-Nager, Hadeel El-Sayed, Marwa Adel Farid, Khadeega Ga'far, Yossra Hamouda, Hagar Hod, Kat Jorge, Isa Kundra, Julia Kushnirsky, Aya Morsi, Monica Abed Ramsis, Sherif Salem, and Heba Youssry. We offer especially hearty gratitude to Audra Sim for her superb and very thoughtful copyediting.

Wendy Lochner at Columbia University Press dedicated herself to getting this book published, offering fruitful suggestions while exhibiting patience with several missed deadlines, and we offer thanks to her. Sarah Hammad—an exceptional former student and research assistant—did superb work contributing to chapter 3. Farida Youssef—another outstanding ex-student and research assistant—went above and beyond, preparing over 1,000 endnotes and catching errors in the text and references along the way. Carly Prowdley, funded by and studying at Grand Valley State University, rendered figures 1.1 and 2.1 and the brain diagrams in appendixes 1 and 2. Other institutions have lent forms of support, especially the American University in Cairo, Georgetown University's Department of Neuroscience, and Humboldt University's Berlin School of Mind and Brain. They are joined by O. P. Jindal Global University, Pusan University, and the University of Washington. For various kinds of additional support, we would also like to thank Arwa Al-Magariaf; Ramy Amin; Amy Carrillo; Sean Collard; Franca DeAngelis; Stuart Dennie; Diane, Peter, and Matthew Dixon; Steve Formaneck; Taha Gebril; Senica Gonzalez; John and Karine Hauser; Betty and Daniel LaBrash; Jeff Langman; Adham Mandour; Mariam Matar; Gord McClennan; Fred Nix; Ian Rennie; Aislinn Rose; Sandy Schulkin; Judy Straut; Susan Straut-Collard; and Günther and Mathilde Struck. Our thanks extend to Pegge Crippen, Marion Hawkins, Rosalind Schulkin, and other family relations, especially Chick Straut, who was a fellow intellectual traveler until he passed away at the end of 2018.

Some of those mentioned have been particularly inspirational on an intellectual level. With this in mind, we dedicate this book to Evan Cameron, Tony Chemero, Mark Johnson, and Tibor Solymosi. We also dedicate it to our immediate families: Bob, Paula, Marc, Najma, and Shannan, and April, Danielle, and Nick.

MIND
ECOLOGIES

INTRODUCTION

A
fter having fallen out of favor, pragmatism is resurging. This is especially so in cognitive science and value theory, two fields increasingly
intertwined. Cognitive science, especially, is undergoing a pragmatic turn away from representational models, with proponents from
competing quarters embracing embodied approaches that recognize the
centrality of aesthetics, emotions, and interests to human experience. This
development is clearly in the spirit of the classical pragmatists, not to mention phenomenologists, who are fellow travelers throughout this book. A
number of neuroscientifically literate philosophers and philosophically literate neuroscientists are approaching comparable conclusions and, accordingly, embracing pragmatism; Antonio Damasio, among others, cites it as
an "anchor" of his thought.[1]

In what follows, we have three goals. The first is to explicate pragmatism, which means looking at it in the context of the history of philosophy,
psychology, and science with the ultimate aim of applying it to contemporary work. Classical pragmatists, of course, recognized the importance of
the brain. However, many pragmatists—including John Dewey, William
James, George Herbert Mead, and C. S. Peirce—also appreciated the role of
active bodies in constituting perception and cognition. A few neurobiologists have picked up on this.[2] They are joined by a larger number of

scientifically and historically informed researchers normally lumped together as philosophers.[3]

Our second goal is relatively modest: to more tightly integrate classic and contemporary views, detailing how actions and less considered bodily functions, such as gustation and digestion, bring perception and cognition into being. We do this as a corrective against the brain-centered outlooks that currently dominate, and to further advance the work that other historically sensitive scholars have already begun. At the same time, and as our book title indicates, we think any reasonably thorough account of mind must be grounded in an understanding of neuroscience.

Classical pragmatists made some of their more impressive breakthroughs when focusing on the arts and emotions, and something similar is occurring today. A third, more ambitious goal of this book, therefore, is to integrate hints from classical pragmatism with contemporary cognitive science and neurobiology to demonstrate that emotional, aesthetic, and interested capacities—what we term affective or valuative life—are at the heart of action, perception, and cognition. In sum, we aim to show that behavior, perception, and cognition, along with anticipatory patterns of emotion, mood, and arousal, are mutually coordinating, often aesthetic, and emphatically co-constituting.

Though framed in neurobiological and cognitive scientific terms, our work draws heavily on historical texts. As the archeologist and philosopher R. G. Collingwood observed, texts answer questions specific to the time in which they were written, and these questions determine much of a text's meaning.[4] This point applies broadly: misunderstood context may cause confusion. It also applies specifically, where the connotations of a statement like "The ring is in the garbage" may vary depending on whether the inciting question was "Where's your wedding ring?" or "Where's the cheap novelty ring you found on the way to the park?"[5] This suggests that we cannot adequately appreciate texts just by reading their words. An important step is looking at the historically specific problems and questions that prompted authors to say what they did. Accordingly, in chapter 1, we offer a historical account of the pragmatic ideas that are critical to our arguments.

In line with Collingwood's approach, chapter 1 specifically considers how pragmatists were provoked by discoveries in biology, by debates between empiricistic and a priorist psychologists, and by developments in scientific methodologies,[6] especially those tied to late nineteenth- and early twentieth-century physics. Further, we examine the efforts of pragmatists to wed old and new ideas—for example, how ancient Greek thought inspired Dewey's concept of experience, which is cutting edge by today's standards.[7] Building on past traditions and the intellectual movements of their day, especially rising experimentalism and evolutionary theory, pragmatists insisted that perception is primarily enacted through doings in the world and the effects that this, in turn, has on organisms. This anticipated and, indeed, influenced J. J. Gibson's landmark ideas about perception.[8]

Pragmatists also foresaw—sometimes in exact detail—understandings advanced in the decades following Gibson by cognitive scientists and neurobiologists. This includes ecologically oriented theorists who emphasize environmentally-embedded bodily actions as bases of perception and cognition. It also includes the recent insistence of Damasio and others that emotion underpins rational decision making. In classical pragmatic literature, this embodied, biologically-based, valuative view is grounded to a significant extent in evolutionary ideas stressing the adaptability of affective capacities. In the case of William James, it is founded in an assimilation of Darwinian thinking into a theory of mind, albeit without the need for accepting the biological theory.

Specifically, James held that ideas can emerge somewhat independently of an environment that either reinforces or extinguishes them afterwards. He also held that valuative capacities such as interests could increase or decrease susceptibility to stimuli received from our surroundings, leading us to abstract from our worlds and rationally connect selected targets of attention in certain ways. In effect, therefore, we register things by cognitively altering and messing with them in ways loosely analogous to experimental scientific methods. The application of experimental frameworks to models of mind is all the more apt because valuative considerations enter both scientific and everyday decision making—for example, an aesthetic and emotional preference for elegance and economy when evidence weighs equally in favor of two competing alternatives. For pragmatists, valuations and beliefs are completed and measured in the context of environmental action. Pragmatists accordingly laid groundwork for the thesis that perception, cognition, and affective life mutually coordinate around doings and undergoings in environments, a process wherein environments are also defined.

Along with phenomenologists such as Maurice Merleau-Ponty,[9] a companion throughout this book, pragmatists introduced a more literal analogue to experimentalism in their account of mind. They did so by arguing that bodily structure and objects encountered limit action, thereby shaping the way we manipulate and alter things, bringing rhythm and form to doings and undergoings and hence to experiences arising out of them. As Dewey reasoned as early as 1896 in a landmark *Psychological Review* article, experience is not simply the world eliciting sensory excitations that are then wired to and interpreted by the brain. Though all of this is involved, experience is an outcome of the way sensory stimuli coordinate with motor activity and thus also the world. In Dewey's terminology, perception is sensorimotor, and is accordingly shaped by immediate movements, but also habits, emotions, and anything else relating to actions. In this way, Dewey and others such as Mead and Merleau-Ponty suggested that bodily action achieves many of the integrative functions traditionally attributed to inner mechanisms of mind.

Many readers will immediately recognize the resonance between these views and embodied, embedded, enacted, and extended theory—what has come to be known as 4E cognitive science. Chapter 2 details this. Mark Rowlands credits Shaun Gallagher for coining the expression "4E."[10] Laid out schematically, the term suggests:

1. That perceptual and cognitive processes are *embodied*—that is, comprised of neural and extraneural bodily structures engaging with the world
2. That perceptual and cognitive functions are *embedded* and scaffolded by structures in the surrounding physical and social environment
3. That perceptual and cognitive processes are *enacted* not only in neural systems, but also as consequences of interactions in the world
4. That perception and cognition *extend* into the world, including that of human technology; for example, notepads deployed as external memory enhancers, or canes that blind people use to engender spatial perception of their surroundings

These four views overlap and largely imply one another, with the example of the cane applying equally to all of them. Arguably, differences are not so much between perspectives as between the individuals championing them. Thus, self-identified enactivists tend to emphasize nonneural mechanisms;

they also largely eschew concepts of inner experience and mental representation, pushing the perceptual and cognitive outside the brain, albeit without denying the latter's importance. Extended theorists like Andy Clark pay comparatively more attention to the role of tools in mental life.[11] While likewise pushing beyond the head, extended-mind proponents are more comfortable inside of it. They talk about internal representations, and their models of mind are more squarely influenced by computer science, often adopting its language.

In line with the pluralism of pragmatism, we do not cling to any single perspective. So while advocating 4E approaches, we are unconcerned with fleshing out differences, and we obviously accept Mark Johnson's suggestion that an emotional "E" be added.[12] Moreover, although we think the theoretical machinery of representation is far too broadly assumed, often adding nothing or just confusing discussions, we grant that it may occasionally capture what occurs in human interactions. Thus, while avoiding such terminology, we do not go to lengths to dismantle it. Instead, our primary critique is of views that suggest that experience is built up inside our heads. These typically invoke the language of representation to defend an outlook that has become utterly mundane in philosophy, psychology, and neuroscience, even if edgy in everyday life and movies such as *The Matrix* (1999) and *Inception* (2010).

In connecting pragmatism to 4E views, and attempting to thereby augment both, we particularly focus on enactivism, which closely aligns with the classic American movement. As Rowlands puts it, enactivists hold that perception and cognition "are constituted in part by the ways in which an organism acts on the world and the ways in which world, as a result, acts back on that organism."[13] This is a word-for-word formulation of what Dewey repeatedly expressed from the late 1890s until the end of his career, but there is very little recognition of pragmatism in landmark enactivist statements. The similarities are perhaps most painfully apparent in Kevin O'Regan and Alva Noë's work.[14] Similarities also show up in treatments by Daniel Hutto and Erik Myin and by Evan Thompson,[15] the latter of whom is one of the movement's founders along with Francisco Varela and Eleanor Rosch.[16]

Though favoring extraneural approaches, we do not—as stated uncontroversially at the outset—bar the nervous system and brain from our account. Chapter 2 accordingly includes discussions about neurobiological factors in bodily synchronization.[17] It additionally looks at how perceptual, habitual, grammatical, semantic, and motor functions, along with probability prediction, are handled by overlapping and indeed sometimes the same brain regions. This lends further credence to the claim that perception, action, and cognition knot together. So too does the fact that pragmatic treatments of perception—in company with phenomenological and Gibsonian interpretations—presaged ideas increasingly important in artificial intelligence (AI) and robotics, particularly the precept that human-like intelligence requires a human-like body in addition to a CPU capable of brain-like functions.[18] Insofar as pragmatic, phenomenological, and Gibsonian approaches stress that perception occurs through total coordinations of bodily capacities, they suggest that multiple modalities always mobilize in actions and hence habits. This provides avenues—underexploited by enactivists—for understanding intermodal perception, which we examine in the context of contemporary experimental and neurobiological work.

Chapter 2 generally considers the body as a synergistic system that falls into coordination around environmental contours such that organic activity structures and constitutes perception and cognition. This embodied position challenges accounts dominant since the early modern era that see the mind or brain primarily as a mechanism that generates internal representations of the external world. Insofar as the world of representation is one of appearance, this epistemological dualism of inner-versus-outer leads directly to skepticism. Challenging this standpoint, embodied views (especially ones grounded in pragmatism and phenomenology) overwhelmingly counter the conclusion—to put it crudely—that the human mind is feeble.

Continuing with the idea of perception and cognition as bodily coordination, chapter 3 explores how synchronized activity among groups of organisms achieves similar functions. Though sometimes stated more than defended, this idea is again introduced in classical pragmatism, especially Dewey's work, which advances a notion of experience as culture that parallels phenomenological concepts of worldhood. In addition to resonating with 4E positions, these views also enrich and are enriched by empirical research on child development that suggests cognition is co-defined by caregivers. Theoretical work in psychology suggests something similar in adult relations, and everyday observation affirms that we remain dependent throughout life, with even simple tasks relying on the collective efforts of many. Social life, in turn, characteristically involves the deployment of emotional capacities, thus accentuating links between action, affect, cognition, and perception—links that persist even when a lone perceiver scans a space for openings and threats. We focus on aesthetic aspects of social cohesion in discussing group activity, drawing on experimental psychology to augment our account of perceptual and cognitive coherence. Emotion is central to almost any study of aesthetics, and it is also part of what unifies human action in group settings, giving additional reason to suppose that affect knits together with action, cognition, and perception.

Neurobiological research supports this view. Broca's area, to consider an example, is classically associated with language capacities. However, in different neural coalitions, it also appears to facilitate movement preparation and action recognition and imitation.[19] It also contributes to music perception, and thus aesthetic experience.[20] Interinnervated with Broca's area, the basal ganglia likewise connects to movement and habit organization and cognitive functions, including the syntax of regular verbs, the probabilistic prediction of events, reward appraisals, and emotional evaluation.[21] This neural coalition points to the knotted character of action, affect, cognition, and perception. Moreover, motor-related neural areas in the brain appear to mobilize when an agent either witnesses another agent moving or engages in a task. Perhaps more interestingly, activation in the former case seems to depend not on observed movements alone, but awareness of goals and their achievability.[22] Together with other evidence, this suggests that recognition of intentions in others is important, highlighting links between standard sensorimotor accounts and social cognition.

After laying out connections between pragmatic, developmental, neurobiological, and experimental work, chapter 3 next examines Gibson's perceptual theory of affordances, along with recent research tying it to aesthetic experience and social life. While emphasizing physical movement, chapter 3 reiterates that our psychological landscape begins as social and remains so throughout life. In other words, movement knits with social life all along, so that the latter is not built up from the former in a reductive way. By emphasizing that things registered by us are consequences of our conduct in primarily shared worlds, we once again challenge the skeptical notion that experience is an exclusively private, subjective phenomenon.

In chapter 4, we attend to the affective or valuative side of embodied cognition. This is a fairly central point in classical pragmatic literature, which frames cognition as knotted with emotional, interested, and aesthetic dimensions. Pragmatists were not exactly original in positing this. However, up until their time, those who saw cognition as affectively infused largely suggested that this degrades the epistemic basis of human thought and beliefs. Pragmatists stood out in arguing that the aesthetic, emotional, and interested sides of cognition enhance its rationality. More specifically, they held that thinking, abstraction, and—indeed—experience fruitfully knit together with emotional, interested, and aesthetic life, bringing us more in touch with what may colloquially be called "reality." This, in turn, aligns with conclusions increasingly accepted in neurobiology, especially since the 1990s—conclusions that collectively suggest that little in the way of differentiated thought and reasoning occurs absent emotions and interests.[23] However, pragmatic and recent accounts also share a common oversight. This is a neglect for ties between emotions and interests in spite of obvious conceptual, experiential, and neurobiological overlap—an oversight that arguably follows from treating emotions as visceral and interests as other than that.[24] Without denying, for instance, that emotions typically have visceral components, there are clear counterexamples to this. There are also cases in which interests are emphatically visceral. We argue that standard separations of emotions and interests are conceptually, experientially, and neurobiologically unwarranted and prevent what might otherwise be a more expansive account of valuative cognition.

In addition to resonating with neurobiology, pragmatic views align with perspectives from psychology and cognitive science. This includes outlooks stressing satisfying closure and coherence[25] and research emphasizing emotional motivations behind inquiry.[26] It also includes experimental work that identifies a meeting of cognition and affect in environmental exploration.[27] We focus on the active, anticipatory, and searching sides of cognition and perception implied in both pragmatic and recent accounts. This encompasses not only literal manipulation and selective gathering that lead to knowledge, but also outcomes achieved through selective attention and emotional weighing. Both imply a kind of cognitive foraging[28] understood as an appetitive process—that is, a driven and active search for what is cognitively satiating. This search, once again, entails doings and undergoings, coordinated in immediate and longer time frames with emotions systematically directing attention and action, which circles back on affective life.

It should be evident that all this applies not only to cognition, but perception too, a topic we return to in chapter 5, once again focusing on affective or valuative dimensions. Though more squarely directed at cognition, James's treatment of interests can be expanded into an account of perception that emphasizes valuations as information-rich ways of being perceptually in touch with the world. This aligns with more recent scholarship in philosophy, neurobiology, and cognitive science. James, in company with Dewey, Gestalt theorists, and phenomenologists—all at their height in roughly the same period—also anticipated Gibson's theory of affordances and, by extension, 4E cognitive science.[29] This is not surprising since Gibson's intellectual lineage leads back to these historical schools.[30] Inasmuch as all these perspectives hold that we perceive according to what our bodies can do in environments, they suggest that our experience is grounded in what is biologically needful and aversive. This, in turn, highlights human perceptual systems as valuative and, more specifically, as organized around attractions, preferences, satisfactions, and aversions. In this sense, they are structured according to aesthetics, too.

In chapter 5, we specifically elaborate on exploratory sensorimotor activity, not just in humans, but in organisms ranging from unicellular life to insects to warm-blooded vertebrates. Connecting these diverse forms of life are the activities of foraging for nourishment and avoidance of hostile elements, which, even in the case of single celled organisms, involve complex strategies. For example, the unicellular species *Physarum polycephalum* coordinates in groups to explore their environments, secreting trails of slime to mark where they have already been, leaving external chemical recordings similar to what extended cognitive scientists in the vein of Clark have described;[31] insects and warm-blooded vertebrates achieve the same with pheromone markers,[32] as do humans by technological means. All of this—especially in the cases of unicellular and insect life—matches what the prominent roboticist and AI researcher Rodney Brooks has proposed: that intelligent behavior is achieved through direct sensorimotor coupling with the world, combined with layered behavioral tendencies that might include random wandering, collision avoidance, approach of distant or certain kinds of objects, and more. On a general level, it affirms

Brooks's dictum that the world is better than any model of it that can be constructed in central processing units or brains.[33]

In articulating these points, we do not go so far as to claim that unicellular and insect life are valuative. However, such organisms exhibit preconditions for valuative and aesthetic experience, realized fully in the case of human beings and arguably a range of other cephalic species as well. An important point we add—building on chapter 4—is that much of this is driven by the interweaving of visceral, neural, and motor activity. Put another way, many affordances link to core drives such as gustation, digestion, and the search for mates or shelter from predators.[34] Exploratory behavior largely moves towards these ends. This is part of the evolutionary past of humans, and according to experimental psychologists such as Rachel and Stephen Kaplan, it provides explanations as to why humans take aesthetic pleasure in exploration and discovery of environments promising reasonable degrees of navigability.[35]

Indeed, while many of us do not face the extreme exigencies of our evolutionary predecessors, we may still find ourselves exploring restaurants and other spaces in search of food or mates, to flee inclement weather, or to escape someone harassing us on the streets. At a more global level, we may find certain spaces emotionally tempting and others threatening, suggesting valuative—which is to say, emotional, interested, and aesthetic—dimensions in perception. Insofar as this connects to exploration and movement, action is knitted in too, along with cognition, since such behavior entails judgment and evaluations of surroundings. Moreover, this account suggests that valuative aspects of surroundings, like perceptual ones, are qualities of interactions in which both extra-organic things and organisms partake.[36] Accordingly, they are not projections of mind onto the world. This is in keeping with the antiskeptical flavor of pragmatism, phenomenology, and Gibsonian psychology.

As the book title suggests, our general aim is to detail how embodiment, neurobiology, and affective aspects merge into an integrated ecology of mind, psychic life, or self. A more specific goal is to show that affect, cognition, motility, perception, and valuations fuse in human life. By virtue of being embodied, and therefore structured around doings and undergoings in the physical and social world, these processes are simultaneously embedded, enacted, and extended. Classic commentators—pragmatic, phenomenological, and Gestalt—stressed the last point in various ways. For instance, they tended to agree that affective life delineates situations in which decisions, perception, and behaviors take shape: we commonly speak of finding ourselves in moods, as opposed to characterizing emotions merely as states discovered inside of us. They further suggested that emotions and interests bring world-changing shifts in attention, and this means alterations in thought and worldly action as well. Dewey more specifically observed that emotions are attached to objects and events, and are only private and internal in cases of breakdown. These views anticipate a variety of more recent work.[37] It should be added, however, that many recent accounts in this vein lack the biological justification we supply. Classic theories necessarily lacked the same because of the comparatively ill-developed state of biological knowledge on which they were based.

In our sixth and final chapter, we offer a philosophical and biological culmination of the view that action, cognition, perception, and affective life do not merely contribute to one another. In the case of human beings, they are essential to one another; they make one another what they are. These endowments function in global contexts, which include bodies, capacities for movement, viscera, and a great deal else synchronizing through total interactions with surroundings to compose the fabric of psychic life. In other words, mind is ecological.

The word "ecology" typically evokes the environment and systems in it. What goes unrecognized in most philosophical treatments of mind is that we have ecologies within us. In one sense this is literal: a variety of species—including humans—host vast populations of microorganisms, which play critical roles in bodily functioning generally, and psychic life specifically.[38] We explore this in our final chapter. As intimated, psychic life can also be said to be ecological insofar as the body, including organs such as brain, viscera, limbs, and more, form mutually dependent, integrated systems that bring about *mind*. In our last chapter, we argue that this in fact undersells the situation when it comes to what is conventionally divided into action, cognition, perception, and emotion. That is, we argue that cognition is emotional, and emotion, cognitive. Perception is likewise cognitive and emotional, and emotion, being perceptual and cognitive, helps us grasp what is occurring in the world, accentuating possible ways of acting to handle issues that arise. This implies motor potentialities, whether or not they are actualized.

In noncephalic creatures such as the earlier mentioned *P. polycephalum*, this similarly holds, albeit obviously without intervention from brains. These organisms explore environments in groups with the help of slime trails left to externally mark where they have been. They can successfully navigate mazes while selecting from diverse ranges of nutrient options and avoiding harmful substances. These behaviors are sensorimotor and hence perceptual in fairly standard ways. They are cognitive, too, insofar as they involve weighted selections, approach, and avoidance according to what is life-promoting or life-diminishing. For just this reason, they might be said to be at least prevaluative. Importantly, a single response—for example, movement towards food—is all of this at once, suggesting that action, cognition, perception, and valuation (or something like it) fuse in even relatively simple instances of life. In human beings, action, cognition, perception, and affective life similarly bind together, but with help from brains in addition to bodies and environments.

Views contrary to this are longstanding and are advanced in the notion of the brain as the seat of human psychic life. This notion persists in standard interpretations of mind that maintain separation between action, cognition, emotion, and perception. Likewise, neuroscientific literature often divides the brain in terms of interpreting versus responding to external events. Certain neural structures may indeed be geared towards interpreting the environment, while others may be more oriented towards generating emotional reactions. However, evidence converges to indicate that different neural structures perform these and other operations together, with individual regions simultaneously executing more than one role. Experimental, neurobiological, and theoretical work on the psychology of aesthetic experience reinforces this point. Aesthetic responses are obviously perceptual, but there is behavioral and neurobiological evidence suggesting that they are simultaneously affective and thus visceral, and, in this way, embodied.[39] Moreover, they are explainable in terms of motor potentialities[40] and have been observed to activate motor areas in the brain.[41] They are simultaneously cognitive insofar as they involve immediate appraisals of situations and what it is possible to do in settings. Some of these points connect aesthetic experience to Gibson's affordance theory. Given that Gibson's work is a development of pragmatism, phenomenology, and Gestalt theory and is a widely accepted antecedent to 4E views, all of this fits within our general framework.[42]

Although we reject modularity, some regions in the brain are fairly specialized. For example, Broca's area, while not operating alone, appears dedicated to speech comprehension and production, with lesions in this neural region associated with language impairment; likewise, the fusiform gyrus is crucial to facial recognition, since damage to it often impedes this ability.[43] Yet facts such as these do not stand as strong evidence that these regions solely handle the aforesaid capacities; nor does this undermine the broadening ecology we seek, since the aforementioned pathologies may be consequences of interrupted pathways. More critically, most brain structures—and perhaps all of them—seem to perform more than one task.[44] Thus, although active during language processing, Broca's area also appears involved in music perception.[45] This is perhaps not the most compelling example, since Broca's area is rather large. However, much the same occurs even at the level of individual neurons, which in some brain regions handle multiple functions, as seen in organisms ranging from roundworms to mammals.[46] Outside the brain, the human body also has specialized appendages and organs, but many of these likewise can perform multiple operations. This is the case with hands, which predominantly grapple with things, but can also be deployed in language expression and comprehension, as when signing and reading braille. Moreover, activities such as signing typically entail a global coordination of posture, gaze, and more, once again highlighting the integrated nature of organic existence. All of this resonates with classical pragmatic views, not to mention phenomenological ones advanced by thinkers in the vein of Merleau-Ponty.

The organization of action and bodily sensibility are replete with cognitive function. This supplies a nonneural explanation as to why life mostly unfolds prereflectively,[47] though we do not completely reject brain-based accounts of the unconscious. We do, however, think they are overplayed because a great deal of what constitutes psychic life in fact happens outside the brain. Evidence suggests that moods, thoughts, and psychiatric conditions are moderated by what goes on in the viscera.[48] Pragmatists, Merleau-Pontian phenomenologists, and 4E cognitive scientists add that a great deal of processing occurs through body-environment interactions. This includes general doings and undergoings in the world that supply integrative fabric for experience. It encompasses more specific body-environment interactions, such as mountains funneling monarch butterflies towards destinations on migratory paths,[49] with light cues and basic magnetic sensitivity providing further guidance.[50] Much the same holds across species, whether in humans navigating less through mentation and more through coordinating with contours of paths; or in *P. polycephalum* doing the same by means of external traces of slime; or in a variety of organisms adapting to physical laws not by performing brain-based calculations but rather by virtue of appendages that behave according to those laws.[51] Along comparable lines, neurobiology and other internal factors are not sufficient for explaining emotions. Neither inner feelings nor nerve firings alone specify emotions independently of worldly contexts; the same tremor might be fear or excitement depending on the environmental situation.[52] In addition to this, theoretical reasoning and tentative evidence suggest that expression and action make particular emotions more distinctly identifiable.[53] This is in line not only with pragmatic and phenomenological standpoints, but also enactive views, all of which suggest emotional experience is part of a sensorimotor loop.

At the same time, none of this suggests ignoring neurobiology, as findings from that field reinforce pragmatic, phenomenological, Gestalt, Gibsonian, and 4E renderings of psychic life. To offer an example, the cerebellum—a rostral structure abutting the brain stem—is involved in both motor coordination and emotional experience along with cognition and perception.[54] This is exactly the kind of finding that the aforementioned traditions would predict, and it points to something nearly everybody accepts: that the brain is always involved in the psychic life of cephalic organisms.

Although 4E theorists know that sensorimotor activity, which is fundamental to many of their accounts, requires a brain to occur in humans, too few attend to the nervous system in detail. Noë, to give one example, asserts that sight substitution devices (see chapter 2 for more detail) do not activate the visual cortex, even while citing evidence that they do in some circumstances and despite there being no dearth of other studies testifying to this at the time that he made the claim.[55] This error might have come from ideological commitment to a tenet that has become something of a slogan, especially in enactive quarters: that we are not our brains. While agreeing with this sentiment, we obviously advocate careful discussions of neurobiology. The earlier illustration of the cerebellum is but one in a long list of neurological cases that align with 4E accounts, and there are a variety of others. To name one, leading experts on phantom limb pain have suggested that the unpleasantness of such cases arises partly from motor signals going out without receiving normal motor feedback, so that sufferers might have the experience of continually clenching and tightening hands.[56] This supplies neurobiological support for what Merleau-Ponty specifically said about the phenomenon.[57] It also generally accords with pragmatic, phenomenological, and 4E standpoints that conceive of perception in terms of sensorimotor loops. Such explanations of phantom limb discomfort, along with current understandings of the cerebellum, offer noncontroversial neurobiological affirmations of 4E accounts.

Pragmatists and, later, Merleau-Ponty insisted that mind is not traceable to any one structure. This is because it arises out of a totality of sensitivities and capacities working in concert with the world, not to mention the microbiome within us, though classic thinkers were obviously not in a position to know this. The reticular formation in the brain stem does not equal consciousness, nor do the amygdala bulbs, nor still the brain and nervous system in their entirety, nor the body. Bodily engagement and practiced patterns that characterize psychic life distribute across brain regions and infuse action as it occurs in the world. This and more constitute the integrated organization of cognition, motility, perception, and affective life. Feelings and emotions are not detached from cognitive and perceptual systems in the brain, and they are not separate from bodily conditions and actions either. These are the points we defend in this book and bring together in the final chapter.

1

LIFE, EXPERIMENTALISM, AND VALUATION

lthough we primarily aim to advance contemporary cognitive science
and neurobiology along pragmatic lines, we begin with a historical
examination. This is, as intimated in the introductory chapter, to
explicate relevant pragmatic ideas. It is also to show it is no accident that
pragmatism remains relevant, for it bears the imprint of major nineteenth-
and twentieth-century scientific shifts that continue to dominate our
understanding of the universe. It also reflects longstanding debates in the
philosophical world, with pragmatists negotiating tensions between
empiricism and post-Kantian rationalism, the legacy of Greek philosophy, and more. Many of these debates have evolved and continued in
psychology, for example, where empiricist versus rationalist debates parallel disputes between behaviorists and cognitive linguists.[1]

Dewey insisted that the very task of "thought is to establish working connections between old and new subject-matters."[2] We attempt to live up to
this charge. In this chapter, we look at intellectual contexts that led pragmatists to emphasize embodiment (taken broadly in the 4E perspective)
along with valuative theories of mind, which is to say that we concentrate
on accounts that regard cognition and perception as emotional, aesthetic,
and interest-based. In the broad view of this book, pragmatism lays the
groundwork for the integrated ecologies of mind that we hope to advance.

Obviously, biology is central to our argument. It is also important in pragmatic philosophy, and among classical pragmatists, James had the most formal biological training. After stints as a student of painting and then chemistry, James studied anatomy and physiology, finally graduating with an MD in 1869. He had the added advantage of attending Lawrence Scientific School at Harvard (at that time a center of the Darwinian debate) and began his career as an anatomy and physiology instructor. Not unexpectedly, therefore, James came to believe that psychology must presuppose a certain amount of brain science, and opened his *Principles of Psychology* with a statement to this effect.[3]

Dewey and especially Mead upheld the same premise, albeit—in keeping with the ecological orientation of nineteenth-century evolutionary views—warning not to replace mind-body dualisms with brain-body ones.[4] These slightly later pragmatists emphasized the brain as an organ for coordinating bodily doings and undergoings in the world. Less obviously, they also saw coordinations of sensory and motor activity as more than mere brain functions, conceiving of them as the basis of human perception and cognition. While forward-thinking, this idea was also emblematic of its time. One reason was that evolutionary theories abounded, with Darwinian and non-Darwinian variants both stressing adaptation as something connected to the body but also to intelligence, thus introducing links between motoricity and mind.[5] Dewey and Mead were therefore not alone in their views. Henry Calderwood,[6] John Hughlings Jackson,[7] Francis Galton,[8] Edward Titchener,[9] Margaret Floy Washburn,[10] and others from that era took comparable positions, often invoking evolutionary theory in formulating what some of them called "sensorimotor" accounts of mind. Chauncey Wright, though still locked in early modern concepts of mind, was an early defender of Darwin. He was influential upon the pragmatists, and also made baby steps towards a motor theory of consciousness.[11] Herbert Spencer—similarly a proponent of evolutionary theory, albeit more Lamarckian than Darwinian—did much the same, even while likewise trapped in early modern empiricist suppositions and despite his ideas being relentlessly dismembered by James.[12] Classical pragmatists were therefore part of a general movement towards embodied views, but, unequivocally, their positions were more developed than those of most of their contemporaries, not to mention many of today's theorists.

In addition to evolutionary themes, pragmatists and especially Dewey assimilated the other two major scientific shifts that rocked the late modern period—namely, relativity and quantum mechanics. This, too, fits the ecological thrust of pragmatism. In the case of relativity, as the name suggests, it fits because the framework holds that determinable properties cannot be expressed outside interrelations, such that specifying mass, physical dimensions, and the like mean taking up a standpoint relative to the object described. Quantum mechanics does the same in its assertion that observing the microscopic world means changing it. One of Dewey's breakthroughs was to realize that this is not a peculiarity of extreme circumstances (e.g., subatomic sizes or high relative velocities). Instead, the notion that we come to know reality by mucking about with it is endemic to experimental methods, which pragmatists assimilated before scientists formalized them. It also characterizes everyday life, where we perceive and know things by mentally and physically altering them, in some cases literally hefting them, pushing into them, feeling their resistance, and otherwise changing them through our observational activities. This insight is central to the classical pragmatists' embodied theories of mind and knowledge,[13] and also key to recent 4E variations.[14]

By developing embodied positions, which are ecological by definition, pragmatists moved beyond standard inner-versus-outer divides that ruled the modern era and that still dominate today. Peirce, James, and Dewey all anticipated ecological psychology,[15] guiding future trendsetters such as Gibson.[16] Dewey and Mead further foresaw—sometimes in exact detail—extraneural accounts advanced by 4E proponents, especially the enactive position of O'Regan and Noë.[17] That the classical pragmatists proposed that bodily action is constitutive of mind is more remarkable in light of recent neuroscientific work, which is at least consistent with the thesis. For instance, the basal ganglia are critical for motor control and the organization of behavior and habits, simultaneously linking to cognitive functions, including linguistic activity, probabilistic prediction of events, and reward appraisal.[18] Also remarkable is the fact that pragmatists identified cognition and perception as emotional and aesthetic. They thereby anticipated recent work in neurobiology, experimental psychology, and philosophy[19] while offering means of expanding and enriching it. This is what we, too, hope to accomplish in this book.

EVOLUTION AND PRAGMATISM

While evolutionary accounts of life date to antiquity, the late eighteenth and early nineteenth centuries saw outpourings of them, the most prominent among them being Charles Darwin and Alfred Russell Wallace's theory of evolution by natural selection. The theory influenced both Dewey and Peirce, though Peirce leaned more on Lamarckism. It also informed pragmatist thought in more general ways that did not require allegiance to the biological components of the theory.[20] Darwin and Wallace's influence on Mead was arguably deeper, and it connected to his embodied account of mind. Mead suggested that perceptual adaptations are, in effect, environmentally scaffolded capacities evolved to enact objects—that is, realize properties of objects through perception-engendering action.[21] For example, the tactile exploration of a beer bottle enacts its roundness and glassy smoothness, and simultaneously the perception of these qualities; yet we are only able to enact these properties and experiences insofar as selective pressures have endowed us with hands and connected organic structures. While Lamarckian and other evolutionary and perhaps even non-evolutionary schemes would imply comparable outcomes, the prominence of Darwin's theory motivated Mead's views.

Darwinism likewise influenced James and is central to his work,[22] but, again, one can accept James's views on mind while rejecting Darwin's biological theory. This is because, despite assimilating Darwinism, the main target of James's argument was not evolutionary. Specifically, James's primary objection was to British empiricist psychologies, and he saw neo-Lamarckism as biologically extending these models of mind.[23] Darwinian frameworks offered alternatives to either view. Simultaneously, and because neo-Kantian *a priorist* schools were rivals to empiricist psychologies, James—with help from Darwinism—developed his alternative in the neo-Kantian vein. However, by emphasizing how interests and emotions (or what might be called personal affectivity) structure our experience and cognition, he deviated from the *a priori* logical schemes of post-Kantian psychologists. At times, he also expressed unmitigated hostility towards traditional Kantianism despite assimilating tenets from it.[24]

To appreciate how and why James synthesized all these outlooks, it is important to recognize that Herbert Spencer was a main target of his attacks. Spencer was not only a British empiricist, but also an evolutionary theorist committed to the neo-Lamarckian notions of direct adaptation and inheritance of acquired traits. Direct adaptation holds that environmental pressures elicit adaptive variations rather than merely reinforcing them, a point illustrated by the overused example of a giraffe's neck elongating by virtue of habitually reaching up to eat leaves. In this scheme, the organism gets the adaptation it needs during its lifetime directly from its activity in its surroundings. Inheritance of acquired traits simply means that characteristics gained in this way are passed to the next generation. By adopting these standpoints, Spencer obviously abandoned the blank-slate position of earlier British empiricists. However, he retained the core empiricist tenet that environments are the prime shaper of minds, extending it only to include both the environments of individuals and those of ancestors. In James's view, therefore, Spencer did what other empiricists did. And as is often the case when people attempt to define philosophical schools, James put it in exaggerated terms: he complained that British empiricists reduced the organism to a passive recipient of sensations, ideas, and dispositions, so that the world molds the mind through "a kind of direct pressure, very much as a seal presses . . . wax into harmony with itself."[25]

When it came to neo-Lamarckism, James detested the idea of the inheritance of acquired traits, particularly its applicability to human psychology, and urged that there was little evidence for it.[26] He concluded that any existing inborn adaptations are legacies of "congenital variations, 'accidental' in the first instance, but then transmitted as a fixed feature."[27] Even more fervently, James protested direct adaptation and the analogous empiricist claim that experiences—understood as impositions of impressing environments—directly mold minds. Impressions, he maintained, could not by themselves do this because things usually interrelate in myriad ways, meaning that multiple instantiations are typically possible in perception.[28] With a Necker image, for instance, we can see different planes as front or back, or, alternatively, we can see the figure as a two-dimensional pattern (figure 1.1). Seeing the figure all ways at once, on the other hand, would render it unpicturable.[29] This is one example among many, and it suggests our experience would be chaos if it were simply a raw outward order impressed on the senses.[30] Moreover, were environment the sole shaper, as stock empiricist accounts suggest, people raised in the

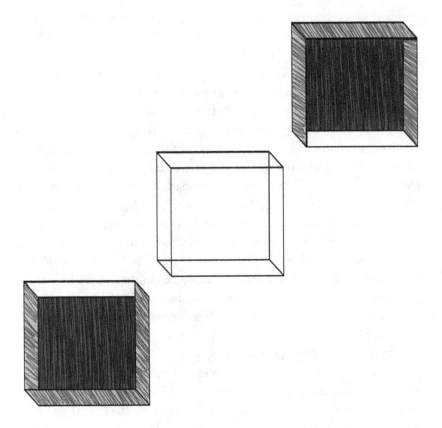

FIGURE 1.1 Necker images

By alternating your focus between the two square planes in the middle figure, you can make either pop forward so that the overall form looks like either the top-right or bottom-left figure.

same circumstances ought to develop identical minds—an outcome not supported by evidence, according to James.[31]

Empiricist accounts further suggested that the strongest beliefs should correspond to our most frequent observations. While granting that this can happen, James cited science as a domain that sometimes progresses "by ignoring conditions which are always present."[32] Thus it was that Galileo established kinematic laws of motion by envisaging marbles rolling over nonexistent frictionless surfaces, or Isaac Newton conceived his laws by reducing celestial bodies to point-like objects. This suggests that outward observation alone does not impress justified beliefs on us. As a result, James proposed the reverse of what empiricists argued: that ideas sometimes precede things noticed.[33]

James's problem, in short, was not with empiricists' emphasis on experience, but their conceptualization of it as the world impinging on a passive mind. He noted that we encounter overwhelming numbers of stimuli, most of which do not enter our experience.[34] This is repeatedly affirmed by researchers after him.[35] Thus, James held that "consciousness is at all times primarily *a selecting agency*" that focuses on "one out of several of the materials so presented to its notice, emphasizing and accentuating that and suppressing as far as possible all the rest."[36] For James, the mind operates on the basis of interests, ideas, and functionally similar mechanisms, working on sense data "very much as a sculptor works on . . . stone,"[37] by which it "*makes* experience" of the world.[38] Insofar as the aforesaid mechanisms influence action, and actions change the world, the world-making power of mind extends to the material conditions of life.[39] James insisted, therefore, "that the knower is not simply a mirror . . . passively reflecting an order that he comes upon . . . existing. The knower is an actor [who registers that] which he helps to create."[40]

James, of course, did not deny that experience sometimes elicits interests. However, without the shaping influence of interests, coherent experience would not exist in the first place, as might be the case if one were absorbing everything simultaneously at a cocktail party. This position relates back to the Darwinian idea of indirect evolution. James lauded the standpoint as triumphantly original for recognizing separate cycles of operation in nature;[41] in other words, he appreciated the theory for highlighting that variations arise for reasons independent of the environmental pressures that select or discourage them. By applying this evolutionary idea on the scale of the individual, James arrived at two interrelated explanations of how the mind can fit the environment without being directly molded by it.

One was that "accidental out-births of spontaneous variation in . . . the excessively instable human brain" spawn new ideas and new ways of drawing relationships between things.[42] Many of these inventions "perish through their worthlessness,"[43] but some help us notice and pull things together intelligibly. The environment reinforces the latter ideas, but this "is the cause of their *preservation*, not that of their production."[44] James's second explanation was that the environment supplies sensory variations, which are reinforced or deemphasized depending on our interests. Thus, during a cocktail party, we do not register every perceptible phenomenon. Instead, our attention narrows, focusing on what interests us and has emotional pull. Necker forms are similar in that focusing attention can flip a plane to front or back or make the figure appear two-dimensional. In most cases, such focusing occurs automatically, and the point is that attention must be narrowed to extract anything coherent. James further suggested that selective interests or attention shape how we rationally put things together.[45] The Necker illustration conveys something like this on a perceptual level.

Aside from the emphasis on multiple cycles of operation, James's account reiterates the nonpassive sides of adaptation, an implication that applies to pragmatic theories of mind. It also characterizes Gestalt and phenomenological outlooks, along with more recent Gibsonian and 4E views, all of which will become critical in this book. Stephen Jay Gould nicely illustrates the active and—in a sense—embodied side of Darwinism that James picked up on.[46] Gould points out that evolution is sometimes driven by both the presence of variation and the actions of organisms. Thus the appearance of a new food only alters the evolution of bill shape if birds have an ability to eat it and an interest in doing so: when they act upon this, the added nutrition allows those better suited to the task to propagate more. James made roughly the same point when joking that successive generations of dogs raised in an art gallery would not evolve an appreciation of painting.[47] They would not because they would be uninterested in the aesthetics of painting in the first place, and hence would not be affected by it.[48] James adapted this reasoning on an ontogenetic level, maintaining that interests dictate what environmental aspects we notice and therefore what features affect us within the span of our own lifetimes.

By modeling a Darwinian account of mind, and especially by emphasizing independent cycles of operation, James combined the empiricist claim that the world structures the mind with the rationalist claim that the mind imposes form on the world.[49] In this reconciliation, and by insisting that factors independent of our experience limit cognition and perception and actively shape how the world appears, James mirrored what Kant had done.[50] He recognized this, claiming in broad strokes to defend "the account which the apriorists give,"[51] that interests are "the real *a priori* . . . in cognition,"[52] and that "interests form a true spontaneity and justify the refusal of *a priori* schools to admit that mind was pure, passive receptivity."[53] However, these quotations also highlight a break from conventional Kantianism—namely, that James replaced the logical *a priori* with valuative mechanisms that nonetheless serve comparable functions by limiting cognition and experience.

This insight, as we will especially argue in chapters 4 and 5, has important but underappreciated ramifications for contemporary work on affective bases of rationality and perception. A missing link in both contemporary and classic literature—to be expanded on later—is the inadequate appreciation of conceptual, experiential, and neurobiological overlap between interests and emotions.[54] Insofar as interests connect to emotion and bodily feelings, James's account also marks a departure from "in the head" approaches to mind. Dewey and Mead would widen this fracture (while retaining Kantian elements) by arguing that possibilities of bodily action limit possibilities of experience.

EXPERIMENTALISM AND PRAGMATISM

The term "scientist" as a designation for those investigating material nature is a recent neologism, coined by William Whewell at the prompting of Samuel Coleridge in the 1830s.[55] Before that, "science," from the Latin *scientia* for "knowledge," meant any corpus of "systematic and orderly thinking about a determinate subject-matter."[56] Over time, it came to mean systematic thinking directed towards a particular end—namely, acquiring knowledge about physical nature. This usually implies physical engagement with investigative targets, whether in laboratories or through telescopes, though exceptions exist.

Although classical pragmatists kept apace of scientific developments and the ones we deal with might be counted as theoretical psychologists, only Peirce and James were scientists in the sense described above. In addition to and because of his training, James joined Louis Agassiz—a scientific celebrity at that time—on an expedition to Brazil. The purpose of Agassiz's expedition was ideological: to discredit evolution, and James derided this agenda.[57] One of James's jobs on this venture was sifting through biological samples, something he found tedious. Ironically, James was never keen on hands-on work, even while extolling practice. Later, he would eagerly hand his psychology laboratory over to Hugo Münsterberg, despite taking pride in having established one of the first in the world. But while never enthusiastic about practical scientific work, James nonetheless had considerable knowledge of it and experience doing it, especially by the standards of his day, and this impacted his views on mind and epistemology.

The scientific achievements of Peirce, who studied chemistry at Lawrence Scientific School at the same time as James, were unequivocally impressive, and this even by today's standards. The United States Coast and Geodetic Survey employed him for over thirty years. During this time he refined the use of pendulums to detect small variations in Earth's gravity. He spent about three years as an assistant at Harvard University's astronomical observatory investigating the shape of the Milky Way and measuring the brightness of stars. He pioneered the expression of the meter as a number of light wavelengths at a determined frequency. A keenly rigorous mind, Peirce also made important contributions to statistics, logic, topology, and algebra, advancing the intellectual tools with which scientists have conducted and communicated analyses.

It therefore comes as no surprise that a great deal of pragmatism and its methods are grounded in science. Peirce's pragmatic definition of the notion of meaning is a case in point.[58] It holds that thought-distinctions are never "so fine as to consist in anything but a possible difference of practice" and that our concept "of anything *is* our idea of its sensible effects."[59] To ascertain the meaning of a concept, therefore, we need only ponder "what effects, which might conceivably have practical bearings, we might conceive the object of our conception to have."[60] Understood thus, a hard object might be conceptualized as one that has the possible effect of scratching other substances. This is known as Peirce's pragmatic maxim, and though it presents a method of getting clear about ideas, it bears the unmistakable imprint of experimental science, where we come to know things by systematically messing around and observing the consequences. Likewise, it gets close to the scientific procedure of operationalizing. Peirce's maxim forms a central thread in pragmatic ideas about knowledge and mind.

Though Peirce's work from the late 1870s is generally extolled as the beginnings of American pragmatism, it did not achieve widespread notice until James began delivering and publishing popular lectures a few decades later. James's variant of pragmatism irritated Peirce, especially insofar as it emphasized the role of individual interests and, hence, personal affectivity in reasoning. However, James retained Peirce's core idea, and his version of pragmatism unequivocally bore the mark of experimental science. As James put it, quoting the future Nobel Laureate chemist Wilhelm Ostwald, "I am accustomed to put questions to my classes in this way: In what respects would the world be different if this alternative or that were true? If I can find nothing that would become different, then the alternative has no sense."[61] In other words, James explained that if rival views have the same consequences, then they mean practically the same thing; and practical meaning, according to James, is the only kind. James elaborated the point with another example from Ostwald:

> Chemists have long wrangled over the inner constitution of certain bodies called "tautomerous." Their properties seemed equally consistent with the notion that an instable hydrogen atom oscillates inside of them, or that they are instable mixtures of two bodies. Controversy raged; but never was decided. "It would never have begun," says Ostwald, "if the combatants had asked themselves what particular experimental fact could have been made different by one or the other view being correct."[62]

Summing up and repeating Peirce's position, James observed that many disputes evaporate when you trace out the concrete consequences of either side. There can be no abstract distinctions that do not express themselves "in concrete fact and in conduct consequent upon that fact, imposed on somebody, somehow, somewhere and somewhen."[63] The lesson James took from Ostwald, quoting him once again, is that "realities influence our practice . . . and that influence is their meaning for us."[64]

Like other pragmatists, James added what most scientists at least tacitly understand: that we influence realities, and so come to know them better. This view is inherent in experimental methods, which again open the world to us by jostling it in systematic ways, and is also part and parcel of James's model of cognition and Kant's before him.[65] In passages exhibiting nascent 4E views, James described "belief" as the "mental state or function of cognising reality,"[66] and "cognition" as an intermediary stage in "what in its totality is a motor phenomenon."[67] He meant that cognition, when confronted by some thing or event, is more concerned with the question of "What is to be done?" than the question of "What is that?"—or as Merleau-Ponty would later say: "Consciousness is in the first place not a matter of 'I think' but of 'I can.' "[68] James further said: "Cognition . . . is incomplete until discharged in act."[69]

James, then, associated belief with action, arguing that "the test of belief is willingness to act,"[70] and that "there is some believing tendency wherever there is willingness to act at all."[71] He meant not only that action measures strength of belief, but also that belief functions to facilitate action. When wavering between contradictory options, unsure what to believe, one hesitates to act, especially if acting carries weighty consequences. With strong belief, however, there arrives "an idea which is inwardly stable, and fills the mind solidly to the exclusion of contradictory ideas. When this is the case, [actions] are apt to follow."[72] On the grounds that beliefs enable and guide action, James proposed that the truth of a belief "is not a stagnant property," but something that happens through "a process of valid-*ation*,"[73] or, in other words, valid-*action*. Belief in atomic particles, for example, has led to scientifically fruitful theorizing and experimentation. Hence, it has led scientists to act in ways benefiting their field. So long as the belief continues to reliably cultivate beneficial or "valid" actions, scientists are apt to continue trusting it.

Based on the intimate connection between action and belief, James further speculated that people, by willing themselves to action, can will themselves into a state of belief.[74] He did not mean, however, that this occurs on a primarily psychological level, since, for him, actions complete beliefs. Furthermore, people cannot capriciously believe whatever they want, for they cannot act however they want.[75] They cannot because the world supplies resistance, as Dewey, Mead and Peirce also said at various times. Most will accordingly find it impossible to act on the belief that they can walk on the Sea of Galilee; maddening to act on the belief that traveling south will get them from Florida to New York; and mortifying to act on the belief that Queen Elizabeth II led the Cuban Revolution. In sum, beliefs are not invariably correct, as many are not even tested and are held merely by virtue of nothing contradictory interfering,[76] but this is not the point. The point, rather, is that the world—including everything from the physical world to the world of already existing beliefs—checks certain actions, and therewith certain beliefs. It also generates evidence affirming or repudiating beliefs, as when putting scientific hypotheses to the test, or when risking small actions with someone we like romantically and discovering whether our feelings are returned.

Dewey, along with Mead, advanced roughly the same view, albeit in a more bodily and less cognitivistic fashion and in ways that squarely anticipated Merleau-Pontyian, Gibsonian, and 4E outlooks. They did this while incorporating Kantian elements, which some readers may take to be completely at odds with pragmatic positions; yet pragmatism, in fact, explicitly draws inspiration from Kant.[77] "Perception," as Dewey suggested in a mix of rationalist and empiricist language, "is an act of the going-out of energy in order to receive."[78] He held the same idea of cognition, as will later be seen. Thus, to adapt illustrations from Mead and Merleau-Ponty that resonate with Dewey,[79] we perceive by reaching out with our hands, exploring, and receiving the form of things. In so doing, we come to realize physical properties such as the glassy smoothness and roundness of a bottle. In this case (and we will later show it applies generally to other modalities such as vision), perception is active; that is, perception is literally an activity. It involves directing our bodies into the world. Bodily structures are perspectives or even biases of sorts,[80] albeit in nonpejorative senses, and robust perception necessitates adaption, just as getting to know people requires adjusting initial preconceptions of them, but without which we would have no place to start. So it is similarly with our hands: we normally do not keep them rigidly flat, and doing so would impoverish experience. Instead, our hands and fingers adjust to the form and texture of the bottle while nonetheless setting limits ahead of time on what can be experienced, and the doings and responses consequently undergone are integrated into the particular encounter.

Dewey's account arose in his ongoing efforts to circumnavigate debates between rationalists and empiricists.[81] Put crudely, these debates were about whether the mind structures the world or whether the reverse holds, and Dewey effected a reconciliation plainly within the Kantian trajectory. Against rationalists, Dewey argued that ways of cognizing worlds follow from ways of inhabiting them, which is to say, from habits.[82] To a significant extent, therefore, habits precede thoughts, a position defended by Peirce, James, and Mead as well. "Ideas . . . are not spontaneously generated. There is no immaculate conception," wrote Dewey. "Reason pure of all influence from prior habit is a fiction."[83] But so too, for him, were the "pure sensations" of empiricists, for they "are alike affected by habits."[84] According to Dewey, empiricists who attack the notion of thought pure from the influence of experience, usually identify experience with sensations impressed upon an empty mind. They therefore replace the theory of unmixed thoughts with that of pure unmixed sensations as the stuff of all conceptions, purposes and beliefs. But distinct and independent sensory qualities, far from being original elements, are the products of a highly skilled analysis. . . . To be able to single out a definitive sensory element in any field is evidence of a high degree of previous training, that is, of well-formed habits. A moderate amount of observation of a child will suffice to reveal that even such gross discriminations as black, white, red, green, are the result of some years of active dealings with things in the course of which habits have been set up. It is not such a simple matter to have a clear-cut sensation. The latter is a sign of training, skill, habit.[85]

In sum, Dewey chided rationalists for not being empiricists—that is, for not recognizing the priority of experience. Yet this is, strangely, also why he assailed empiricists. "Our ideas," he wrote, "truly depend on experience, but so do our sensations. And the experience upon which they both depend is the operation of habits."[86]

While critical of both rationalism and empiricism, Dewey sympathized somewhat more with the rationalistic view that the world conforms to the structure of mind; that it is because of this that the world is an object of possible knowledge; and that it is by virtue of sharing the same structure that minds come to have similar experiences of the world, making it an object of shared knowledge. However, rationalists proposed that the world conforms to the structure of mind either because the mind imposes rational structure on the world or because the world is an expression of the rational mind of God. Dewey accounted for the conformity in a much less esoteric way: "The world is subject-matter for knowledge, because the mind"—or what Dewey sometimes called the "body-mind"—"has developed *in* that world."[87] Even though this sounds like British empiricism, Dewey went on to explain that the body-mind itself participates in and contributes to the patterns of acting, interrelating, and habits that make the structure of worlds. As Dewey said elsewhere, "habits"—and therewith the self or body-mind—"incorporate an environment within themselves," and in this sense conform to it. Habits also bring the environment into conformity with themselves ("they are adjustments *of* the environment, not merely to it"[88]), much like Kant's Copernican Revolution. It is to be expected, therefore, that the "body-mind . . . will . . . find some of its structures to be concordant and congenial with nature, and some phases of nature with itself."[89]

Here, bodily structure and things encountered limit actions and hence experience, supplying an analogue to the Kantian *a priori*.[90] Although not conventionally logical, the limits can be nearly as unyielding. Thus while we can roll beer bottles between our palms, the same action and, hence, experience is impossible with cinder blocks. Coherence-bringing activity, moreover, is typically subtler than gross movements of reaching and handling. As Dewey observed, experience is organized—prior to intervention from internal, mental mechanisms—by "adaptive courses of action, habits, active functions, connections of doing and undergoing" and "sensorimotor co-ordinations."[91] He reasoned that this meant even a presumably non-conscious organism such as an amoeba had preconditions of experience. It has them because of its bodily organization and the structures of things it encounters, along with activity-shaping organic demands. This combination ensures enactment of systematic patterns of doing and undergoing that form bases for experience, a point that more recent commentators also invoke unicellular life to illustrate.[92]

Dewey was arguably the most historically oriented of the original pragmatists (though Peirce's knowledge was also vast), and he identified his embodied position with ancient Greek views.[93] His account of embodiment also highlights links between ancient thought and James's concept of experience, though he did not state this and James was likely unaware. Dewey wrote that "sensation and perception," according to ancient Greek accounts, supply experience with "pertinent materials, but [do] not themselves constitute it." Experience arises with the addition of "retention . . . and when a common factor in the multitude of felt and perceived cases detache[s] itself so as to become available in judgment and exertion."[94] Here, Dewey summarized Aristotle, who, in his *Posterior Analytics*, observed that animals have a "discriminative capacity which is called sense-perception." Aristotle explained that "the sense-perception comes to persist" in memory, and when "such persistence is frequently repeated," there "develops a power of systematizing" and integrating individual memories: "So out of sense-perception comes to be what we call memory, and out of frequently repeated memories of the same thing develops experience."[95] This Greek account resonates with the pragmatic assertion that experience is not sensation, as well as with James's argument that experience is instead what is left over after our concerns have chiseled it into coherent form.

While mostly accepting what James said, Dewey added that experiential shaping also occurs through bodily engagement, where activities of a certain form engender similarly structured experiences. This occurs when fingers coordinate around the contours of a bottle so that rhythms of interaction integrate into an experience. It happens globally and socially, such as when activities and hence experiences cohere around children or spouses, a point conveyed in Dewey's concept of experience as culture.[96] Dewey's account fits ancient etymologies. "Experience" derives "from the Greek . . . 'empiria' (ἐμπειρία), and the Latin word 'experiential,' or 'experimentum.'" *Experiri* means "try, to put to the test."[97] *Empeiros* means being experienced or practiced in an activity, based on *peira*—trial or attempt. For the Greeks, wrote Dewey, experience "signified a store of practical wisdom, a fund of insights useful in conducting the affairs of life."[98] Thus, in mounting his enactive view, Dewey remarked that "the key to the matter" is curiously found in the ancient Greek notion that "experience was itself a product of experience."[99] In other words, having experience means being experienced. Recognizing that Plato and Aristotle distinguished experience or *empeiria* from art or *tekhnē*,[100] Dewey switched from a past-tense historical narrative to a present-tense voice in his summative conclusion—"experience is equivalent to art,"[101] with "art" understood more in the practical than fine art sense, though Dewey did not separate the two. So conceived, "experience is exemplified in the discrimination and skill of the good carpenter, pilot, physician, captain-at-arms."[102] In Dewey's scheme, then, experience is a mode of skilled coping, again precisely the position of contemporary enactivists, especially O'Regan and Noë.[103]

Ancient distinctions between experience and art are worth just a little more elaboration, since they bear on Dewey's position. In *Metaphysics*, Aristotle articulated the difference with an example from medicine.[104] He explained that one may have observed occasions when patients suffering an ailment improved after receiving a specific treatment, and thus know, as a matter of experience, that a good result was realized in these cases. Art comes about when a universal judgment is abstracted from many experiences. Notice here that experience implies some skill. Art implies higher skill. This aligns with Dewey's and more recent enactive accounts of experience. Notice, moreover, that the English words "experience" and "expert" share the same Latin root, *experiri*. Expertise implies skill, and the Greek adjective *empeiros* can connote those having skill and expertise, while *empeiria* means experience. The point is that for Dewey and some ancients, experience implies skilled coping, which is consonant with enactivism.[105]

At the same time, root words of "empirical" have sometimes been used to denote quack physicians. Plato's *Laws* helps resolve this apparent inconsistency. There, he described slave physicians "who gain their professional knowledge by watching their masters and obeying their directions in an empirical fashion, not in the scientific way."[106] They thereby acquire habitual routines, as opposed to reasoned knowledge. Routines may give them a knack and skill for treating patients, and thus the appearance of knowledge of medical arts. Yet as Plato's *Gorgias* reiterates, "routine" or "knack" is "no art,"[107] for by itself it supplies "no principle in virtue of which it offers what it does, nor [can it] explain the nature thereof, and consequently is unable to point to the cause of each thing."[108] Thus, as Dewey explained, "empirical" in this ancient Greek sense "does not mean 'connected with experiment,' but rather crude and unrational."[109] Consequently, empirical knowledge meant that which was "accumulated by a multitude of past instances without intelligent insight into the principles of any of them. To say that medicine was empirical meant that it was not scientific"[110]—science, again, understood in the older sense that connotes systematic and orderly bodies of knowledge. Empirical medicine would here be enslavement to "a mode of practice based upon accumulated observations of diseases and of remedies used more or less at random."[111] By contrast, "experimental science," in the late modern innovation of the term, "means the possibility of using past experiences as the servant, not the master, of mind."[112]

Dewey consequently did not have Plato in mind when equating experience to art, for he explicitly said that his ancient predecessor identified experience with "established customs formed . . . by repetition and blind rule of thumb."[113] According to Dewey, experience for Plato "meant habituation, or the conservation of the net product of a lot of past chance trials,"[114] and hence could also mean "enslavement to the past."[115] Notice that the experimentalism of pragmatists ran contrary to such notions, keeping instead with the late modern understanding of experience. Notice also that Dewey did not understand habits as blind reflexes. On the contrary, he wrote that a "sensitive habit grows more varied, more adaptable by practice and use."[116] That is, while fruitfully depending on routine instilled through past activity, and while there would be no experience in Dewey's sense without habits that enable rhythmic coordination, habits can be intelligent, inventive, adaptive, and responsive. Such is evident when a skilled musician creatively enacts ingrained training, whether by adjusting to an unfamiliar guitar neck or through spontaneous innovations during a performance. Dewey therefore did not use the word "experience" in the same way as Plato did, to denote blind routine; indeed, he argued that experience evaporates and things withdraw from notice when actions become mechanical.[117] However, this does not diminish his openly acknowledged debt to Plato's concept of experience as habit, along with his concept of experience as trial and error, or, in other words, unsystematic experimentalism. In effect, Dewey's concept is Plato's rendered more artful.

Dewey's concept of experience correspondingly implies creative adaptive habits and artful experimentalism. Because of this, it was close to his concept of knowledge, which was also based in experimentalism. Dewey wrote that "to maintain, to expand, adequate function is [the] business" of life,[118] and "the business of organic adaptation involved in all knowing [is] to make a *certain* difference in reality."[119] For Dewey, living and knowing both involved "co-operative and readjusted changes in the cosmic medium."[120] Here the word "cosmic" corresponded roughly to the ancient Greek word *kosmos*, which might be translated as "orderly arrangement" or "system," and Dewey's statement expressed what physicists of his day had just come to recognize: that illuminating a system so that it can be seen and known actually disturbs the system.[121] In quantum physics this is emphatically so, for as Dewey noted, at least one photon of light "is required to make, say, an electron visible," and the collision between the photon and the electron "displaces to some extent the object observed."[122] The lesson Dewey drew is that knowing—like experience—is a "kind of interaction that goes on within the world,"[123] so that the knowledge is "a product in which the act of observation," which is an act of rearrangement, "plays a necessary role."[124]

In this regard, quantum mechanics is commonsensical, for we frequently observe things better by changing or manipulating them. When "trying to make out the nature of a confused and unfamiliar object," as Dewey elaborated, "we perform various acts with a view to establishing a new relationship to it, such as will bring to light qualities which will aid in understanding it. We turn it over, bring it into a better light, rattle and shake it, thump, push and press it, and so on."[125] Or if we cannot directly jostle an object, we "deliberately alter the conditions under which we observe [it], which is the same thing in principle."[126] Thus, inquiring—like perceiving—is an act of going out in order to receive. This means, on the one hand, that we garner observations from the world in return for acting upon it; and, on the other, that what we receive is of little worth when we fail to act within certain limits. If we are using ill-suited instruments, ignoring constraints of materials and banging about randomly, we elicit a series of observable effects, but the relationship between them will likely appear haphazard and meaningless.[127]

These points reiterate Dewey's basic rationale for characterizing "knowing" in terms of art and building. Carpentry is an art, and a carpenter a builder, and what makes someone a builder, according to Dewey, "is the fact that he notes things not just as objects in themselves, but with reference to what he wants to do to them and with them; to the end he has in mind."[128] Put otherwise, the end the carpenter has in view and the objects he encounters limit possibilities of action. The end, if it is to be achieved, limits what materials can be used; and the materials used limit how the carpenter can deploy his skills. Consequently, the carpenter acquires the habit of seeing things in terms of possible uses:

> Fitness to effect certain special changes that he wishes to see accomplished is what concerns him in the wood and stone and iron which he observes. His attention is directed to the changes they undergo and the changes they make other things undergo so that he may select that combination of changes which will yield him his desired result.[129]

So when calling pine "a soft wood," a carpenter indicates it is easy to sand and hammer nails into, but also easy to scratch and thus inappropriate for kitchen floors. From his standpoint, things and their properties "*are* what they can do and what can be done with them,"[130] and it "is only by . . . processes of active manipulation of things in order to realize his purpose that he discovers what the properties of things are."[131]

In these various examples, properties appear not only in consequence of what the subject does, but in consequence of "changes [that things] make other things undergo."[132] Dewey's point, therefore, echoes Peirce's pragmatic maxim, which defines an object of conception according to its possible effects on other things or those of other things on it; "hardness" identifies a substance resistant to being scratched and liable to scratch other things. A similar notion appears in James's insistence that willingness to act measures strength of belief, and that actions generate effects that either confirm or disconfirm beliefs. C. I. Lewis noted generally that the "appearance of an object is conditioned also on other objects."[133] Mead expressed the idea more specifically in terms of spatial experience when he said that the "ultimate fact" of it "is that of the effective occupation of . . . by the physical individual, both in the experience of resistance to what invades his place and in the advance to occupy other places, together with the sense of boundary which comes with the tactile surface experience."[134] All this suggests that properties only appear to us in relational or ecological contexts, so that, as Friedrich Nietzsche said in the spirit of Peirce's pragmatic maxim, "the properties of a thing are effects on other 'things': if one removes other 'things,' then a thing has no properties."[135]

The earthy red of Mars, to consider an illustration, is conditioned not only on the presence of the planet with its abundance of iron oxide compounds. It also requires a being that can see it and light that is selectively reflected as rusty red. Moreover, the color depends on relative velocity, since rapidly approaching objects are shifted to higher frequencies of light and receding ones to lower frequencies. Thus, even before a perceiver shows up, the color is already an effect of interrelations.[136] Perceived color is, of course, typically regarded as a secondary property, or in other words, as the way the mind represents the object, as opposed to the way the object really is. Put otherwise, color is conventionally regarded as a perceived outcome of a relationship between an organism with certain sensitivities and an object. However, notice that many properties are products of relationships, such as the dimensions of an object and its mass being dependent on relative velocity, a point Dewey made in reference to Einstein's theory and experimental methods more generally.[137] It is likewise with the solidity of objects, which, in the case of neutrinos, almost never manifests. The list goes on. Most hesitate to dismiss properties such as length, mass, and solidity as merely secondary, even though they are not determinable outside of relationships. Given this, pragmatic thinking suggests that it is capricious to claim that color, smell, and so forth are secondary and subjective merely because an organism is involved, especially compared to other properties such as length, which are conventionally deemed primary, yet also require relationships and observers to be expressed.[138] As Mead summed up, there "is an uncritical tendency to identify the organism with a so-called 'consciousness,' to make it in some sense subjective as over against an objective world of things." This is arbitrary because "the organism is a part of the physical world we are explaining."[139]

It is with something like this in mind that Dewey insisted upon perceived qualities as not being mere internal representations. They are "never . . . 'in' the organism at all," he wrote; they are "always . . . qualities"—or effects— "of interactions in which both extra-organic things and organisms partake."[140] Thus, as Dewey elaborated in another work, qualities such as "red, or far and near or hard and soft, or big and little" are undeniably relative.[141] They are so in the literal sense of involving "a relation between organism and environment," but, according to Dewey, this is no basis for "proclamations of the agnostic 'relativity' of knowledge."[142] Instead, it is "an argument for the ultimately practical value of distinctions."[143] As Dewey put it, this means "they are *differences* made in what things would have been without organic behavior—differences made not by 'consciousness' or 'mind,' but by the organism as the active center of a system of activities."[144]

The combination of this system of activities and the active organism can be thought of as a situation. If we are in a kitchen, our situation includes us, and likely floors, walls, chairs, counters, sink, and stove, and perhaps also a cat sniffing at our feet. These constituents delineate possibilities of action and organization. If we remove, say, the table, then we can traverse the space it once occupied, but cannot lean our elbows where we once did; nor can the tablecloth continue to rest a meter and a half above the floor. Limits also vary depending on the active center. The cat is one center; a person another. Because the two come to the kitchen with different capacities, they confront different constraints: they are faced, as the expression goes, with different situations. We can therefore imagine they perceive their situations differently, yet "perceiving situations differently" is here equivalent to "perceiving objectively different situations." This is because the varying restraints are not confined to heads or brains of the cat and person; they, like the table and sink, are able to do different things. It is accordingly no mere variation in mental outlook that differentiates the experience of sipping a glass of milk from that of lapping from a dish while on all fours. It is a difference in action. If the person caresses a table leg while the cat sharpens its claws on it, they perform different actions that realize different properties as effects. Easy sweeps of fingertips realize a smooth lacquered finish, while the cat's claws realize the sinewy toughness of wood, all of which is to say: perception here involves acting and making differences in the world, as opposed to merely representing it.[145]

Brains and sensory systems, of course, make differences too, and Mead advocated something like this when he observed that an "organism determines in some sense its own environment by its sensitivity." He argued that "the only environment to which the organism can react is one that its sensitivity reveals," meaning that "the sort of environment that can exist for the organism, then, is one that the organism in some sense determines."[146] Mead's views, along with Dewey's, resonate with Jakob von Uexküll's concept of *umwelt*.[147] *Umwelt* literally translates as "around-world." In everyday German, it connotes "surroundings" or "environment." Uexküll's concept means something like "organism-centered world." It points to the fact that while different organisms occupy the same physical spaces, their different nervous systems leave them sensitive and responsive to different things. For example, bats inhabit a rich world of echolocation, whereas this sphere does not exist for us in any significant way. Though highly sympathetic to this view, we would caution against interpretations that reduce different *umwelten* to different inner experiences or representations of the world. We do not go so far as to deny such differences exist. However, we maintain that different organisms literally inhabit different worlds because they enact different responses to different things, generating different consequences. Differences are therefore not merely or even primarily "in the heads" of organisms. This view is defended in various ways by Dewey, Mead, and later Gibson and a host of 4E researchers along with Merleau-Ponty and Heidegger, all of whom are companions in this book.

AFFECTIVITY AND PRAGMATISM

Affectivity—taken broadly to connote interests, emotions, moods and aesthetics—was prominent in classical pragmatism, especially the work of James and Dewey. In this, they were hardly unique. Thinkers dating from Sigmund Freud and Friedrich Nietzsche to David Hume and Spinoza all the way back to Plato have similarly appreciated that affectivity dominates everyday life. However, James and Dewey stand out in their rejection of affectivity as an epistemological contaminant, arguing instead that it brings us perceptually and cognitively into contact with the world. In this, they anticipated Damasio, among others. Yet they did so while countering Damasio's emphasis on inner life for reasons grounded in their experimentalism. This led them to anticipate affordance theory and treat affectivity as an *umwelt* or world-altering lens.

Let us begin by considering affective sides of pragmatic experimentalism. We have already seen that pragmatists assimilated scientific methods in general and specific ways. As is standard with scientists—this has not been discussed so far—pragmatists were falliblists. That is, they recognized that evidence was never enough to adjudicate beliefs, which is why basic research methods classes consistently stress that theories are never absolutely verified and that, typically, more than one theory can match a data set. James made just this claim in *Pragmatism*.[148] He pointedly subtitled the book "A New Name for Some Old Ways of Thinking," arguing that while evidence is critical, additional non-evidential criteria had long been employed to adjudicate beliefs. These criteria are, in fact, those that scientists continue to rely on today—namely, elegance, economy, and sense-making power, which fit with other accepted beliefs and even appeal to emotional preference. Giving such significance to emotional preference appears unscientific and seemed so to James's contemporaries, especially irking Peirce, though he, too, saw a place for affectivity insofar as the irritation of uncertainty motivates inquiry.[149] Moreover, all things being equal, the aforementioned criteria tend to align with emotional preferences. We emotionally value economic beliefs—ones requiring fewer assumptions and offering ease of application— even though parsimony is not a law of nature. We similarly value ones that do not contest established knowledge. That which lacks sense-making power frustrates and pushes us away. We find elegance aesthetically attractive and therefore emotionally satisfying. We also incline emotionally towards beliefs that fit evidence: first, because this is usually more adaptive, and, second, because humans desire knowledge, as philosophers since ancient times have asserted. Even the decision not to decide until evidence dictates choice is based on emotional inclination.[150] James added that these standards are not lax, but extremely difficult to meet, as scientists well know.

As an illustration, consider debates about the structure of the solar system. The heliocentric view dominates today, and the story many science instructors offer is that it vanquished the geocentric model because it more closely fit observations. But the story is misleading. The Ptolemaic system in fact predicted planetary movements better than Copernicus's, which has circular rather than elliptical orbits. Furthermore, when Galileo compared the two and presented Copernicus's as better, he left out a third model. This was one proposed by Johannes Kepler's teacher, Tycho Brahe. It had Earth at the center, orbited by the sun, with other planets going around the sun. Where observation is concerned, Tycho's model fits evidence just as well as Copernicus's does. A little activity will demonstrate why. Imagine inserting your left index finger through the hole of a DVD and moving your right index finger around the circular edge while keeping your left index finger stationary. Next, picture keeping your right finger stationary and using the left index finger to roll the edge of the DVD around it. Though the spatial relations between the fingers remains the same in the two instances—everywhere equidistant both times—the right finger appears to circle the left in the first instance, and the left the right in the second (figure 1.2).[151]

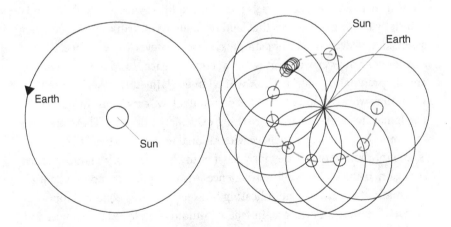

FIGURE 1.2 Depictions of the relationship between the earth and the sun using a DVD
In the first case, a stationary finger through the central hole represents the sun, while the other finger moving around the DVD edge stands in for the earth. In the second case, the finger representing the earth is stationary, while the finger representing the sun rolls the DVD around the first finger, tracing a circular path. Notice that the spatial relationships between the two fingers remain the same in the two instances, i.e., always equidistant.

The reversibility implied in the DVD illustration applies to ellipses and other shapes, and holds regardless of the center of motion in the bounded figure.[152] If updated with appropriate elliptical orbits, either of the models just described (heliocentric or geocentric) would account for the observed motions of planets, for why Venus appears largest during its crescent phase, and so on. However, the geocentric model would not fit James's pragmatic criteria. With two centers of motion rather than one, it is less economical, especially since the model requires background stars to undulate back and forth annually to account for the observed data. Having two centers of motion also seems less elegant. It does not fit with other accepted beliefs in physics and, on the whole, has less sense-making power. Notice, then, that while both geocentric and heliocentric models are evidentially backed, the primary deciding factors between the two are valuative. Yet, valuative interpretations—including emotions, interests, aesthetics, concerns, moods, and so on—are not opposed to evidence. On the contrary, they can be ways of making evidence stand out; in other words, the identification of facts depends on interested valuations of the world. We will revisit and defend this view later in this and future chapters.

The stock criticism of pragmatic, valuative criteria, of course, is that they allow us to choose whatever feels good and right, as with an alcoholic preferentially leaning on studies that laud the benefits of drinking. As has been asserted, however, we cannot simply choose whatever we want because we cannot act however we want without consequences, and evidence is key among pragmatic criteria. Thus, while James claimed that beliefs are true when they work, he added that this does not allow "capricious impunity."[153] For James, to work "means something extremely difficult." There are narrow limits on what we can believe with rational warrant, so that "the squeeze is so tight that there is little loose play for any hypothesis."[154] James's criteria, again, include observational evidence. Also included are non-evidential, pragmatic standards for adjudicating between theories equally supported by data. These standards are, in fact, commonly employed in science, and any working scientist knows that they are not easily achieved.

James's treatment of logic followed a course similar to that of his account of pragmatic decision-making, insofar as the latter involved affectivity. His 1884 article, "On Some Omissions of Introspective Psychology," was the basis for his famous 1890 chapter, "The Stream of Thought."[155] In these pieces, James suggested that feelings underlie logical connectors. In other words, whereas standard logic reduces the statement "P but Q" to the conjunction "P and Q," the two have different phenomenological meanings, and these meanings are engendered by different feelings.[156] "But" has a feeling of uncertainty, and the portion of the statement following "but" often has the rising intonation of a question. "And" is less ambiguous and more matter of fact. Thus, in contrast to the firmness of saying "I'll go to the party and won't have fun," Johnson argues:

> To feel James's "but" is to feel the quality of a situation as a kind of hesitancy or qualification of something asserted or proposed. When you think "I may go to the party, but I won't have fun," you are expressing some unsatisfactory qualification of your anticipated situation. You are feeling that if your situation should develop in a certain way (i.e., if you go to the party), then there will follow a certain unresolved quality of your situation as it has developed to that point in time (i.e., you feel the disease of not having fun). In addition to this feeling of hesitancy, "but" also marks a feeling of conjunction. In the example above, the not-having-fun is tightly connected to having attended the party. To say "x but y" asserts that both x and y are taken together, but y is taken (or given) with some hesitancy.[157]

These observations leave formal logic untouched, since "P and Q" and "P but Q" continue to equally imply the truth of both conjuncts. However, the above does highlight the role that feelings play in meaning, ordinary communication, and the linguistic synthesis of thought.

A key to James's valuative accounts of mind and knowledge is his concept of "selective interests."[158] It asserts that interests direct attention, carving and atomizing the world and making it more manageable, thereby facilitating human cognition. In this regard, James's position came close to the neurobiologically backed positions of Damasio and others over a century later.[159] James's central thesis was that people cannot cognitively handle everything at once; accordingly, they focus on one bit of reality to the exclusion of others, whether in the sciences or everyday life. James further suggested that interests and, therefore, affectivity form the heart of concept formation. Thus, it might be that a mechanic values and hence conceives oil as combustible or lubricating, whereas a carpenter might be interested in oil as a darkener of wood, cognizing it in this way instead.[160] James's concept of selective interests is a variation on Peirce's pragmatic maxim that introduces personal affectivity into the mix.[161] Yet it is not any less grounded in the world, for the object of attention—in this case, oil—is valued, and thus noted for producing different effects that are unequivocally there.[162]

Without devolving into unmitigated subjectivism, James accordingly argued that an object's "essence"—that is, the key set of features that make an object what it is—"varies with the end we have in view."[163] Hence, essence is nothing more than those key properties that are

> so *important for my interests* that in comparison with [them] I may neglect the rest. . . . The properties which are important vary from man to man and from hour to hour. . . . But many objects of daily use—as paper, ink, butter, horse-car—have properties of such constant unwavering importance, and have such stereotyped names, that we end by believing that to conceive them in those ways is to conceive them in the only true way. Those are no truer ways of conceiving them than any others; they are only more important ways, more frequently serviceable ways.[164]

As James insisted through his career, a concept "is a teleological instrument" and "a partial aspect of a thing which *for our purpose* we regard as its essential aspect."[165]

Again, in arguing that personal interests guide thinking, James did not mean that thought is "just in one's head." Rather, he meant that subjective interests direct attention to realities in the experienceable world that are germane to pursuits concerning the conscious agent; and that they hone conceptual instruments so as to handle and negotiate these realities. Instead of reducing conceptual meaning to "images in one's head," therefore, James equated it primarily to functions in human activities, writing that "if [concepts] have any use they have that amount of meaning."[166] James thought this to be true of relatively concrete concepts such as "circle" and "man," and truer of highly abstract ones. "There are concepts," he observed, in which

> the image-part . . . is so faint that their whole value seems to be functional. "God," "cause," "number," "substance," "soul," for example, suggest no definite picture; and their significance seems to consist entirely in their

tendency, in the further turn which they may give to our action or our thought. We cannot rest in the contemplation of their form, as we can in that of a "circle" or a "man"; we must pass beyond.[167]

By functionally or pragmatically reading metaphysically charged concepts, James broke with many empirically and analytically oriented thinkers before and after him who have dismissed some such concepts as nonsense. Such concepts, James insisted, "bring new values into our perceptual life, they reanimate our wills, and make our action turn upon new points of emphasis."[168] Consequently, they have effects in our activities and thus also our experience, including sensible experience. The outcome is that they have sense or meaning.[169]

The aforequoted passages simultaneously suggest that valuations and concepts point beyond the immediate perceptual order, something that chapter 4 attends to in the context of recent work on mind. James maintained that without temporally-oriented cognitive tools, "we should live simply 'getting' each successive moment of experience, as the sessile sea anemone ... receives whatever nourishment the wash of waves may bring."[170] Valuations and concepts allow us to "go in quest of the absent, meet the remote";[171] we connect experiences from different times and places, project forward, look behind. "All these," wrote James, "are ways of *handling* the perceptual flux and *meeting* distant parts of it; and as far as this primary function of conception goes, we can only conclude it to be ... a faculty superadded to our barely perceptual consciousness for its use in practically adapting us to a larger environment than that of which brutes take account."[172] For example, if we gaze on unexploited oil sands in Alberta and ponder the jobs, energy, pollution and political strife stored within, we anticipate interests or ends—or what might be loosely called "abstract affordances"—that exist in the broader human world, but not those in sensations received from the current, physical environment. Moreover, the notions we employ—"energy," "pollution," and so forth—indicate future consequences and, more specifically, concerns about the object of conception (in this case, the sands). Thus, by emphasizing aspects of the object important to our interests, the concept actually postulates future ends—say, greater energy security and a cleaner environment—around which future actions, experiences, and worlds are likely to organize.[173]

Interests relate to aesthetics insofar as interests chisel our world into coherent form. James identified loosely aesthetic qualities in reasoning[174] using his affectively charged variant of Peirce's pragmatic maxim, where concepts are defined by personally valued effects. James also described how emotion helps distinguish between rational and irrational beliefs, nudging us toward the former. Thought-obstructing inconsistencies feel irritating, for example, like gridlocked traffic. Emotions such as this can push us to resolve and thereby escape inconsistencies, and the ensuing flow into rational comprehension, like a movie resolution, comes with feelings of "relief and pleasure."[175] Extreme complexity similarly agitates while excessive simplicity stultifies, so we seek an aesthetic balance between economy and oversimplification.[176] These accounts are somewhat idealized insofar as more recent investigations—for example, now classic cognitive dissonance studies—have found that people sometimes suppress inconsistency by focusing on distractors, or by simply adopting evidentially unwarranted standpoints that accord with extant beliefs.[177] However, some of the same literature indicates that the stress of inconsistency and extreme complexity motivate us to process information and update our understanding. This is in line with James's argument, and it makes sense insofar as it is unlikely that evolution gave us emotional and cognitive systems that overwhelmingly push us out of touch with things.

Dewey's account of reasoning, though not identical, went in similar directions insofar as he viewed it as an aesthetic process that builds and culminates, and thereby coheres like music or narratives. He asserted that all knowledge is "instrumental." This terminology was arguably unfortunate, but he meant by it that knowledge is a work(ing) of art, and he often used examples from science to reinforce the claim. However, he did not thereby insinuate science (here understood in the contemporary sense) as the only path to knowledge. Dewey emphasized this when he remarked that thinking, which is not always scientific, "is pre-eminently an art; knowledge and propositions which are the products of thinking, are works of art, as much so as statuary and symphonies."[178] This meant that thinking—like perception—entails building. Since we build with materials at hand, for example, using already available ideas, thinking is reconstructive. Accordingly, "if defective materials are employed or if they are put together carelessly and awkwardly, the result is defective,"[179] casting "a fog which obscures."[180] In the case of well-executed construction, however, the working of thought integrates and illuminates: "Every successive stage of thinking [becomes] a conclusion in which the meaning of what has produced it is condensed; and it is no sooner stated than it is a light radiating to other things."[181]

The account of emotional life delivered in Dewey's late period is especially relevant to this book; it connects to ideas offered by Merleau-Ponty roughly a decade after, and, as will be detailed later, to those of Heidegger and a range of more recent commentators as well.[182] We often regard emotions as quintessential examples of "inner phenomena," which is to say, phenomena accessible only through introspection. Dewey thought this misleading[183] for reasons Merleau-Ponty would echo in a piece he published near the end of Dewey's life.[184] There, Merleau-Ponty observed that "introspection gives . . . almost nothing."[185] If we try to study anger or love from inner observation alone, we "will find little to describe: a few pangs, a few heart-throbs—in short, trite agitations" that convey little about what these emotions mean.[186]

A first point to note is that we understand the significance of emotions such as anger or love by grasping them in relation to objects and events. An overwhelming anger connected to nothing in the world constitutes a kind of breakdown or disorder in the sense that it is felt for no reason. As Dewey added, however, the emotion "demands something beyond itself to which to attach itself" in order to persist.[187] Hence, pathologically angry people perceive insults where none occurred, generating "delusion in lack of something real."[188] To say that emotions are not purely private and that they demand objects is to say what is obvious: emotions are manners of comporting ourselves towards people, things, and events, ways of interacting and being in the world. They are, to be sure, something we feel, but they are also something we do and live.[189]

A second point, accordingly, is that emotions are constituted through life situations. We say emotions are "in" us. Yet we also speak of ourselves as "being in" emotional moods,[190] and at times this means being in the midst of a situation. The merry-making occurring at a Christmas party, for example, constitutes a situation with a public emotional atmosphere. Some attendees, it is true, may not be merry, as they could be suffering from depression or linking the holidays to painful past experiences. The claim, however, is not that given situations demand specific feelings, but that situations have public moods, which we can sense even if unable to partake in them.[191]

A third point, then, is that emotion is "a mode of sense."[192] Much as colors are qualities of things, an emotion "is *to* or *from* or *about* something objective."[193] Dewey characterized emotional qualities as openings and closures, invitations to approach and avoid,[194] and thus as something approximating what Gibson would call affordances. Thus, when one is tired, a hill looks steeper and less accessible and is emotionally deflating.[195] An alley might likewise appear emotionally inviting or threatening. Dewey noted that we regularly ascribe emotional qualities to the world, speaking of situations as threatening, depressing, intolerable, and triumphant.[196] He further insisted that the emotionality of situations does not follow from subjective impressions projected out, but rather characterizes interactions in which organic and extra-organic things partake. Earlier examples of hands grappling with various objects showed how perceived qualities appear through a mutual participation of subject and object. The earlier illustrations from quantum mechanics and relativity, along with the discussion of the earthy red of Mars, demonstrated a comparable point. Dewey maintained that the emotional qualities of lived situations likewise build, develop, and come into appearance through "an interpenetration of self with objective conditions."[197] He did not thereby deny private feelings. Instead, what Dewey asserted was that "love," "fear," and such words "do not express elements or forces which are psychic or mental *in their first intention*."[198] What we think of as inward movements "are working adaptations of personal capacities with environmental forces."[199] "They denote *ways of behavior*" through which love and fear, in part, come to be known as love and fear.[200]

In *Art as Experience*, Dewey elaborated that varieties of emotion are not any more subjective than other things we perceive. His ideas about perception, particularly of art, reinforce this view. Anticipating Merleau-Ponty, Gibson, and enactivists, Dewey urged that subject and object build up in the same continuing operation.[201] Thus, fingers caressing rough-cut wood realize coarseness as a perceptual effect, with the activity simultaneously actualizing potentialities of the hand. Dewey's approach to art was similar: it is not a painting or photograph "as a *picture* that causes certain aesthetic effects 'in us.' The picture is *itself* a total *effect* brought about by the interaction of external and organic causes."[202] This indicates that the picture only exists for entities with capacities to see it as such, and "its 'beauty,' which ... in being an intrinsic part of the total effect, belongs to the picture just as much as do the rest of its properties."[203] An art critic, recognizing emotionality in aesthetic characteristics, might look at an Ansell Adams photograph and describe a serene mountainscape and turbulent clouds; and, under Dewey's rationale, might say that these emotional attributes, as aspects of the total effect, are as real as the picture. People encountering serene mountains and turbulent clouds in person might conclude the same.[204]

As with experience more generally, then, Dewey pushed emotions outside the organism, offering what is in effect an extended account in the 4E sense that also encompasses enactive, embodied, and embedded understandings. But he did this while not denying emotions are something we feel, even if regarding private experience as somewhat incidental.[205] We will return to Dewey's treatment of emotion near the end of the book, connecting it and James's account of selective interests to Gibson's theory of affordances, while endeavoring to integrate all of them within a biological framework.

LOOKING FORWARD, LOOKING BACK

We have offered a historical account in this chapter to clarify pragmatic ideas important to bodily and affective accounts of mind. Readers versed in 4E traditions and especially enactivism will already recognize this. Those familiar with embodied phenomenology, Gibsonian psychology, and Gestalt traditions will likewise appreciate resonances between these schools and pragmatism. Before going on to examine pragmatism in the context of contemporary cognitive and neurobiological sciences, we want to look back on history once more, but with the intention of progressing forward. In doing so, we follow pragmatists who looked back to Kant and further to the ancient Greeks to integrate older views with the sciences, psychologies, and philosophies of their day.

One point we have not discussed regarding the relation between pragmatism and the past, which in turn connects to the contemporary scene, is its appropriation of the Kantian insight about the interpretive nature of acts through which we encounter the world. In the case of Kant, the insight was grounded in his inversion of the way Western philosophers had conceived of the mind's relation to the world. As he asserted, a common standpoint up until his day was that we know the world because our cognition conforms to it, mirroring its structure.[206] Citing problems with this view, Kant pursued an alternative, "namely that we can cognize of things *a priori* only what we ourselves have put into them."[207] In other words, we can only grasp what is brought into accordance with the structure of cognition, as determined ahead of time by what logically must be so for beings inhabiting space and time as we do. Kant described his approach as analogous to that of Copernicus, "who, when he did not make good progress in the explanation of the celestial motions when he assumed that the entire celestial host revolves around the observer [i.e., the earth], tried to see if he might not have greater success if he made the [earth] revolve and left [the sun and other] stars at rest."[208] In this sort of case, cognition is not made to conform to independent celestial patterns, but rather the reverse. The thought that the sun is at rest relative to other objects in the theoretical scheme literally changes how we must picture the paths of planetary objects if we are to picture them coherently at all. Our cognition thereby pulls objects into an arrangement, makes them appear in conformity with it.

Kant proposed that something similar occurs on a more universal level, arguing that *a priori* limits—in effect, logic—constrains interpretive encounters with the world. More than this, he asserted that people have knowledge and coherent experience only insofar as the world is brought into conformity with these limits.[209] Kant accordingly maintained that the mind actively constitutes knowledge and experience. James advanced an analogous thesis, as we have seen, albeit substituting traditional logical mechanisms for affective ones. Dewey did this too. However, in company with Mead, he went further, introducing organism-environment transactional frameworks as additional substitutes. Again, the hand and things encountered limit possibilities of action and, hence, experiences. Habits and affairs handled have the same limiting outcome.

That Kant and pragmatists shared this commonality led them to adopt comparable, though by no means identical, approaches to metaphysics—metaphysics being a field directed at the conditions under which anything can be said to have "reality." Drawing insight from the emerging experimental sciences, Kant insisted that registering reality means actively manipulating it.[210] This indicated that mind not only acts to impose form on reality, thereby rearranging it, but that it must act so in order to coherently grasp anything at all. The mind manages this by working within logical constraints. In other words, it operates according to *a priori* limits that prescribe how reality—the phenomenal world—is configured, put together, and synthesized, thus dictating how reality shows up in perceiving, experiencing, and thinking.[211] This synthesis or joining together is an interpretive act in which things unite with others according to *a priori* or logical conceptual forms. Moreover, acts by which things are so united are, in effect, acts of judgment—that is, processes in which certain affirmations are made about things. Kant contended, accordingly, that the human experience of reality is necessarily judgmental: any consciously registered reality is always already constituted through interpretive and hence judgmental acts of mind. If this is so of perception, then it follows that perception is cognitive. This is one of the less controversial claims defended in this book, and pragmatic positions affirm it.

The thesis was in fact tacitly introduced when we described the hand's activities coordinating around shape and smoothness, integrating into a kinesthetic experience of a bottle; and when we further suggested the hand as a kind of bias or perspective thrust onto the world, albeit in such a way that it adjusts to accommodate things encountered, just as most of us hopefully do with initial biases we assign onto other people. In other words, we suggested the hand as an interpretive framework—a cognitive means by which we perceive the world. James's account of interests and affect (developed further by Dewey and in many ways advanced by Gestalt theorists, mirrored by Gibson, and reinforced by recent experimental work) is similar and will be thoroughly explored in chapters 4 and 5. Interests, emotions, and the like create bias by directing attention, and thus providing an interpretive and synthesizing framework through which we come to experience the world, and without which, as James compellingly argued, encounters would be chaos. This suggests, furthermore, that affective engagements are interpretive and cognitive, and even conventional psychological and neurobiological language speaks of them as appraisals. However, affect is also cognitive insofar as it is overwhelmingly action oriented, and actions, again, occur within the interpretive frameworks of body-environment transactions. Since we will also maintain that cognition is rooted in motor activity, it follows—or so we will argue—that it is perceptive. Insofar as motor systems neurologically integrate with affective ones, we will argue also that cognition is emotional—a point again substantiated by a host of biological evidence to be discussed, as well as conceptual arguments, again taken up in later chapters. The reader who was not already convinced of these positions before beginning will likely not be persuaded at this early point, and understandably so. Nonetheless, in this first historical section, we have already sketched a preliminary defense for a set of theses outlined in the introduction to this book: that cognition is affective or valuative and vice versa, and that perception is cognitive and emotional and vice versa. All are embodied.

2

PRAGMATISM AND EMBODIED COGNITIVE SCIENCE

Following James's and Dewey's suggestion that progress occurs by wedding old and new, the rest of our book merges classical pragmatism with contemporary neurobiological and cognitive sciences and with phenomenology and consonant movements, using each to develop the others further. As already detailed, pragmatists circumnavigated debates between empiricists and *a priorists* by showing that active bodies perform integrative operations traditionally attributed to "inner" mechanisms. Heidegger and especially Merleau-Ponty did the same (the latter coming extremely close to Dewey's ideas),[1] though these continental thinkers were unversed in the philosophy of their American counterparts and vice versa. This is to say: all held mind emerges from body-environment transactions. James emphasized affective, quasi-visceral, action-oriented integrative mechanisms. Dewey, Mead, Merleau-Ponty, and Heidegger furthered this by offering more thoroughly integrated bodily views. This clearly anticipates a range of 4E approaches, and phenomenological ideas have long been acknowledged in that quarter.[2] Pragmatic contributions are less appreciated, though increasingly recognized.[3]

Pragmatists most squarely anticipated enactivism of all the 4E movements, particularly the later variants of O'Regan, Noë, Hutto, and Myin. Enactivism was also anticipated by Merleau-Ponty and to some extent Heidegger, more obviously so since the movement's founders drew openly on them. Together, these classic figures also foresaw an idea adopted by some AI and robotics researchers: that human-like intelligence requires human-like bodies in addition to CPUs capable of brain-like functions.[4] This provides correctives to older mainstream views. Many might assume, for example, that monarch butterflies—famous for their multigenerational migrations—possess complex internal cognitive maps. However, the butterflies have pinhead-sized brains, and much of their behavior is explained by interactions of their bodies with environments, such as mountain ranges funneling them towards their destination.[5] To be sure, they could not do this without a nervous system.[6] But, as with other organisms and indeed recent robotic constructions, a great deal of processing occurs through the exploitation of organism-environment dynamics, with only sparse control exerted by the brain or CPU equivalent.[7]

Alva Noë, Louise Barrett, and others have therefore suggested that we err when locating the seat of human subjectivity inside our heads. This is why Noë titled one of his books "Out of Our Heads."[8] Indeed, something comparable to monarch butterflies holds for human beings: surgeons, elite dancers, and other experts report minimal reflection when performing, their activities depending on practiced habits.[9] Here, complicated nervous activity in the brain and elsewhere is involved, without which the incorporation and deployment of practiced habits could not occur.[10] At the same time, much of the activity underlying perception and cognition occurs outside the brain. It depends on the body, things encountered, and tools used—for instance, the way different tissues resist the scalpel's edge, or the contours over which a performer dances, as in Gene Kelly's famous (1952) *Singin' in the Rain* scene. Thus here, too, in the words of the founders of enactivism—Varela, Thompson, and Rosch—cognition is not the representation of a pregiven world . . . but rather the enactment of a world and a mind on the basis of a history of the variety of actions that a being in the world performs."[11]

Noë's aforementioned book, whose subtitle promises to demonstrate "Why You Are Not Your Brain," goes on to state: "Consciousness does not happen in the brain."[12] Though we agree with the sentiment, we worry that such rhetoric—while striking in its extremity—distracts from the importance of neurobiology. We therefore frame our story in more moderate terms, or, one might say, pragmatic ones. We take neuroscience seriously but avoid overselling it, and we highlight extraneural aspects of perception and cognition. Likewise, while we think a great deal can be explained without recourse to the language of representation and computation, we do not see a need to completely eschew such concepts, nor in fact does Noë.[13]

A point we especially pursue in this chapter is one emphasized by Dewey and repeated by Mead and Merleau-Ponty: that the body is not a collection of adjacent organs, but rather a synergistic system that falls into synchrony by coordinating around worldly contours. From these pragmatic and phenomenological standpoints, which resonate with 4E views, this brings about sensorimotor organization that constitutes perception—as when fingers (which could in principle move this way or that) cohere around explorations of a bottle, integrating movement and sensation into perception. As recent 4E researchers and especially enactivists have done, Dewey and Mead developed this into a general theory of perception and cognition.

In this chapter, we map similarities between classic pragmatic ideas and 4E positions, focusing on enactivism. We also attend to kinships between pragmatism and embodied phenomenology. Importantly, we discuss how these earlier schools can add to recent embodied views—for instance, with a richly developed concept of habit as "organic memory" helping to explain intermodal perception, or how emphasizing continuities with the past highlights intellectual resources largely unexploited by 4E researchers. In the spirit of James, Dewey, and Mead, moreover, who endeavored to take the brain into account, we begin by elaborating on neuromechanisms in bodily coordination—something that Andy Clark, Richard Menary, and Evan Thompson consider in limited detail,[14] but that enactivists such as Noë, Hutto, and Myin tend to pass over with too little regard.

BRAIN AS SIMULATOR AND ALTERNATIVE VIEWS

Ideas about brain and mind have varied according to the technological zeitgeists of their times. John Daugman eloquently summarizes:

> The water technology of antiquity (fountains, pumps, water clocks) underlies ... the Greek pneumatic concept of soul (*pneuma*) and the Roman physician Galen's theory of the four humours; the clock work mechanisms proliferating during the Enlightenment are ticking with seminal influence inside La Mettrie's *L'Homme Machine* (1784); Victorian pressurized steam engines and hydraulic machines are churning

underneath Freud's hydraulic construction of the unconscious and its libidinal economy; the arrival of the telegraph provided Helmholtz with his basic neural metaphor, as did the relay circuits and solenoids for Hebb's theory of memory; and so on.[15]

A dominant metaphor since the mid-twentieth century has, of course, been the computer, though it is under pressure from embodied quarters. Interestingly, and in spite of the technology's newness, basic mind-as-computer models—especially popularized versions—re-express what has been agreed upon by figures who otherwise stand on different sides of the philosophical spectrum: that the mind or brain generates or receives representations of the world. Since representations are symbolic stand-ins for reality that leave us removed from it, this position has fueled skepticism, even in cases where thinkers have attempted to challenge it, as Descartes did.

The writings of cognitive scientist Donald Hoffman offers a recent example of the metaphor and its accompanying skepticism.[16] He compares perception to Microsoft Windows and comparable computer operating system interfaces that leave the user unaware of their underlying electronic processes. Such programs guide operations of the mouse and keyboard, allowing us to trigger complicated cascades inside the computer that remain hidden. Indeed, Hoffman stresses that reports about causal chains of electric impulses, magnetic fields, and other such processes would be overwhelming and irrelevant to tasks, and consequently a paralyzing hindrance. Hoffman reasons that the same holds for perception: that our experience is not of what is really going on, but rather a representational framework that allows us to handle tasks. In Hoffman's words, "perceived space and time are simply the desktop of the perceptual interface of *Homo sapiens*. Objects, with their colors, shapes, textures, and motions, are simply the icons of our space-time desktop."[17]

Hoffman's reasoning as to how we ended up this way mirrors his account of computer interfaces: he argues that perceptual systems cost organisms valuable calories and that natural selection weeds out unnecessary energy consumption. As with computer interfaces, Hoffman maintains that veridical percepts of the environment are not only expensive in terms of energy and time; they are also irrelevant because the capacity to get things done— "utility, not truth"—is what matters.[18] Summing up, Hoffman explains that the reason we rarely jump off cliffs or in front of cars is the same as that for why we avoid carelessly dragging files to the trash. Though "the shape and color of the file icon do not resemble anything about the true file," dragging "the icon to the trash" could cost "many hours of work."[19] Consequently, we take icons seriously but not literally. The same holds for perception. We take it seriously despite its lack of veridicality because not doing so is costly—at least this is what Hoffman contends.

The thinking of Richard Dawkins provides another example of mind as computer as well as of the skepticism-inducing scientism that Dewey was trying to escape. While something of a caricature (even though Dawkins is in earnest), it offers an illustrative case precisely because it is a caricature of mind as representing machine that typifies modern-era views that persist today. Dawkins writes in a 2006 work intended to cast doubt on everyday perception and cognition: "The human brain runs first-class simulation software" so that "eyes don't present to . . . brains a faithful photograph of what is out there."[20] Instead, "brains construct a continuously updated model: updated by coded pulses chattering along the optic nerve."[21] Dawkins repeatedly emphasizes a digital, pulsing on-off picture that characterizes computer coding, and to some extent captures the fact that action potentials either fire or not, thereby signaling through axon-dendrite, neuronal synapses. However, brains are not desktop computers and neurons do not always behave in this on-off manner. In some cases, they communicate directly with other neurons, which is to say, without axons, and hence without action potentials.[22]

As with Hoffman's desktop theory, Dawkins's mind-as-simulator account asserts that the brain simulates experiences of what is not actually occurring in the world. For instance, Dawkins mentions separate harmonics that he claims the brain synthesizes into the brassy or reedy timbres of trumpets or clarinets. His story about cognition follows suit. "Each of us builds, inside our head, a model of the world in which we find ourselves,"[23] he says, adding that we are constrained by our evolutionary past. Consequently, the challenge is to move beyond "the minimal model of the world" that "our ancestors needed in order to survive."[24] Because brains are "onboard computers, evolved to help us survive in a world . . . where the objects that mattered to our survival were neither very large nor very small" and "where things either stood still or moved slowly compared with the speed of light,"[25] we struggle to cognize vast cosmic distances or quantum peculiarities, to picture atoms or electron zones, and so on.

Though insisting that everyday perceptions and thoughts are erroneous, Dawkins nonetheless insists that something approaching accurate representational knowledge is possible. Here, too, he invokes the concept of "simulation software." Only in this instance he argues that brains can construct tolerably good models of the universe because scientific methodology, technological instruments, and mathematics widen the slit of what he calls our "mental burka"—for example, allowing us to detect and understand portions of the electromagnetic spectrum outside the visual range.[26] Human brains, according to this reasoning, "turn out to be powerful enough to accommodate a much richer world model than the mediocre utilitarian one that our ancestors needed to survive."[27] We can thereby come to grasp the universe through what he calls "a model-building enterprise,"[28] and in this way appreciate that we are vastly mistaken in our everyday experience of the world, or so Dawkins argues.

Again, Dawkins's account of mind is somewhat superficial and does not address the nuances and differences between the ideas of, say, Jerry Fodor, Hilary Putman, and Fred Dretske, nor does it approach their depth or sophistication. Nonetheless, it reflects academic positions that have gained traction, and it is also popularly circulated by movies such as *The Matrix* (1999) or by psychology and philosophy professors who present the mind or brain as world-representing machines. And, despite being framed in the language of neuro- and computer science, the account offered by Dawkins, like that of Hoffman's, parallels early modern outlooks.

Dawkins and Hoffman's views preserve early modern outlooks most obviously by emphasizing an inner-outer divide. Though many take this divide for granted today, it has not always been so readily accepted. Hubert Dreyfus, for example, observes that the Homeric Greeks regarded inner experience as exceptional,[29] as evidenced by Homer considering "it one of Odysseus's cleverest tricks that he could cry inwardly while his eyes remained [dry]." Dreyfus further notes that there are few if any other references to inner experience in Homer's works. Along these lines, Dewey claimed that "inner experience" was a modern-era discovery,[30] whereas to the ancient Greeks, "experience was the outcome of accumulation of practical acts, sufferings and perception gradually built up into the skill of the carpenter, shoemaker, pilot, farmer, general, and politician."[31] Having experience meant being experienced: "There was nothing merely personal or subjective about it."[32] This was the model that Dewey adopted, anticipating enactivists such as O'Regan and Noë who, in their landmark 2001 article, proposed that perception is *"the activity of exploring the environment in ways mediated by knowledge of the relevant sensorimotor contingencies."*[33] For them, "seeing is a way of acting" that "rests on know-how, the possession of skills."[34] According to this position, which is exactly Dewey's, a great deal of perception occurs—borrowing once more from the title of Noë's 2009 book—"out of our heads."

A related way in which Hoffman and Dawkins's stances resemble early modern views is their suggestion that representations mediate mental contact with the world. By definition, representations are not things themselves, but appearances or semblances of them. Thus, while materialistic and therefore antagonistic towards mind-body dualism, outlooks in the vein of Hoffman and Dawkins imply a kind of "epistemological dualism" that is in line with psychology from the early modern period onwards. The framing of Dawkins and Hoffman's examples, despite occasional claims to the contrary, also assumes that there is more or less only one reality; and, by extension, that there is only one truth on any given matter—something that even Descartes and Hume agreed upon despite differences elsewhere. Both accounts consequently assume that a world is really as it is, and that our representations of it, by virtue of being representations, only give us the world as it appears.

However, as discussed last chapter—and this connects to pragmatic theories of mind and will have relevance through this book—most properties and even so-called primary ones such as length and mass are not determinable outside of interrelations, and worldly interaction affects what is observed. Even scientific facts are products of human actions, as when experiments make light manifest as either a wave or a particle depending on the setup used. It is, as the late physicist David Finkelstein put it polemically, as if we are in an age of "non-objective physics."[35] According to Finkelstein, Werner Heisenberg pioneered "quantum theory in the same city and decade in which Kandinsky coined the phrase 'non-objective art,'" and one may speculate that he "borrowed from Kandinsky when he called quantum theory 'non-objective physics.'" Whereas "classical physics . . . represses the observer and the *act* of observation and talks naively about 'things as they are,'" Finkelstein argues that a "main idea of quantum theory is to talk about *what you do*, not about 'things as they are.'"[36] As discussed, this applies elsewhere and outside of science proper. Thus, brushing hands over polished wood realizes smoothness. Rapping knuckles on it brings out hardness, though caressing also does so to some extent. As we will begin to show later in this chapter, what applies to kinesthetic-tactile experience holds generally for perception and indeed cognition.

Finkelstein's account repeats the experimentalist ideas of pragmatists, especially as expressed by James and Dewey,[37] both of whom saw a deep resonance between their theories of mind and what was going on in the sciences of their day. What classical pragmatists added—and what Finkelstein doubtless believed, despite his use of "non-objective"—is that the described state of affairs does not make results arbitrary or unreal. Certain experiments unexceptionally make light manifest as waves, others as particles, so there is nothing haphazard here. And while there are probabilistic aspects to quantum mechanics, just as in everyday life, these probabilities can be given definite expression. That encountered realities are products of actions and effects actualized in the world does not (and this is critical) undermine veracity. As Dewey would say, echoing Peirce's pragmatic maxim: reality is that of things bumping into each other, of alterations introduced to states of affairs. Dewey further insisted that changing reality does not, as realists have charged, pervert it. It simply introduces changes that were not previously there, and it is how the universe shows itself to us.[38] These scientific views, as already argued, enter into pragmatic accounts of mind and resonate with recent 4E interpretations, especially enactive renderings.

Pragmatic theories of mind, influenced by the foregoing ideas from physics, therefore suggest that human beings are akin to "experimental setups." Although the "detectors" and "physical architecture" of the specific setup differs from that, say, of Clinton Davisson and Lester Germer's 1927 double slit experiment, human beings, in their everyday perceptive capacities, bring worlds into light by the actions and effects they undergo. Dewey made just this comparison in his *Essays on Experimental Logic* with regards to cognition, but it applies equally to perception.[39] Thinking, he wrote,

> is not an event going on exclusively within the cortex or the cortex and vocal organs. It involves the explorations by which relevant data are procured and the physical analyses by which they are refined and made precise; it comprises the readings by which information is got hold of, the words which are experimented with, and the calculations by which

the significance of entertained conceptions or hypotheses is elaborated. Hands and feet, apparatus and appliances of all kinds are as much a part of it as changes in the brain. Since these physical operations (including the cerebral events) and equipments are a part of thinking, thinking is mental, not because of a peculiar stuff which enters into it or of peculiar non-natural activities which constitute it, but because of what physical acts and appliances do the distinctive purpose for which they are employed and the distinctive results which they accomplish.[40]

Applying this view to perception, which Dewey did repeatedly, leads to basic conclusions that Clark and Noë arrived at roughly a hundred years later. Clark opens his 2008 *Supersizing the Mind* with portions of the aforequoted passage (albeit with no further mention of Dewey). Although Noë makes no reference to pragmatism in his early landmark statements on enactivism,[41] he exactly repeats Dewey's thesis when he argues that "perception is not a process in the brain,"[42] but rather "a mode of skillful"—or one could say, experimental—"exploration of the world."[43]

To say that perception does not occur in the brain is not to claim the brain is uninvolved. Dewey did emphasize its importance, in fact, more than Noë, and in the vein of Clark and Thompson.[44] Nonetheless, Dewey anticipated what he termed brain-body dualism, writing over a century ago: "The advance of physiology and the psychology associated with it have shown the connection of mental activity with that of the nervous system. Too often recognition of connection has stopped short at this point; the older dualism of soul and body has been replaced by that of the brain and the rest of the body."[45] In effect, Dewey anticipated what Tibor Solymosi has called "neuro-hype" and "neurophiles."[46] Rather than diminishing the brain, Dewey was advancing the view, as Merleau-Ponty put it some years later, that the "body is not a collection of adjacent organs, but a synergic system, all the functions of which are exercised and linked together in the general action of being in the world."[47] Like Merleau-Ponty, Dewey held that we only perceive and cognize through bodily sensitivities and capacities coordinating into joint action. This includes real-time coordinations and also past ones, the latter bestowing habits that structure our interactions with things and hence our perception of them. Critically, capacities and sensitivities coordinate not on their own, but by synchronizing around environmental contours—points taken up in detail later. Dewey, to be sure, granted that the nervous system is essential, but added that it

> is only a specialized mechanism for keeping all bodily activities working together. Instead of being isolated from them, as an organ of knowing from organs of motor response, it is the organ by which they interact responsively with one another. The brain is essentially an organ for effecting the reciprocal adjustment to each other of the stimuli received from the environment and responses directed upon it.[48]

Crucially, moreover, "adjusting is reciprocal; the brain not only enables organic activity . . . to bear upon any object of the environment in response to a sensory stimulation, but this response also determines what the next stimulus will be."[49]

As O'Regan and Noë have observed, "seeing involves testing the changes that occur through eye, body, and attention movements,"[50] and the brain plays a role in these adjustments and in determining the stimuli received. Walking is similar, and it is perceptive. When striding and pressing feet into a trail, our bodies undergo reciprocal adjustments: our gait, posture, and gaze modify to the sandy softness, rocky hardness, or the knarred unevenness of roots, which show up in patterns of activity that integrate into our kinesthetic-visual experience. It is a truism to say that the brain plays a role. As physician and physiologist Kaoru Takakusaki explains, somatosensory, visual, and vestibular brain areas activate along with a host of postural reflexes that range from head-eye coordination to the accompanying alignment of body structures.[51] Descending pathways from the brain stem to the spinal cord mediate all of this, with spinal locomotor structures and the reticulospinal tract in the brain stem contributing too. Basal ganglia and the cerebellum also enable automatic adjustments. Less familiar circumstances involving more overt cognitive processes see heavier activation of the temporoparietal cortex along with the involvement of motor cortices in anticipatory adjustments.

Dewey was not alone in emphasizing the brain, since James and numerous others from his period also did so. From today's standpoint, such emphasis is commonplace but oversold in its explanatory power,[52] especially because neuroscience is in its relative infancy as a field. Even Dewey's emphasis on motor mechanisms, as discussed last chapter, was emblematic of his time, albeit more developed than that of most of his contemporaries. A few passages from Calderwood, Jackson, Galton, Titchener, and Washburn—some of whom pragmatists were aware of—are worth examining. In 1909, Titchener proposed: "Meaning is, originally, kinaesthesis; the organism faces the situation by some bodily attitude."[53] Shortly after, Washburn, in her 1916 *Movement and Mental Imagery: Outlines of a Motor Theory of the Complexer Mental Processes*, hypothesized that "the whole of the inner life is correlated with and dependent upon bodily movement."[54] Over two decades earlier, in 1883, Galton described an "incipient motor sense, not of the eyeballs only but of the muscles generally," which he related to mental imagery.[55] Calderwood's work, while not going so far as to advance a sensorimotor account of mind, was also rife with such terminology. Calderwood was an ordained pastor and philosopher interested in explaining human psychology and physiology in the context of evolution. He spoke incessantly about sensorimotor coordinations in his 1879 *The Relations of Mind and Brain*, citing numerous contemporaries, including Jackson, who did likewise. He also displayed remarkable knowledge of neuroanatomy, using much of the nomenclature that we deploy today and, indeed, attributing functions that remain generally accepted. Around the same time, Jackson—a neurologist concerned with the evolutionary organization of the nervous system, among much else—wrote: "The whole nervous system is a sensori-motor mechanism, a co-ordinating system from top to bottom. . . . The highest centres are only exceedingly complex, and special sensori-motor nervous arrangements representing, or co-ordinating in ways we shortly go on to show, the whole organism."[56] Jackson's emphasis on "the whole organism" coordinating, while inclined more towards hierarchical organization and the concept of representation than Dewey, Mead, Merleau-Ponty, Thompson, and Noë, nonetheless resonates with all these thinkers' views. It connects to avenues that Dewey, Mead, and Merleau-Ponty open for understanding intermodal perception, which are points we turn to shortly.

Pragmatic views and those of Merleau-Ponty (who worked during Dewey's late period and began his career as Mead's life was ending) resonate with neuroscientific developments. This is especially in light of research that indicates deep neurological connections between motor function and cognition and suggests the same between motor function and perception. Broca's area classically associates with language capacities. However, in different neural coalitions, it also turns out to be involved in movement preparation, action recognition, and movement imitation, as well as music perception.[57] Dewey would have appreciated this last twist, given his motor account of aesthetics. Broca's area is also interinnervated with the basal ganglia. The basal ganglia, in turn, are associated with movement and habit organization and cognitive functions, including language production, the syntax of regular verbs, the prediction of probabilistic events, reward appraisals, and the emotional weighing of options.[58] Neurochemistry suggests similar connections. Dopamine pathways, for example, underlie functions ranging from probabilistic reasoning to incorporation of motor patterns, and Parkinson's patients, who have depleted levels of dopamine, suffer problems with ordinary syntax and movement.[59] Dopamine neurons are related to learning when encountering novel happenings: dopamine neurons in the frontal cortex and striatum nuclei in the basal ganglia become active in anticipation of such events. Dopamine and these structures do not activate separately, but rather in coalition with other transmitters and regions in the central and peripheral nervous system as the extraneural body operates in its surroundings. Taken together, and considering other examples, this indicates neurobiological support for the thesis that perception, cognition, and action are codependent.

Other studies testify to connections between action and perception. One that Varela, Thompson, and Rosch cite[60]—albeit overstating results—was conducted by Richard Held and Alan Hein.[61] The experiment kept kittens mostly in the dark, with pairs only briefly exposed to light each day. One in each pair was harnessed to a mechanical arm attached to a central hub that allowed it to move in circles. The arm extended to the opposite side, where it had a box confining a second kitten that moved in the same pattern as the first, hence undergoing similar visual stimuli, but without autonomous exploration. Later tests suggested that kittens in the second condition developed impaired vision. Varela, Thompson, and Rosch have made much of this, declaring somewhat inaccurately that kittens in the second condition "behaved as if they were blind,"[62] citing this as an affirmation of their view that perceptual capacities emerge through sensorimotor exploration. Jesse Prinz, however, has pointed out that the kittens were in fact able to negotiate obstacles, albeit clumsily.[63] He accordingly argues that the study does not show the kittens had impaired vision, but rather that their coordination suffered from lack of practice.

Other studies can be added to the story, specifically ones on tactile-vision and echolocation-vision substitution devices, and this is an additional example enactivists favor.[64] While again leaned on overmuch, it does support their position and, by extension, that of their pragmatic and phenomenological forerunners, especially when considered in light of available neuroscience. Paul Bach-y-Rita pioneered tactile variants of these devices,[65] which used head-mounted cameras to deliver stimulation to participants via electrical current on their tongues or vibrations on their backs. For these devices to function effectively, participants must actively explore their environments; given this, participants acquire an analogue to vision in a relatively short time. Moreover, just as we do not experience the visual world as being on our retinas, but rather as before our eyes, participants experience objects in front of them, not on their tongues or backs where the stimulation is received. They can make discriminations, perceive pathways, numbers of objects, and the like. This suggests that action is preeminent in perception. More than this, it is not specific sense organs but rather sensorimotor coordinations—in this case, ones that combine to negotiate space—that determine whether or not perception is vision-like. This is a key enactivist thesis, and one that Dewey, Mead, and Merleau-Ponty all would have appreciated.

Noë, who emphasizes this research in his 2004 *Action in Perception*, was strangely insistent at that time that "there is no involvement of . . . the visual cortex,"[66] even while immediately citing and discussing preliminary evidence to the contrary in an endnote. Evidence has since affirmed that the visual cortex is active in at least early blind participants (participants who became blind early in life),[67] as, in fact, did earlier work on echolocation-vision sensory substitution devices.[68] But whereas visual areas are activated in early blind participants, this does not seem to occur in seeing participants.

There are a number of ways in which sensory substitution devices and their activation of visual brain areas strengthen the enactivist case, especially as originally articulated by Varela, Thompson, and Rosch.[69] Notice, to begin with, that regardless of whether visual areas activate, participants learn an analogue of sight by actively exploring with these devices. This affirms that the nature of the sensorimotor coordination determines the perceptual experience. Moreover, that visual brain areas in blind participants become active fits the very point that Varela, Thompson, and Rosch invoke the kitten study to reinforce.[70] The explanation as to why involves a number of steps. First, it appears that brain regions traditionally associated with vision are—like the eyes—especially appointed to handle spatial information, albeit with significant multifunctionality.[71] Thus, occipital areas mobilize not only when we see objects, but also when people—both blind and sighted—touch and handle them,[72] with the activity providing spatial information about shape, size, and position. Second, differences in brain activation between blind and seeing participants using sensory substitution devices suggest that perceptual and related neurobiological capacities develop through sensorimotor training. It seems, in short, that the early blind have greater practiced habits of negotiating space by means of touch and sound. Their visual cortices accordingly develop to interpret spatial aspects registered through nonvisual sensory modalities.[73] Visual regions in the brain consequently activate when people negotiate a space using sensory substitution devices.

Bach-y-Rita has in fact suggested that such neurologic activity occurs with sensory substitution devices. He postulates similarities between the employment of these apparatuses and braille reading,[74] citing studies indicating that the latter mobilizes occipital regions (parts of the visual cortex).[75] Some evidence suggests that reorganization can occur rapidly. Researchers have found occipital regions activating in braille reading (also a task involving spatial patterns) in blindfolded participants after about five days. Occipital regions likewise activate in blind participants doing the same task.[76] Further, it may be that such activation follows from the absence of the normal burden of handling conventionally visual information, since these regions become less active during braille reading when blindfolds are removed.[77] The take-home point, at any rate, is that perception is sensorimotor coordination that takes some practice, and sensorimotor practices exert organizing influences over brain and bodily function (which we, of course, do not wish to separate).

While classical pragmatists did not have the resources to investigate the functioning of living brains, these sorts of outcomes are consistent with their motor theories of cognition, particularly those of Dewey and Mead. In his 1920 *Reconstruction in Philosophy*, for example, Dewey wrote:

> The organism acts in accordance with its own structure, simple or complex, upon its surroundings. As a consequence the changes produced in the

environment react upon the organism and its activities. The living creature undergoes, suffers, the consequences of its own behavior. This close connection between doing and suffering or undergoing forms what we call experience. . . . [S]uppose a busy infant puts his finger in the fire; the doing is random, aimless, without intention or reflection. But something happens in consequence. The child undergoes heat, he suffers pain. The doing and undergoing, the reaching and the burn, are connected. One comes to suggest and mean the other. Then there is experience in a vital and significant sense.[78]

Aside from offering a good phenomenology and not making the empiricist error of assuming that parts analyzed out of wholes are building blocks of experiential situations, this orientation was in line with findings that were to come. It suggested that motor control is embedded in perceptual systems and vice versa, and it hinted the same for cognition—a position Dewey explicitly developed elsewhere.

Dewey, in particular, criticized the reflex arc concept inherited from Descartes, which advocates the still widely accepted idea that stimuli come first, followed by mental activity, then responses. Dewey's problem with this was that stimuli only show up as such in the contexts of overall motor responses, which, as a coordination of motor and sensory activity, form perception and also the basis of cognition. Dewey emphasized, moreover, that there are myriad stimuli in any given instance; our perception of the stimuli and, hence, the ones we notice, are determined by what we do, meaning that stimuli only manifest emphatically after responses. In his seminal 1896 "The Reflex Arc Concept in Psychology," Dewey put it thus: "The real beginning is with the act of seeing; it is looking, and not a sensation of light."[79] He expanded:

Upon analysis, we find that we begin not with a sensory stimulus, but with a sensori-motor co-ordination, the optical-ocular, and that in a certain sense it is the movement which is primary, and the sensation which is secondary, the movement of body, head and eye muscles determining the quality of what is experienced. . . .

Now if this act, the seeing stimulates another act, the reaching, it is because both of these acts fall within a larger co-ordination; because seeing and grasping have been so often bound together to reinforce each other, to help each other out, that each may be considered practically a subordinate member of a bigger co-ordination.[80]

This article was voted by leading American psychologists the most important in the first fifty years of the *Psychological Review*.[81] This—like the article—seems a prescient stroke, given that the piece exactly anticipates ideas that enactivists would advance roughly a hundred years after its publication. It also seems that the insight was rapidly lost to most working within the field of psychology, who largely continued with standard stimulus-response models.

ELABORATING CLASSIC AND CONTEMPORARY EMBODIED VIEWS

We have seen in pragmatic accounts—particularly the embodied ones of Dewey and also Mead—that perception is an outward pushing against and accommodation to what is received. Merleau-Ponty likewise defended this view, offering beautifully elaborated descriptions attesting to it. Like pragmatists, not to mention the Gestalt psychologists who deeply influenced him, Merleau-Ponty was antagonistic towards British empiricist psychologies. Among other reasons, this was because they theorized the experiential field as built up from base units akin to atoms or pixels, a proposition that also exasperated classical pragmatists. This does not match our experience of things, and the outlook also conflicts with the science of Merleau-Ponty's day, not to mention ours.

As with Dewey, Merleau-Ponty maintained that situations are ontologically and epistemically primary. This is to say, situations are where things first exist and appear to us. Merleau-Ponty connected his views to those of Gestalt psychologists, who similarly maintained that the whole precedes the parts and is not reducible to them. Merleau-Ponty had no shortage of examples drawn from that school. Many of the illustrations that Gestalt psychologists employed indicated that perceived phenomena could not be built up from parts, as in the case of figures with missing segments in the bounding lines that we nevertheless see as triangles, circles, and the like (figure 2.1).[82]

Merleau-Ponty similarly noted that visual perception does not reflect the lack of homogeneity in distributions of rod and different kinds of cone cells,[83] though this should follow if perception is built up from impressions

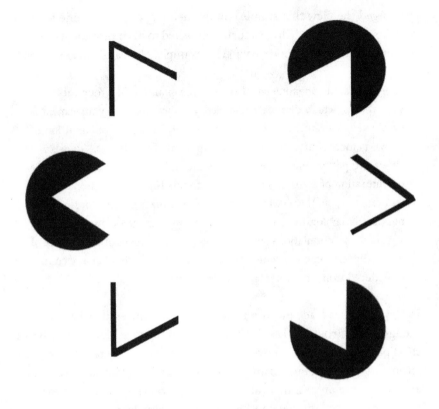

FIGURE 2.1 Gestalt illustration

FIGURE 2.2 Blind spot test

Close your right eye and put your nose to the page between the dot and X, then slowly pull the book away while focusing on the X. At some point, the dot should disappear.

impinging the eye. Nor do we have gaps in vision despite the absence of photoreceptors at the junction of the eye and optic nerve, another fact that should be otherwise if empiricist psychologies hold (figure 2.2).

Working with Gestalt psychology and against hard behaviorism, which in some ways is British empiricism made bodily, Merleau-Ponty further challenged the view that simple conditioned responses combine to form complex movements.[84] This, in turn, connected to his motor theory of perception. Speaking of a wood carving, for example, Merleau-Ponty explained:

> There are tactile phenomena, alleged tactile qualities, like roughness and smoothness, which disappear completely if the exploratory movement is eliminated. Movement and time are not only an objective condition of knowing touch, but a phenomenal component of tactile data. They bring about the patterning of tactile phenomena, just as light shows up the configuration of a visible surface. Smoothness is not a collection of similar pressures, but the way in which a surface utilizes the time occupied by our tactile exploration or modulates the movement of our hand. The style of these modulations particularizes so many modes of appearance of the tactile phenomenon, which are not reducible to each other and cannot be deduced from an elementary tactile sensation.[85]

Dewey and Mead advanced a nearly identical anti-empiricist position, though rarely articulated as richly and attempting to achieve so much in a short passage—as, in the above quotation, simultaneously rejecting empiricism and behaviorism, affirming a bodily variant of Gestalt psychology, and advancing an extraneural account of perception in the context of a more general account that considers nervous physiology.

This outlook, which had significant traction in a variety of quarters in the late nineteenth and early twentieth centuries, is repeated by contemporary enactivists, most of whom are aware of Merleau-Ponty because the founders of the movement drew on him, albeit while virtually ignoring pragmatists. For instance, Erik Myin and Jan Degenaar observe that "tactile feelings of hardness or softness are determined by particular patterns of experiences one has when engaging in such activities as squishing a sponge or pushing a brick wall."[86] Using an example of a bottle, Noë expresses a similar point, essentially repeating Merleau-Ponty's aforequoted illustration of the carving, albeit emphasizing engagement through understanding to an extent not typically seen in the work of Merleau-Ponty or radical enactivists such as Hutto and Myin.[87] "The bottle as a whole is present to you," writes Noë, "not because you now represent the sense of having an internal model of it, but in the sense that you now understand the way in which it structures and controls your

movements, and so your sensory stimulation." Thus, "content of your tactile experience is enacted by your exploratory movements."[88] This is also a
squarely pragmatic point. As Mead put it in a Deweyan vein, "the perception by the hand is also one that involves such movement in the exploratory processes of hand and fingers and the movements of the skin."[89]
Accordingly, "perception has in it . . . all the elements of an act—the
stimulation, the response represented by the attitude, and the ultimate
experience which follows upon the reaction."[90] In other words, perception
is sensorimotor.

Interestingly, views advanced by Dewey and Merleau-Ponty and later by
enactivists mix elements of pre-modern and modern philosophy, though
not all proponents show awareness of this. We have already talked about
this, but elaboration would serve the current discussion. Note, to begin with,
that shaping influences that objects have, say, on the activities of the hand
and therewith experience does not mean that the body and more generally
the human subject yield like clay. The body is constrained in its capacities
and possibilities of action. So far this sounds like rationalism made bodily.
Notice, however, that actions in the world—acquired habits and skills
that Dewey, Merleau-Ponty, and enactivists link to other perceptual modes
such as vision—are rather unlike traditional rational structures, are primary. In pre-modern philosophy, habit was often equated with experience,
and in everyday language we say people are experienced when they are
skilled (possessing an accumulated history of doing). Thus, along empiricist
lines, experience has priority. Dewey indeed included that word—
experience—in numerous book titles, and Merleau-Ponty correspondingly
titled one of his better-known essays "The Primacy of Perception and Its
Philosophical Consequences." On the whole, however, Dewey's views were
a bodily variant of the post-Kantian rationalist variety, just as Merleau-
Ponty's arguably were.[91] As Dewey summed up, this time stressing the a
priorist idea that we actively shape our worlds, "the organism brings with
it through its own structure, native and acquired, forces that play a part in
the interaction. The self acts as well as undergoes, and its undergoings are
not impressions stamped upon an inert wax but depend upon the way the
organism reacts and responds."[92] With minute variations and combinations
of action, and with the particularities of individual subjects and objects
encountered, possible actions are unlimited in number. Yet they are not
unbounded in scope. Fingers cannot spin like drills: subjects enter each

encounter endowed with certain potentialities and consequently cannot enact whatever pattern they wish.

Thus, Dewey rejected the empiricist notion that the subject waits "passive and inert for something to impress itself upon it from without."[93] Other classical pragmatists, Merleau-Ponty, and enactivists have made the same assertion. Noë, for instance, has remarked that "perception cannot be represented in terms of merely passive, and internal, processes" since it is constituted through outward action.[94] At the same time, Deweyan, Merleau-Pontian, and enactivist views challenge the position that subjects "project" perspectives outwards, as if onto an empty screen. To repeat what is becoming a leitmotif, we cannot perceive or even think whatever we want because we cannot do whatever we want. Hence, while the hand projects onto objects by reaching out and pushing into them, it does not make things completely in its own image. In projecting out, it meets the countervailing press of objects and therewith things limiting its actions and adoptable positions. So whereas one can roll bottles between one's palms, the same action and—therefore—same experience is impossible with cinderblocks. In this instance, it is therefore a misnomer to say that what appears in experience is merely subjective, and the same, as will soon be seen, applies elsewhere. It is a misnomer because the possibilities of action delimiting particular experiences are not merely conditioned upon the subject (and much less on internal mental structures alone), but also upon objects encountered, and the experience is largely constituted through actions in the world.

Seen accordingly, perceived qualities are qualities of interactions that occur in the world rather than mere subjective representations of it.[95] Phenomenologically and conceptually, glassy smoothness is a property of a bottle that we handle; yet, in the current context it is inseparable from the hand's movements, save analytically. The property is inseparable because it is, in Dewey's phraseology, a consequence undergone in the course of an interaction—a consequence undergone when fingertips sweep over a surface that does not bite flesh. "We speak of perception *and* its object," Dewey remarked. "But perception and *its* object are built up and completed in one and the same continuing operation."[96]

Here one perhaps wants to plead that the bottle *really is smooth*, and not just when touched—we can see its smoothness, for example. True enough, and this is affirmed by the experimentalism of pragmatists. However, this is still to say it really is smooth within the context of particular interactions; or, to put it another way, the smoothness is an effect of certain manipulations, and the effect is real.[97] This is even so when the bottle meets the eye. Mead, among numerous others, pointed out that seeing involves the movement of eyes, their convergence and focusing of lenses.[98] It also necessitates rapid microsaccadic movements, without which visually perceived objects fade or completely disappear.[99] At the same time, events in the mind and presumably the brain exert subtle control, such as pupil dilation changing when people imagine shapes of different brightness.[100] By virtue of these changes and indeed the mere fact that the lens changes light, seeing entails changing conditions under which things are observed.[101] Moreover, while our brain is a coordinating organ, as Dewey insisted nearly 100 years ago, motor responses are not fully determined by the brain. Among much else, motor activity is also "mediated by the design of muscles and tendons, their degrees of flexibility, their geometric relationships to other muscles and joints, and their prior history of activation."[102]

Critically, therefore, "not just the visual apparatus" becomes active, "but the whole organism," as Dewey went on to say in 1934,[103] echoing Jackson's observations from about fifty years earlier.[104] Though researchers separate "the optical apparatus . . . in anatomical dissection, it never *functions* in isolation," Dewey explained. "It operates in connection with the hand in reaching for things and in exploring their surface, in guiding manipulation of things, in directing locomotion."[105] Enactivists repeat the point: "Seeing involves testing the changes that occur through eye, body, and attention movements"[106] and therefore "activity on the part of the animal as a whole."[107] The sight of a brimming beer cooler invites outstretched arms, grasping, twisting of caps, opening of mouths, tilting back of heads, gulping, and more, all of which characterizes the experience. It is true that we sometimes look without grabbing. Yet it is also true that we spend most of our waking life handling and ambulating, which means coordinating actions around objects and settings; moreover, our eyes participate in most of this. Consequently, Deweyan and enactivist positions, along with those of Merleau-Ponty, suggest that we learn to see in such terms.

All of these positions further insist that visual experiences—like the earlier examples of tactile perception—are constituted through action. This point has in fact already been introduced with the kitten and sensory substitution device examples, and Myin and Degenaar offer further explanation:

> Seeing a scene or an object is, in the sensorimotor approach, comparable to feeling a surface or object, where the experience is of the whole surface or object, despite the fact that momentary tactile stimulation is limited to the fingertips making contact only at particular places. . . .
>
> For example, tactile feelings of hardness or softness are determined by particular patterns of experiences one has when engaging in such activities as squishing a sponge or pushing a brick wall. Similarly, that experiences of seeing differ as a class from experiences of hearing is due, according to the sensorimotor theory, to patterns of sensorimotor contingencies specific to vision and audition, such as that in seeing, but not in hearing, stimulation from a particular source stops when one turns one's head sideways, or closes one's eyes.[108]

Similarly, Noë summarizes:

> Like touch, vision is *active*. . . . You and your eyes move around the scene the way you move your hands around the bottle. As in touch, the content of visual experience is not given all at once. We gain content by looking around just as we gain tactile content by moving our hands. You enact your perceptual content, through the activity of skillful looking.[109]

Comparable are the following remarks by Dewey:

> As we manipulate, we touch and feel, as we look, we see; as we listen, we hear. . . . The eye attends and reports the consequence of what is done. Because of this intimate connection, subsequent doing is cumulative and not a matter of caprice nor yet of routine. In an emphatic artistic-esthetic experience [and therefore a skillfully enacted encounter], the relation is so close that it controls simultaneously both the doing and the perception.[110]

Because of this, Dewey and O'Regan and Noë have all rejected camera snapshot models of vision, arguing instead that perception develops in time with unfolding actions.[111]

Enactivists have maintained that visual perception, along with other modalities, "is constituted by the exercise of a range of sensorimotor skills,"[112] a claim substantiated by the examples of sensory substitution devices discussed earlier. There, it will be recalled, the nature of the sensorimotor coordination rather than the specific sensory modality determined the nature of perception, and for this to happen, participants had to actively explore and thus acquire a degree of practiced skill. Experiments with glasses that invert retinal images reinforce a similar point because, according to O'Regan and Noë, they show, first, that individuals wearing such distorting lenses see relatively normally after five or six days; and, second, that this improvement is limited when head movement is restricted.[113]

This once again highlights the centrality of action and exploration in perception. It further offers some elucidation of why those wearing distorting glasses adjust to see mugs, cars, and the like in normal ways while the same might not occur with reading license plates. O'Regan and Noë explain:

> Reading alphabetic characters involves a subspecies of behavior connected with reading, judging laterality involves another, independent, subspecies of behavior, namely, reaching. An observer adapting to an inverted world will in the course of adaptation only be able to progressively probe subsets of the sensorimotor contingencies that characterize his or her new visual world; and so inconsistencies and contradictions may easily arise between "islands" of visuo-motor behavior.[114]

In addition to this—and commonsensically—enactivists maintain that the organism's history of embodied habits of interacting impact its perceptual encounters. Habits are rarely jettisoned overnight, and specialized ones may be more stubbornly fixed. This reiterates Dewey's suggestion that the ability to perceive is a matter of "training, skill, habit."[115] Dewey also added, in almost Merleau-Pontian and Heideggerian language, that "through habits formed in intercourse with the world, we also in-habit the world. It becomes a home and the home is part of our every experience."[116] Merleau-Ponty likewise wrote that "the human body, with its habits which weave round it a human environment, has running through it a movement towards the world itself" and that "habit is both motor and perceptual."[117]

The emphasis on habits may seem to have echoes of empiricist psychology. This is only to the extent that habits are acquired through past dealings, which is to say, environmental experiences. Pragmatic, phenomenological, and enactivist accounts all depart from empiricism by emphasizing that habits are not merely impressed from without, but learned through active engagement—engagement that in fact begins prenatally.[118] This does not, to be sure, obviate the fact that we sometimes look without overtly acting. But none of this jeopardizes the claim that we learn to perceive through coordinating sensory and motor capacities and that we overwhelmingly perceive in terms of action. As Dewey put it along Gibsonian lines, knowing "what to look for and how to see it is an affair of readiness on the part of motor equipment."[119] Thus, we see hallways as spaces for movements, bottles as things we can handle, regardless of whether or not we exploit these possibilities of action. When we see shape and line, we see "ways in which things act upon one another and upon us; the ways in which, when objects act together, they reinforce and interfere."[120] Evidence attests to this,[121] and we will especially review it in future chapters that focus on Gibson and affective aspects of perception in the context of pragmatism and neurobiology.

The point we want to advance for current purposes is one that the earlier beer-drinking illustration reinforced: that perception is sensorimotor, meaning it occurs through synchronizations of global assemblies of sensitivities and activities. Looking motivates touching and vice versa; looking also leads us to crane our head to hear. Eyes, ears, and other modalities work in concert with motor capacities, as when a siren turns our head, or limbs collaboratively work pedals and a steering wheel to keep a car on the road seen ahead. Counting change, perusing magazines, typing, swaying to music, leaning in for a conversation, crinkling our face and turning to locate a foul odor and then withdrawing—in all these cases, as Dewey observed, "motor and sensory structure form a single apparatus and effect a single function."[122] If an object stimulates only one organ, say, the eye, then "experience is thin and poor." However, "when the tendency to turn the eyes and head is absorbed into a multitude of other impulses and it and they become members of *a single act*," then "perception"—as opposed to "some specialized reaction"—"occurs."[123]

This position resonates with that of Merleau-Ponty and was repeated by Gibson fifty years later, unsurprisingly, since his intellectual lineage leads back to the pragmatists and phenomenologists. Gibson observed, for example, that

> a man on a mountaintop who turns around completely, taking several seconds for the act, keeps his eyes anchored to the dual ambient array for the whole period except for a small part of the time, totaling only a fraction of a second, during which the jumps have occurred. This is what happens in "looking around," and the result is a vivid perception of the whole environment. [. . .]
>
> The head turning and the eye turning are concurrent movements of a single act. The active turning of the head involves the opposite turning of the eyes in much the same way that the contracting of the extensor muscles of a limb involves the relaxing of the flexor muscles. The neck muscles and the eye muscles are innervated at the same time, reciprocally.[124]

In short, wrote Gibson, this perceptive act "is not a reflex to a stimulus but a coordination."[125]

INTERMODAL PERCEPTION

Noë argues that perception—visual and otherwise—is "a mode of activity on the part of the *whole* animal" whereby the organism coordinates with its surroundings.[126] This means that how "things look, smell, sound, or feel (etc.) depends, in complicated but systematic ways, on one's movement."[127] Noë's theory further "proposes not that perceiving is *for* acting, but rather that perceiving is constituted by the exercise of a range of sensorimotor skills."[128] Nonetheless, Noë does not deny that perceiving is for acting, and his position implies this, just not as a primary explanatory mechanism. Dewey, while arguing that perceptual organs are vehicles for action,[129] anticipated Noë in maintaining that perception demands joint action and coordination around environmental contours. This view offers insight into intermodal perception—the notion, for example, that in perceiving color we simultaneously *see* temperatures, textures, and more. Though elaborated on by classic embodied theorists such as Dewey and Merleau-Ponty, this is an avenue of insight largely untapped by recent 4E theorists.

Dewey's *Art as Experience* expanded this line of inquiry to give reasons for the intermodal nature of everyday perception. Dewey wrote:

Nothing is perceived except when different senses work in relation with one another except when the energy of one "center" is communicated to others, and then new modes of motor responses are incited which in turn stir up new sensory activities. Unless these various sensory-motor energies are coordinated with one another there is no perceived scene or object. But equally there is none when—by a condition impossible to fulfill in fact—a single sense alone is operative. If the eye is the organ primarily active, then the color quality is affected by qualities of other senses overtly active in earlier experiences. In this way it is affected with a history [or in other words, habits]; there is an object with a past. And the impulsion of the motor elements which are involved effects an extension into the future, since it gets ready for what is to come and in a way predicts what is to happen.[130]

It appears, moreover, that such is the case from very early stages. For example, one study showed olfactory-visual-motor synchrony, where infants exhibited more oral activity when stimulated by the sight and smell of their mother's breasts. Less activity occurred when olfaction and vision were stimulated separately.[131]

While enactivists, echoing Dewey, emphasize that perception occurs as total coordinations of sensitivities and capacities, they tend to give only passing detail—if even that—to the intermodal implications of this position.[132] From Dewey's standpoint, along with Merleau-Ponty's, perception includes traditional categories of sense but also occurs through other modes, for motor, intellectual, and emotional capacities are also involved.[133] Consider the experience of eating popcorn and how it mobilizes tongue, jaw, hands, and eyes along with ears that hear the crunch and skin that feels the warmth, all while forming a part of cultural rituals and functioning as a socially integrative medium within movie theaters. This explains why a liquid popcorn stimulating the same excitations on taste receptors has little appeal. This also helps account for Massimiliano Zampini and Charles Spence's finding that chips seem extra crispy when people hear crackling,[134] for food is crispy not solely because it fragments easily, but also because it has a certain a look, sound, and manner of mobilizing the jaw and tongue—an overall way of synchronizing sensitivities and actions. As Dewey wrote in an example about painting, which could apply to eating, jogging, skiing, sexual encounters, and numerous other activities, "an integration is effected in the total set of organic responses; eye-activities arouse muscular activities which in turn do not merely harmonize with and support eye-activities, but . . . evoke further experiences," such as by directing posture, gaze, and attention and altering behavior rhythms. "Moreover," Dewey added, "as in every adequate union of sensory and motor actions, the background of visceral, circulatory, respiratory, functions is also consonantly called into action."[135]

Dewey's most extensive elaborations on the intermodal nature of perception are in works talking about aesthetic experience—emphasizing it as an everyday phenomenon and citing engagement with fine art as a characteristic example. The focus on aesthetic exemplars of intermodal experience makes sense, since, by Dewey's definition, it is here that joint integrations are fullest. Merleau-Ponty argued likewise, writing that

> Cézanne declared that a picture contains within itself even the smell of the landscape. He meant that the arrangement of colour on the thing (and in the work of art, if it catches the thing in its entirety) signifies by itself all the responses which would be elicited through an examination by the remaining senses; that a thing would not have this colour had it not also this shape, these tactile properties, this resonance, this odour, and that the thing is the absolute fullness which [a person's] undivided existence projects before itself.[136]

While arguably an overstatement, this characterizes what happens on occasion,[137] and experiments show, for example, that color affects the taste and scent of wine.[138] Merleau-Ponty clarified by prefacing the aforequoted passage with the following:

> If a phenomenon—for example, a reflection or light gust of wind—strikes only one of my senses, it is a mere phantom, and it will come near to real existence only if, by some chance, it becomes capable of speaking to my other senses, as does the wind when, for example, it blows and can be seen in the tumult it causes in the surrounding countryside.[139]

The use of the word "phantom" is instructive: a pure, isolated *quale*—say, of yellow-orange—approaches, for instance, the phantom haloing sensations that sometimes accompany migraines. By contrast, seeing a candle flame—even if only in a painting—means perceiving a warm, waxy-smelling flicker with a soft emotional resonance (though experiences will, of course, vary).

In plainer language, we customarily encounter things together, and through habits of past dealings, we have intermodal experiences. This happens when we see a candle flame's warmth in a painting because of our past experiences with similarly colored things that are hot. The studies on wine fit this, for adding dyes only affected taste and olfaction in trained or semitrained evaluators, who had considerable past experience. Lending further reinforcement are studies finding that multimodal neurons activate for both somatosensory engagements of the hand and for visual events occurring near it, combined with the fact that experiments employ primates *trained* to pick up tools.[140] Work on multisensory neurons similarly indicates that crossmodal activation typically entails some training.[141]

Intermodal perceptual habits, however, are arguably not just instilled through multiple simultaneous sensory impingements. In virtually all encounters, motor activity is involved too. Thus, heat and cold, for example, are never purely thermal. Fireplace heat is a crackling, flickering, spicy, soothing, gentle, orange-yellow warmth that invites us to turn our heads towards it; and cold packing snow is a moldable, trickling frigidness that coats the hand and drips through fingers handling it. Some may object that these examples synthesize what are basic qualities into compound forms, and that they consequently use language to contrive what is never directly experienced from what is initially there. The reverse is actually so. To speak of a pure and unadulterated warmth or cold is to describe what is perhaps endured in pathology, but never perceived in worldly things and events.[142] To feel a pure, isolated heat in one's hand—a heat unconnected to anything the hand is doing or touching—is to suffer illness, such as, for instance, nerve damage. Isolated sensory excitations are not perception.

As Merleau-Ponty summed up, "the senses intercommunicate by opening on to the structure of the thing,"[143] and they open by virtue of bodies—which also include brains—synchronizing around the contours of our world and objects in it.[144] By means of such arguments, Merleau-Ponty and Dewey once again rebuffed the thesis that perception is built up out of sense packets, their ideas aligning again with enactivists (though the latter do not emphatically target British empiricist psychologies since they are more focused on contemporary matters). Dewey and Merleau-Ponty's positions—along with related Gestalt, Gibsonian, and 4E standpoints—also challenge the notion that perceptual experience builds up through the assembly of separate sensory modalities. This misguided assumption is tacitly promoted, even if not explicitly endorsed, by research methods that overwhelmingly look at each modality in isolation.

It is difficult to supply final neurobiological arguments that attest to sensorimotor positions and their intermodal implications. This is because neurobiology is enormously complicated and in relative infancy as a field, with rapidly shifting conclusions. That said, available evidence converges to support the picture advanced here. To begin with, evidence weighs heavily against modularity—the position that the brain has functionally discrete structures dedicated to fairly specific tasks. This is not to deny any cephalic specialization, but to assert that it is more akin to specialization in the rest of the body, and perhaps even less specialized than we imagine other body parts to be. Eyes, eardrums, hands, and so forth are specialized, but they only work in concert with the rest of the body. Moreover, multiple uses are often possible. Hands are used not only for grappling, but also for language expression and comprehension, as in the case of signing and braille reading. The anus, likewise, is critical to waste elimination, but sometimes becomes a sexual vehicle. In the case of brain anatomy, specialization seems even less pronounced. The examples of Broca's area discussed earlier, along with the basal ganglia and the occipital lobes processing tactile information when visual input is removed, stand as a few in a host of available testaments to this possibility. Additionally, neuroimaging studies indicate that cortical pathways once thought to be dedicated to single sensory modalities are actually in reciprocally modulating relationships with others.[145] Moreover, motor areas mobilize when people observe actions.[146]

A specifically relevant experiment had people first listen to words, and then view lips silently mouthing them, and demonstrated comparable activation in auditory cortical sites on both occasions.[147] Again, because it is hard to map intervening processes, this is not conclusive evidence for Dewey and Merleau-Ponty's intermodal accounts (which 4E positions also imply). However, it fits the thesis that we tend to encounter sights and sounds together and habitually connect them, so that there is a sense that we see sounds and hear sights.[148] Furthermore, since at least the late 1800s, researchers have proposed integrated sensory and motor areas in the brain.[149] At an extremely local level, multisensory neurons have been found in the superior colliculus—a midbrain structure—of various species. These neurons can integrate information received through different sensory modalities, where part of the neuronal coordination is elicited by synchronized movement of eyes and ears when interacting with the world.[150] As mentioned some pages back (and in line with pragmatic and enactivist outlooks), neuronal multisensory capacities are acquired through experience,[151] albeit sometimes over extremely short time frames.[152] The differences in brain activity between early blind and seeing participants in sensory substitution experiments discussed earlier provide additional support for the sensorimotor thesis, which, if followed through, implies intermodal perception. Evidence accordingly attests at least tentatively to Dewey's thesis that "nothing is perceived" unless capacities "work in relation with one another."[153]

A host of illustrations with everyday appeal lend further reinforcement. Research on people-environment relations indicates that the overall perceptual milieu affects how we encounter details: a waterfall that sounds pleasant in the context of a park becomes irritating when recorded and played out of context, and is sometimes mistaken for traffic.[154] Studies on "self-vection"—the motion that passengers sometimes experience when stationary and an adjacent vehicle moves—demonstrate intercommunication between the eye, motor-body, and vestibular system, as well as the sensorimotor nature of perception: enclosed in a room in which the walls and ceiling move back and forth, people feel as if the floor is moving, and they sway to maintain balance.[155] Merleau-Ponty detailed a range of additional everyday examples:

> One sees the hardness and brittleness of glass, and when, with a tinkling sound, it breaks, this sound is conveyed by the visible glass. One sees the springiness of steel, the ductility of red-hot steel, the hardness of a plane blade, the softness of shavings. The form of objects is not their geometrical shape: it stands in a certain relation to their specific nature, and appeals to all our other senses as well as sight. The form of a fold in linen or cotton shows us the resilience or dryness of the fibre, the coldness or

warmth of the material. . . . In the jerk of the twig from which a bird has just flown, we read its flexibility or elasticity, and it is thus that a branch of an apple-tree or a birch are immediately distinguishable. One sees the weight of a block of cast iron which sinks in the sand, the fluidity of water and the viscosity of syrup. In the same way, I hear the hardness and unevenness of cobbles in the rattle of a carriage, and we speak appropriately of a "soft," "dull" or "sharp" sound.[156]

Dewey likewise commented that our eyes see "the liquidity of water, the coldness of ice, the solidity of rocks, the bareness of trees in winter," and that "it is certain that other qualities than those of the eye are conspicuous in controlling perception."[157] "Any sensuous quality tends . . . to spread and fuse," he added.[158]

Again, central to both Dewey and Merleau-Ponty's positions—and implied but not emphasized by enactive views—is that the body is a mediating fabric through which all of this integrates. As an illustration from the history of philosophy of what can happen when the body does not coordinate, we may consider Merleau-Ponty's analysis of a perceptual aberration that Aristotle described in his *Dreams*.[159] Sometimes called "Aristotle's illusion," the anomaly is one where one object feels like two when placed between crossed fingertips. Here, wrote Merleau-Ponty, "the synthesis of . . . tactile perceptions in one single object is impossible" because "the crossing of the fingers, being a movement which has to be imposed on them," interferes with their "motor possibilities" (figure 2.3). What results is that "the right face of the middle finger and the left face of the index cannot *combine in joint* exploration of the object."[160] This reiterates Merleau-Ponty and Dewey's emphasis on perception as a total coordination, which enactivists such as Noë also repeat, albeit without highlighting the intermodal implications that follow.[161]

Other perceptual illusions are similar. Consider the McGurk effect: the finding that dubbing the sound /ba/ onto lip movements for /ga/ results in most people hearing /da/.[162] In this instance (and along Deweyan, Merleau-Pontian, and enactivist lines), subjects confront conflicting stimuli that pull sensitivities and capacities out of synchrony. Seen accordingly, the McGurk effect is not a consequence of something being wrong with subjects, but of something amiss in the situation in which they are placed. At the same time—and quite obviously—the effect cannot simply be explained

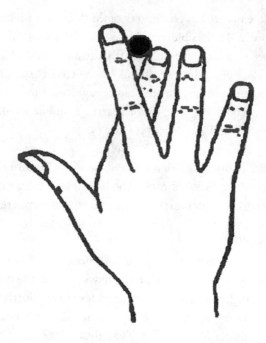

FIGURE 2.3 Aristotle's illusion

Adapted from R. Woodworth, *Psychology: A Study of Mental Life* (New York: Henry Holt, 1921), fig. 68.

in extraneural terms. While remarkably prevalent, it is not universal, and differences in neurobiological activity correlate with susceptibility. Specifically, susceptible participants have increased activation in the left superior temporal sulcus region—an area in the cortex involved in multisensory processing[163]—as compared to those not experiencing the effect.[164] This is a reminder of the need for an integrated understanding of psychic life—in other words, one that takes neurobiological aspects into account along with extraneural ones and the general environment.

This approach accords with the experimentalism of pragmatists. It does not deny that we sometimes perceive and cognize in erroneous ways, yet it highlights that many overestimate the extent to which we do. In the case of the McGurk effect, brain-based differences point to the fact that while human beings come with similar biological endowments and hence experiences of the world, not everybody has precisely the same "experimental setup," as we put it some pages back. Different setups in science realize different observable effects, just as different organic arrangements do. This was demonstrated earlier with examples of wave-particle duality and a cat's claws realizing sinewy toughness versus our fingertips actualizing lacquered smoothness of wood. Without denying the importance of the brain, we argue that a key consideration, once again, is that a great deal of perceptual and cognitive activity—whether in organisms or scientific practices—occurs outside of the head.

RESUSCITATING PERCEPTION'S EPISTEMIC WORTH

Dawkins's example of timbre discussed earlier is even more misleading than commonplace interpretations of the McGurk effect, which insinuate that we are duped on questionable grounds. Dawkins's claim that we are deceived when perceiving brassy sounds of trumpets rather than the separate harmonics follows only from his stipulating—without justification—that an accurate way of registering is to have the soundscape teased apart. Dewey, Merleau-Ponty, enactivism, and indeed contemporary physics, by contrast, have all suggested that most properties only show up in the context of interrelationships in the first place, and that humans are involved in the interrelations in the case of perception and, for that matter, scientific inquiry.[165] For such reasons and others that we will elaborate upon, it makes no more sense to assert that we are deceived when perceiving brassy timbres than it does to claim that we are deceived when registering the smoothness of a bottle we are handling or, indeed, detecting separate harmonics with technical instruments. All these results follow from specific action-scenarios. If judged according to specific scenarios generating them, there is no basis for concluding that we are duped when we register them.[166] This is experimentalism all over again.

Dawkins, it will be remembered, states that there is "simulation software in the brain", and that we do not see "the unvarnished real world but a *model*" built "inside our head."[167] Hoffman, also pushing the idea of simulation software, argues that the perceptual system is geared towards facilitating interface with the world by means of nonveridical representations.[168] Like so many other observers today and throughout history, both cite perceived hues as examples. Dawkins calls these hues "internal labels" with "no intrinsic connection with lights of particular wavelengths,"[169] the latter of which he takes as genuine reality. For Dawkins, perceived hues are "tools" used to construct a "model of external reality" that mark "important distinctions in the outside world."[170] An animal's "world-representing software" is adapted to its particular "way of life," and Dawkins accordingly reasons "that bats may 'see' colour with their ears." After all, "the world-model that a bat needs . . . must surely be similar to the model that a swallow needs."[171] Bats may accordingly use hues "as internal labels for some useful aspect of echoes" so that "the nature of the model is governed by how it is to be *used* rather than by the sensory modality involved."[172]

We take roughly half of this speculative account to be correct. Enactivist theories and, specifically, sensory substitution devices affirm our experience of the world as governed to a significant degree by action in it, so that even conventionally nonvisual sensory modalities can become visual in practice. This is in line with what Dawkins says, minus the dichotomizing of genuine and represented reality that he insinuates with perceived hues. We suggest—and this is a point Dawkins and Hoffman miss—that perceptual experience is not merely constructed in the head, but enacted in interrelations. To requote Dewey, perceived qualities are conceived poorly if reduced to internal representations because they are also "qualities of interactions in which both extra-organic things and organisms partake."[173] This is obviously so for tactile qualities such as smoothness, yet no less so for other kinds of properties. Enactivist accounts attest to this. Also reinforcing this is the notion that interrelationships are epistemically and ontologically primary—that is, they are where things and events with specifiable, perceivable, and knowable qualities initially and actually exist. Color, length, and mass all depend on velocity relative to the point of observation, which is to say, they are effects showing up in the context of interrelations. Consequently, before the perceiver arrives, properties are already effects; and if effects count as "real," and are not arbitrarily deemed "unreal" because a perceiver participates in the interrelations, then there is no reason to grant them less reality than wavelengths of light, which also show up as outcomes of specific relations and, indeed, actions (such as when instruments of detection introduce changes to the phenomena observed).[174]

Consider another example of Dawkins's views on human perception. In this passage he informs readers:

Science has taught us, against all evolved intuition, that apparently solid things like crystals and rocks are really composed almost entirely of empty space. . . . So why do rocks look and feel solid and hard and impenetrable?. . .

Our brains have evolved to help our bodies find their way around the world on the scale at which those bodies operate. We never evolved to navigate the world of atoms. If we had, our brains probably *would* perceive rocks as full of empty space. Rocks feel hard and impenetrable to our hands because our hands can't penetrate them. The reason they can't penetrate them is unconnected with the sizes and separations of the particles that constitute matter. Instead, it has to do with the force fields that are associated with those widely spaced particles in "solid" matter. It is useful for our brains to *construct* notions like solidity and impenetrability, because such notions help us to navigate our bodies through [the] world. . . .[175]

Read pragmatically, this line of reasoning reduces to the claim that perception reflects the situation or relationship we objectively find ourselves acting in, and if we were to find ourselves in another, our perception would reflect it.[176] Relative to our bodies, rocks are, objectively speaking, impenetrable. So in this instance perception is not duped. Also misleading is Dawkins's suggestion that we perceive rocks as solid primarily because of "evolved intuition." This is an overuse of theoretical machinery, specifically Darwinism, and it oversells the degree to which evolved and, in this sense, hardwired representations determine our view of things. After all, if physical laws changed to allow hands to pass through rocks, we would perceive this. Action, in short, is fundamental. This is core to pragmatic, Merleau-Pontian, and 4E views. Furthermore, it is by acting on materials in the world—by altering the situations in which they are observed—that scientists have reached the conclusion that material objects are mostly composed of space, which nevertheless does not contradict our experience of their solidity.

The error in Dawkins's interpretation recalls teachers who sometimes charge, when introducing physics, that Democritus misrepresented reality by claiming that atoms are indivisible. In making this claim, they neglect that *atomas* meant "indivisible," and that it was modern-era scientists who misapplied the term when they initially but wrongly concluded that what they called "atoms" fit this description.[177] Dawkins similarly errs by insinuating that we mistakenly call objects "solid" and "impenetrable," for we rarely use these terms to describe atomic arrangements. Rather, we use them to indicate what we can do with things and what things can do to us— something Dawkins concedes without following the point to its pragmatic conclusion. Solidly built chairs bear our weight. Impenetrably solid fog impedes vision and movement. Unlike liquid or gas, we can handle solid ice, walk on it, or risk falling through it.[178] George Lakoff and Mark Johnson have made comparable points,[179] as has Merleau-Ponty,[180] and none of these linguistic uses of the word "solid" contradicts claims or even says anything about the atomic structure of matter.

As with Dawkins and James before him, Dewey acknowledged a debt to evolutionary biology. Moreover, like Dawkins, who suggests that concepts are tools for worldly adaptation[181] and that we do not see "the unvarnished real world,"[182] Dewey saw concepts as instruments.[183] Consequently, he agreed that they are not "bare transcript[s] or duplicate[s] of some finished . . . arrangement pre-existing in nature,"[184] for copying reality is not a chief function of instruments. Rather, instruments help us negotiate and *change* realities—in other words, there is no finished arrangement to represent.[185] Dawkins highlights some of this. Yet he also stresses that world-models are built "inside our head" and fails to convey,[186] as Dewey stated, that an instrumental classification "does not commit us to the notion that classes are . . . purely mental" or "merely nominal."[187] The latter is true because conceptual classes are realized in "objective action,"[188] as when creating the distinction between hard and soft wood by putting different kinds to different uses.

To sum up, Dewey would agree that adaptability is key, but add that there is no preexisting truth to represent, a point Hoffman acknowledges more readily than Dawkins does while citing quantum mechanics as a case in point.[189] Hoffman's discussions, however, are confused in that he suggested that there is no preexisting truth while citing standard examples of objective reality—for example, wavelengths of light—in order to show that our perception is adaptive but not accurate. Hoffman argues that perceptual theories should be reoriented from "*categories of the objective world*" to "*categories of an organism's perceptual world*,"[190] unaware that these can be the same, as last chapter's example of the cat versus human illustrated. There, we saw that a cat might perceive sinewy toughness of wood by digging its claws in, while the caress of our hand might realize its lacquered smoothness. Here, objectively different things are done and objectively different outcomes happen. For this reason, humans and cats live in different worlds, with different possibilities of action. Consequently, they have different perceptual worlds, but this is not a result of mere mental or even neurobiological variations; they follow, rather, from different outcomes actualized through action—which is to say, objectively different worlds.

Differences in experience are necessarily less across human populations insofar as individuals share similar bodies. However, bodily differences exist, and they help explain perceptual variation.[191] Studies of Gibson's affordance theory, which holds that we perceive in terms of what our environments and bodies allow and disallow, testify to the point. Researchers have found that people with longer arms perceive objects to be closer.[192] In terms of lived experience, "closeness" means "within one's sphere of immediate manipulability." Consequently, objects appear closer when one uses a tool to extend reach.[193] Comparable studies find that the perceived size of gaps varies with jumping ability;[194] that capacity to balance on a beam increases its perceived width,[195] or what might be called its "walkability"; and that ability to pass through an aperture or duck under an obstacle similarly increases its perceived size or height.[196] People likewise perceive hills to be steeper or more distant when they are fatigued, burdened with heavy backpacks, out of shape, or suffering from low blood sugar.[197] A conventional understanding—for example, that of Dawkins or Hoffman, but one might add Plato, Descartes, Hume and many others—would take this as evidence of the nonveridical nature of perception. A pragmatic interpretation, by contrast, emphasizes that the aforesaid variations follow from people's objective capacities to do different things in given environments, and that perception accurately registers these differences.

Hoffman maintains that adaptability trumps truth[198] and specifically challenges enactive theories, seemingly unaware that truth and fitness are typically not in conflict when judged from an action-based standpoint. In advancing this position, he also retains fairly traditional ideas of truth as a correspondence of mental representations to outward facts, even while denying the importance of veridical representations in living processes. As we will increasingly see, classical pragmatists, along with Merleau-Ponty, downplayed conceptions of truth as veridical representations of reality. Such conceptions also align with enactivist positions, but are usually not drawn out there because proponents focus on mind and less on related epistemological implications. At any rate, all these embodied views suggest, first, that much of what we call cognition goes on outside the head; second, that the primary job of perception and cognition is not to represent the world, but instead to adapt, which means acting; and, third, that things only have properties in relation to one another and that humans are part of the relation, making it nonsensical to conclude that we misperceive when we register the solidity of the chair on which we are sitting or hear the brassy timbre of a trumpet. This is experimentalism yet again.

EXTENDING COGNITION OUTSIDE THE CORTEX

For decades, psychologists have argued that "as compared with consciously controlled cognition, . . . nonconscious information-acquisition processes are incomparably faster and structurally more sophisticated."[199] The supposition has usually been that this is because of hidden operations in vast neural recesses, which almost surely goes some way towards accounting for the breadth and complexity of unconscious or prereflective processes. As such, most of today's brain researchers question the assumption that cognition is exclusively cortical. That is, they doubt that it is confined to the cerebral cortex, that outer layer of folded tissue popularly regarded as the least evolutionarily ancient and as the seat of consciousness and rationality. Parkinson's patients, for example, have impaired functioning in subcortical motor regions and degraded cognitive capacities in tasks such as facial recognition.[200] Impoverished dopamine levels relate to these impairments as well as compromised motor capacity in such patients. Studies showed marked improvement in both movement and cognitive tasks when dopamine-elevating medication was administered.[201] Though once again not the final word on the topic, this not only indicates cognitive function in brain regions outside the cortex, but also connects it to action.[202]

Neuroscientific interpretations of what Mihaly Csikszentmihalyi calls "flow" experiences[203] also fit with this. In flow experiences, the self disappears into work or performances; we have something approaching natural highs during which we find ourselves merging with our surroundings and performing with overwhelming confidence and competence and an enhanced sense of wellbeing. Expert surgeons and chess players are among those who report such experiences. Moreover, though engaged in archetypically cognitive tasks, they report performing with little conscious awareness.[204] For such reasons, Arne Dietrich makes a speculative case that the frontal cortex is suppressed while activity in subcortical regions becomes more pronounced during flow states.[205] While possible, this conclusion also carries the Victorian supposition that the cortex handles higher cognitive functions and keeps irrational or basic impulses in check—a thesis that empirical evidence weighs against, as we argue in later chapters.

At the same time, the story we have offered in this chapter (by extension, also the story that might be offered by pragmatists, embodied phenomenologists, and 4E proponents) pushes cognition farther out of the cortex. In addition to recruiting subcortical regions, cognitive processes are enacted through bodily structures outside the central nervous system. As Clark, among others, proposes, bodily actions are "among the means by which certain computational and representation operations are implemented" and consequently "not realized in the neural system alone but in the whole embodied system located in the world."[206] If this is so, then it makes sense that a great deal of cognition occurs prereflectively—that is, on the assumption that the conscious reflection necessitates a brain or something functionally like it.

In line with this, and again ascribing the idea's genesis to the ancients while anticipating developments in AI, robotics, and cognitive science, Dewey urged something comparable to Clark, albeit without emphasis on representation. Rejecting the empiricist reduction of psychic life into basic parts, Dewey urged that with embodied accounts, "the necessity ceases for the elaborate Kantian and Post-Kantian machinery of *a priori* concepts and categories to synthesize the alleged stuff of experience."[207] Yet he did not deny that Kantian processes occur. More to the point, he advocated bodily variants of them, writing that experience coheres because of "adaptive courses of action, habits, active functions, connections of doing and undergoing; sensori-motor co-ordinations. Experience carries principles of connection and organization within itself."[208] Dewey reasoned that much of this even holds for an amoeba. Its life depends on interactions that are necessarily limited and, hence, structured because its locomotive powers, capacity to ingest and expel, its shape, size, and things it encounters all constrain what it does. For this reason, its activities have "organization," "continuity in time," and "reference to its surroundings,"[209] producing coherence and even the basic appearance of intentionality. Hence, as Noë was to put it, "the life of the bacterium is not hidden within it." Rather, its life "is a dynamic in which the bacterium, in its environmental situation, participates. And so it is for consciousness, more generally."[210] This does not mean that embodied life and movement alone supply consciousness,[211] but it does suggest that they are preconditions for having coherent experience and that they structure that experience.

Decades later, the artificial intelligence communities began arriving at similar realizations. In 1969, Herbert Simon—despite his commitment to internal symbolic representation models—developed a well-known ant example where the complexity of the ant's movement is a function of the complexity of surfaces over which it strides. Dewey and Simon's observations about the amoeba and ant connect to recent work by the roboticist John Long, who has built on the ideas of Rodney Brooks to develop simple aquatic "tadpole robots" or "Tadros." Long used them to model evolutionary processes. Specifically, he varied tail stiffness and had the Tadros vie for food (in this case, light), and incorporated the features of the more successful competitors into the next generation. Long reports that his Tadros developed better feeding than earlier generations and "got smarter." Crucially, however, "they did so by evolving their bodies, not their brains."[212] Without claiming that his robots are Harvard-bound, Long insists that they have intelligence by virtue of their goal-directed, autonomous, and embodied behavior.[213] Moreover, in line with Noë, who maintains that "thought arises only when the whole animal is dynamically engaged with the environment,"[214] Long predicts that human-like AI will require not only a CPU capable of functions resembling the brain, but also a human-like body.[215] Thompson similarly urges that a hypothetical Cartesian brain-in-vat experiment would only work if the vat were a body. He adds that the brain would have to be integrated into the body such that it ceases to be a mere container[216] and that it would have to be socially embedded[217]—actively situated in ecological contexts.[218] Terrance Fong and colleagues express ideas comparable to this (along with Long's views and by extension those of Brooks, not to mention the experimentalism of pragmatists) when they write: "Embodiment is grounded in the relationship between a system and its environment. The more a robot can perturb an environment, and be perturbed by it, the more it is embodied."[219]

These observations are instructive for a number of reasons. First, they reiterate that intelligence involves more than brains, albeit without denying their importance in the case of humans. Second, the analysis suggests, along Clark's lines,[220] that bodies perform computational and hence quintessentially cognitive operations. In the case of Tadros, their bodies and especially tails solve difficult physical problems, just as knee joints do when striding over terrain.[221] Long writes: "In response to the tail's coupled internal and external force computations, the body, to which the tail is attached, undergoes the yaw wobbles—recoil and turning maneuvers." The body thus calculates and performs patterns of "acceleration that interact to produce the overall motion of the Tadro according to Newton's laws of motion."[222] Long consequently suggests that active bodies are, by their nature, cognitive structures.

The embodied side of calculation manifests neurobiologically. For instance, basal ganglia link to statistical computation,[223] syntactical function,[224] and prediction in auditory perception[225] while also being critical in the organization of movement.[226] Other studies find that motor tasks activate the same brain regions as the task of reading action words describing the operations.[227] This highlights the connectedness of motor and cognitive function, as does everyday life where gesture and language integrate, a point highlighted by Merleau-Ponty[228] and more recent researchers.[229] Given the story we have told so far, motor function is integrated with perceptual function. Thus, to the extent that cognition is integrated with motor function, it should also knit with perception.

There are many illustrations of this point. To mention some, experiments show that imagining tasks—a conventionally cognitive function—to some extent mobilizes the same neuronal regions (areas involved in perceptual-motor activity) as executing the actions.[230] Performing tasks and perceiving others carrying them out likewise activate many of the same brain regions and specialized mirror neurons.[231] At least this is so in nonhuman primates, which studies have focused on because of the invasive nature of test procedures that insert electrodes into the brain to measure local excitation of individual neurons. That said, functional magnetic resonance imaging (fMRI) does indicate a mirroring effect in humans. Specifically, it shows the inferior frontal cortex and superior parietal lobe activating when people perform an action or see another doing so.[232] Other studies confirm the general point, finding that the frontal motor cortex activates either when executing a task or when watching someone else do so.[233] Similarly, when people look at entities and imagine them, overlapping neuronal regions mobilize, as shown in positron-emission tomography (PET) scans.[234] Here one might speculate that acts of imagination, which are archetypical cases of cognition, are virtual interactions with environments.[235] However, virtual worlds at some point require actual worlds to supply models, even if distorted. Worlds, in turn, are not detached entities, as the earlier cat-versus-human example illustrated. There, we saw that worlds show up to perception largely as a matter of what organisms can do, and the foregoing discussion suggests comparable integrations of cognition and action.

Taken together and with the rest of this chapter, this discussion delivers the key message that mind does not trace to any one structure; rather, it arises out of a totality of sensitivities and capacities working in concert with the world. The brain does not equal mind, nor the entire nervous system, nor still the body. Bodily contact and practiced patterns are distributed in brain regions and also in human action in the world, and together they structure thought and perception. Cognition, perception, and action therefore appear integrated, and later we will see that emotions, interests, and moods also enter this union. These are points raised in the last chapter, and are more thoroughly defended in this one through consideration of classic claims in light of recent work and by beginning to offer empirical support. This chapter also tightened the ecological account of mind by examining intermodal perception. It further introduced habit as another bodily variant of the Kantian *a priori*. Dewey aptly summed up this last point by observing: "The medium of habit filters all the material that reaches our perception and thought. The filter is not, however, chemically pure. It is a reagent which adds new qualities and rearranges what is received."[236]

One point not introduced so far—though it will be elaborated upon in later chapters—is that abstract functions such as language deployment also follow the experimentalism emphasized throughout this book. That is to say, language is used to systematically change the world, not only to represent it. Ludwig Wittgenstein hinted at this in his later work, insisting that words are not, for the most part, defined by reference to things, but by their use.[237] He similarly urged that knowing is not merely an internal process, but a capacity to do something, a point reflected in language where the words "can" and "know" are grammatically parallel.[238] Dewey was arguably more emphatic in this outlook, urging that "to be a tool, or to be used as a means for a consequence, is to have and endow with meaning, language, being the tool of tools, is the cherishing mother of all significance."[239] More recently, Clark has elaborated similar views. He points out that speech often helps us execute tasks that would otherwise be insurmountable,[240] as when a computer expert talks a neophyte through a problem, or when adults do comparably with children, and children then talk themselves through tasks, even if silently, when caregivers are gone. When we engage in conversation with others, we fall into rhythmic synchronizations, and mobilizations of our bodies ensue. Language involves a kind of pushing toward the world, and the world pushes back. This connects to pragmatic and more recent models of cognition and perception, which emphasize loops of doing and effects undergone in consequence of action. Language is emphatically social—something not discussed in this chapter, but raised in the next. There, we will examine not only language, but also the social-embodied nature of perception, cognition, and affect.

3

SOCIAL COHESION, EXPERIENCE, AND AESTHETICS

A ristotle prominently characterized humans as social animals, more so than bees and other organisms that seem to instinctually aggregate. He claimed this because human linguistic and cognitive capacities allow for the formation of social and political concepts such as "public good" and "justice."[1] Though some of Aristotle's assumptions can be interrogated (for example, his assertion that only humans have language, or that conceptual thinking requires it), it is difficult to dispute the interdependence of social life and normally developing human cognition.

Pragmatists—particularly Dewey and Mead and to a lesser extent Peirce—recognized this overwhelming interdependence, and accordingly avoided epistemologies of isolated individualism that make experience essentially private.[2] As discussed in the last two chapters, Dewey repeatedly drew on Greek thought in defending his account of experience, which he equated not only to skilled and public practices, but to culture.[3] This idea was also put forth in ancient sources (see chapters 1 and 2).

Such a view flows easily out of the emphasis placed on body-environment synchronizations by somatic thinkers, whether pragmatic, phenomenological, or 4E. This is because synchronizations overwhelmingly occur in cultural contexts and follow from socially inculcated skills. This chapter develops this position, arguing that body-environment interactions that form experience are public and socially embedded and have similar if not identical integrative outcomes to sensorimotor coordinations. As skills, social sensorimotor coordinations orient towards problem solving, and thus the cognitive. Being social, they also involve emotional attunement. This emphasizes how sensorimotor activity is affectively and cognitively laden, and the same applies, for example, even when a lone hiker scans opportunities and dangers in a desolate vista.

In advancing this case and focusing on social dimensions, we integrate classic philosophic views with empirical pictures from behavioral and brain sciences that likewise focus on interpersonal coordinations as defining human psychic life. We further point out that social cohesion has aesthetic and thus affective aspects that are important for perceptual and cognitive coherence, especially in early development.[4] After linking pragmatic and developmental scholarship, we consider Gibson's landmark perceptual theory of affordances along with recent research connecting it to aesthetic experience and social life. We do this to fill in and expand upon the pragmatic idea of experience as culture, which closely parallels the phenomenological concept of worldhood. These positions be reinforced with evidence suggesting that our psychological landscape overwhelmingly begins as social and remains so throughout life. This is not to deny that physical movement is at play from day one, but to assert that movement is intertwined with social life all along, and that the latter is not reducible to or built up from the former in a quasi-empiricist way. The social world, like the brute physical one, invites and resists action and is consequently experienced in terms of affordances.

Perhaps most importantly, given the main themes of this book, this chapter argues that emotional and sensorimotor social coordinations are replete with cognition and vice versa. The neurosciences support this view, though once again without providing the final word on the issue. Areas of the brain that include the prefrontal cortex and temporal lobes along with the amygdala bulbs and basal ganglia appear crucial in regulating social action.[5] They also regulate cognitive, emotional, and perceptual functions, as discussed last chapter. Regions such as the basal ganglia, to offer an illustration, not only perform cognitive, motor, and perceptual functions;[6] they simultaneously contribute to reward appraisal.[7] Reward appraisal relates to attention, emotion, and perception, as well as motor and cognitive regulation of decision making and action selection. Based on this reasoning, one would likewise expect amygdala structures—classically regarded as emotion areas—to handle motor, cognitive, and perceptual tasks. A combination of human and nonhuman lesion studies (which we do not take to be conclusive, since they could merely indicate disrupted pathways) are at least consistent with this picture. Sufferers exhibit, for instance, inappropriate ingestion of nonfood objects and increased exploration of unsafe things.[8]

Connecting this back to the social side, it may be that some of these same regions of the brain mobilize when discerning others as intentional actors, though much of the evidence comes from experiments with monkeys.[9] Studies on primates further suggest that a significant amount of visual neural resources orient toward socially meaningful activity.[10] This undermines distinctions between cognition and perception as well as affect, insofar as the latter relates to sociality. In social contexts, all these faculties obviously knit with action.

Another key lesson—one shared by pragmatists and ancients and more recent evolutionary and neurobiological researchers—is that humans are social animals. More than just social, however, the accounts integrated in this chapter suggest experience and knowledge are shared. This contrasts with the epistemology of isolated individualism embedded in modern concepts of private experience. Dewey argued that the feeling of being isolated within our own private sphere of subjectivity is symptomatic of a kind of pathology or breakdown,[11] a position that James,[12] Nietzsche, and Heidegger also defended.[13] For Dewey, aesthetic experience is the opposite of such breakdown. The bulk of scholarship on aesthetics, however, appears insensitive to this. As if parroting a classical liberal ethos, it overwhelmingly adopts the model of the lone perceiver engaging with works of fine art or beautiful and sublime settings. This is in spite of the fact that throughout most of history, not to mention everyday life, aesthetic experience has been overwhelmingly shared in rituals, memorable feasts, formal and informal celebrations, romantic episodes, and other social gatherings. The notion of mind as an essentially isolated bubble pervades today's cognitive sciences in spite of nearly insurmountable evidence to the contrary. With a growing but inexplicably small number of adherents, we advocate finally moving beyond this early modern vestige.

ACTION AS GROUP COORDINATION

We have discussed at length the idea of bodily coordination as experience, arguing that capacities and sensitivities integrate into coherent perception and cognition by virtue of synchronizing around things in the world. This happens when our stride patterns differently depending on whether we press into icy cold snow at −20 degrees Celsius or into slush at higher temperatures. These different conditions resist us in different ways, moderating gait and balance. In consequence of interactions, moreover, different effects are realized—for example, a crunching chalkboard scrape that expresses frigid snow, or the diamond glitter that caresses our squint on sunlit days, both characteristically absent at warmer temperatures. Well-trodden trails in times of thaw often become too icy to travel safely. We consequently see and regard them as threatening and therefore as obstacles, and most of us navigate accordingly. In short, the world manifests in emotional, cognitive, motor, and perceptual ways according to what our bodies and environments allow and how the two together constrain and open possibilities. Acting within these constraints and openings literally generates properties we encounter in the world. In pragmatic language—which almost perfectly matches enactive terminology—these properties are registered as consequences of our sensorimotor coordinations with our surroundings.

By virtue of their interactive nature, such encounters often involve tools and other people, with similar results. Bicycles are tools. They respond to our actions and we to theirs, which opens possibilities for movement while also constraining it.[14] This constraining, doing, and undergoing of consequences of our actions form the heart of what we—following Dewey, Mead, and Merleau-Ponty—understand experience to be. A similar example is a blind person navigating with a cane. Here, too, movements are constrained and opened by the body, environment, and tool deployed, so that the cane's tip forms "an area of sensitivity, extending the scope and active radius of touch, and providing a parallel to sight."[15] Introducing two or more organisms changes specific dynamics, but similarities persist. Consider dancing with a partner. This may include certain leaning motions simultaneously constrained and made possible by the presence of a partner and things in one's surroundings such as stairs, handrails, or chairs. It may also encompass emotionally and sexually expressive dimensions and quite a bit more, again both opened and limited by the presence of partners. From this interplay of action, integrated and indeed intermodal experience results.

An important point to note in all this is the pervasiveness of group coordination, which is the norm throughout the biological realm, including the nonhuman one. Thus, migrating birds interlock in patterns opened and constrained by collision avoidance, velocity matching, and proximity maintenance.[16] When in V-formations, wings of birds flap in phase with individuals ahead of them, maximizing upwash capture. In contrast, when each bird is immediately behind another bird in streamwise position, the pattern is antiphase, which may reduce the adverse effects of downwash.[17] Dung beetles similarly coordinate, interacting with gravitational forces, friction, and one another to roll dung into balls very rapidly over significant distances.[18] In some cases, integration is encompassing. For example, a speculative case has been made that "bacteria, in addition to carrying out their individual and localized team activities, together form a planetary entity of communicating and cooperating microbes, an entity that . . . is both genetically and functionally a true superorganism."[19] On a less grand but perhaps more compelling scale, the Portuguese man o' war is a symbiosis of multiple organisms that function and appear as a single jellyfish-like creature. Lichen is another composite organism. Sometimes resembling moss, it emerges from the combining of cyanobacteria or algae among filaments of fungi to exploit chemical and physical aspects of the environment.

These examples demonstrate that what happens on the level of individual bodies reoccurs on group levels, with systems of organisms synchronizing around one another and their surroundings. Humans are no exception when they join in group activity, whether in back-and-forth banter, reciprocal gestures, religious ceremonies, dining rituals, greetings, or coordinated city life,[20] not to mention sporting activities and concerts, which involve synchronization of performers and interplay with audiences. Moving from the everyday to the laboratory, studies affirm that people tend to match one another's gestural activities, postures, intonation, vocabulary, and syntax.[21] Evidence suggests that such coordinations facilitate or indicate social cohesion. One researcher, for example, found that postures of patients and therapists tended to converge and that this correlated positively with rapport.[22] Others have found that sympathetic facial coordination—presumably carried out unconsciously—facilitates recognition of emotions in others. Conversely, ability to discern expression degrades (albeit with variation between genders sometimes found) when capacity to coordinate facial expressions with others is limited, for instance, due to Botox injections or because participants are asked to hold a pen in their mouths.[23]

Chemero, who is heavily influenced by both pragmatism and phenomenology, has conducted intriguing work on human group coordination. In one of his studies,[24] participants engaged in sheep-herding tasks, basically a video game where two people controlled cursors to repel dots, with the aim of keeping the dots inside a circle. They had to do this without speaking. A successful trial—lasting 60 seconds—required confinement of the dots for 70 percent of the last 45 seconds of the task. In early trials, participants tended to work individually, repelling dots on their own side that strayed farthest from the center. This strategy failed. However, after repeated attempts, most participants interlocked in oscillatory patterns with their partners—either following one another's movements or else going in opposite directions. In short, they coordinated with one another instead of working expressly to block the sheep (though the latter factored into the activity, since the goal was to keep the sheep penned), and this strategy proved successful.

Group coordination often occurs on a more global level. John Steinbeck captured this in Heideggerian terms, describing how group activity coordinates into worlds, which are phenomenologically and pragmatically equivalent to experience:

> In the daylight [Great Depression migrants] scuttled like bugs to the westward; and as the dark caught them, they clustered like bugs near to shelter and to water. . . . [T]hey huddled together; they talked together; they shared their lives, their food, and the things they hoped for in the new country. Thus it might be that one family camped near a spring, and another camped for the spring and for company, and a third because two families had pioneered the place and found it good. And when the sun went down, perhaps twenty families and twenty cars were there.
>
> In the evening a strange thing happened: the twenty families became one family, the children were the children of all. . . . In the evening,

sitting about the fires, the twenty were one. They grew to be units of the camps, units of the evenings and the nights. A guitar unwrapped from a blanket and tuned—and the songs, which were all of the people, were sung in the nights. Men sang the words, and women hummed the tunes.

Every night a world created, complete with furniture—friends made and enemies established; a world complete with braggarts and with cowards, with quiet men, with humble men, with kindly men. Every night relationships that make a world, established.[25]

As in other scenarios we have discussed, Steinbeck emphasized the insep-arability of the physical and social world. He added: "A certain physical pattern is needed for the building of a world."[26] This might include "water, a river bank, a stream, a spring, or even a faucet unguarded. And there is needed enough flat land to pitch the tents, a little brush or wood to build the fires."[27]

Dewey—and, following him, Mead—likewise emphasized experience as cultural activity. As with his sensorimotor account, Dewey drew on ancient sources here,[28] which sometimes equated experience to custom and culture. This equation fits with sensorimotor accounts in that learning customs and cultures means acquiring skills—a point also reflected in ancient thinking for reasons covered in earlier chapters. Along these lines (and anticipating a central enactivist thesis), Dewey stressed that having experience meant being experienced. Since skills are overwhelmingly acquired in social contexts, Dewey took this as evidence of the cultural nature of experience. The idea that psychic life is cultural is accordingly not new. However, while cultural aspects of experience have long been recognized, Dewey thought the social side understated in his day due to modern thinkers largely reducing experience to that which hit the senses.

Descartes, in an oft-cited personal correspondence, suggested that nerves, muscles, and even hands can serve memory.[29] Descartes, moreover, was regarded as the father of physiological psychology by no less than T. H. Huxley.[30] Thus, we do not want to overstate the case and caricaturize past traditions. Nonetheless (and Descartes's foregoing assertion notwithstanding), modern thinkers—including Descartes—emphasized sensations and thoughts as represented in the head, and scholars overwhelmingly took experience this way in Dewey's time, as they largely do today. This is one reason that Dewey contemplated changing the name of his 1925 *Experience and Nature* to "Culture and Nature": because people took experience as something internal and therefore other than what Dewey understood it to be.[31] Though the substitution may sound peculiar, we often interchange the words "experience," "culture," and "world," as when speaking of collegiate culture, the collegiate world, or the collegiate experience.

Barbra Smuts's recollection of integrating into the world or culture of baboons captures Dewey's sentiments.[32] It also characterizes the phenomenological concept of worldhood, though Heidegger would likely have been uncomfortable with this application of the term (he seems to have thought nonhuman organisms lack the metaphysical proclivities that allow for worlds). Smuts writes:

When speaking about this process [of engaging with baboons . . .], I've used the accepted scientific term, "habituation." The word implies that the baboons adapted to me . . . But in reality, the reverse is closer to the truth. The baboons remained themselves, doing what they always did in the world they had always lived in. I, on the other hand, in the process of gaining their trust, changed almost everything about me, including the way I walked and sat, the way I held my body, and the way I used my eyes and voice. I was learning a whole new way of being in the world—the way of the baboon. I was not literally moving like a baboon—my very different morphology prevented that—but rather I was responding to the cues that baboons use to indicate their emotions, motivations and intentions to one another, and I was gradually learning to send such signals back to them. As a result, instead of avoiding me when I got too close, they started giving me very deliberate dirty looks, which made me move away. This may sound like a small shift, but in fact it signalled a profound change from being treated as an object that elicited a unilateral response (avoidance), to being recognized as a subject with whom they could communicate. Over time they treated me more and more as a social being like themselves, subject to the demands and rewards of relationship. This meant that I sometimes had to be willing to give more weight to their demands (e.g., a signal to "get lost!") than to my desire to collect data. But it also meant that I was increasingly often welcomed into their midst, not as a barely-tolerated intruder but as a casual acquaintance or even, on occasion, a familiar friend.[33]

As Heidegger might have said if he better appreciated nonhuman life, the baboon community became home for Smuts, a place of familiarity and everyday dwelling. In a passage quoted in the last chapter, Dewey aptly expressed the notion thus: "Through habits formed in intercourse with the world, we also in-habit the world. It becomes a home and the home is part of our every experience."[34]

While advancing a sensorimotor account (and following many of his contemporaries in labeling it so—see chapter 1), Dewey added that first experiences are overwhelmingly social and continue to be through life. Dewey's outlook not only resembled that of his compatriot Mead; it also resonated with ideas of the Russian developmental psychologist Lev Vygotsky—popular among 4E researchers—who stressed the blurring of individual and collective life. Acknowledging a "mystery as to why any such thing as being conscious should exist," Dewey stressed that, given that it does,

> there is no mystery in its being connected with what it is connected with. That is to say, if an activity which is an interaction of various factors, or a grouped activity, comes to consciousness it seems natural that it should take the form of an emotion, belief or purpose that reflects the interaction, that it should be an "our" consciousness or a "my" consciousness. And by this is meant both that it will be shared by those who are implicated in the associative custom, or more or less alike in them all, and that it will be felt or thought to concern others as well as one's self. . . . In short, the primary facts . . . center about collective habit, custom. . . . The problem of social psychology is not how either individual or collective mind forms social groups and customs, but how different customs, established interacting arrangements, form and nurture different minds.[35]

Dewey added that "each person begins a helpless, dependent creature"[36] and that "the contacts of the little child with nature are mediated by other persons. Mother and nurse, father and older children, determine what experiences the child shall have."[37] In short, caregivers modify what the child does and undergoes, and the child in fact does the same to its caregivers, forming a social sensorimotor loop.

Empirical support for this outlook can be drawn from a number of quarters, one being developmental psychology. First, while not wishing to overemphasize the point, we argue that imitative capacities are omnipresent; and though not sufficient for the social patterns of doing and undergoing in Dewey's sense, they facilitate them. Numerous studies find that infants and even neonates imitate facial gestures.[38] Such capacities are important in early interactions, which are again more than just imitative. A caregiver may smile, and the baby may reciprocate and perhaps add a playful vocalization, provoking a new response from the caregiver that continues the cycle. On the neural side, moreover, evidence links social organization to action, perception, and cognition, and therefore also sensorimotor patterns of doing and undergoing. Consider the basal ganglia once more. As discussed earlier in this chapter and the previous one, these subcortical nuclei deal with probabilistic reasoning, syntax, reward appraisal, movement organization, prediction in auditory perception, and more. Movement-enabling dopamine pathways innervate the substantia nigra, the putman, and caudate nucleus. These same regions of the basal ganglia appear to facilitate prosocial synchronization,[39] or, in other words, group coordination.

Though probably overemphasized, mirror neurons—also discussed in the previous chapter—likewise speak to capacities for imitation and more. While some mirror neurons in monkeys fire when executed and perceived tasks are the same, others fire when performed and observed actions have the same goal, even if reached by different methods.[40] Specific cases in which mirror neurons activate or fail to do so are also suggestive. Thus, regardless of whether a monkey reaches into a peanut bag or sees another do so, mirror neurons fire, but not if the actors or observers know the bag is empty.[41] Some mirror neurons respond to both visual and auditory aspects of actions,[42] and some specifically to communicative behaviors such as lip smacking.[43] Insofar as mirror neurons mobilize when either performing goals or recognizing like intentions in others, and insofar as communicative aspects of cognition appear to involve them, this already suggests social dimensions beyond mere imitation. Reinforcing the overall picture is the fact that experiencing distress and witnessing others undergoing it mobilizes some of the same neural regions.[44]

Though drawing on research attesting to the importance of imitation, Michael Tomasello and Colwyn Trevarthen likewise argue that social cohesion between caregivers and infants involves more than this. Both have conducted studies suggesting that caregivers and infants bind into coupled units such that they mutually define one another's psychological terrain.[45] Trevarthen's research specifically highlights how infants and parents enter into rhythmic back-and-forth exchanges, as opposed to mere patterns of imitation. Neonate movements and gaze coordinate with caregivers, with purposeful, affectively charged, provocation-response interplay occurring throughout their interactions. Crucially, moreover, babies do not merely respond; they also engage in behaviors that are "seductive" insofar as they tempt and invite caregivers into interactions.[46]

Such coordinations can be subtle and involve changes in heart rate as well as overt movements and vocalizations. Specifically, studies show that increased heart rates accompany imitative actions in newborns and that decreased heart rates occur with provoking behaviors, which researchers interpret respectively as indexes of emotional arousal and signs of attentional concentration.[47] Together, this indicates environmental attunement centered on caregivers. Infants become avoidant and distressed at inappropriate behaviors, as when confronted by a blank face or even friendly and playful expressions that are out of sync with the particular engagement.[48] Clearly, then, infants are not mere imitators (though it is doubtful many ever thought they were). Instead, they are sensitive agents who respond to and elicit actions.

Such social interactions fit the general pattern of sensorimotor loops described by enactivists, pragmatists, and to some extent embodied phenomenologists. Put otherwise, they fall within the purview of what Dewey thought of as connections of doing and undergoing, albeit ones considerably more involved than rolling bottles between the palms or negotiating icy trails. In these intersubjective scenarios, an infant may push into its social environment by making eye contact. A parent may push back by smiling, which might provoke a coo, which in turn draws a vocalization from the caregiver. Parent and child thereby engage in mutually provoking and receptive patterns of doing and undergoing, integrating into coupled units. Reiterating this and backing it up with decades of studies, Trevarthen emphasizes that infants are not passive recipients of stimuli and indeed are not wholly governed by the actions of others, since they are also elicitors of behaviors in the dyad.

This exemplifies a key pragmatic claim, one obvious to biologists and in everyday observation: that organisms change their environments. This appears unexceptionally true. Existence, growth, expansion, and even death, whether in human beings, frogs, ants, trees, clams, or amoebae, all introduce significant alterations to surrounding environments. Dewey centrally maintained this in his *Reconstruction in Philosophy*, which might also have been titled "Reconstruction in Life," urging that organisms act on and in their environments.[49] They select and ingest food and form sheltering bodily structures such as shells, bones, and membranes. They also build protective edifices and otherwise cultivate ecosystems that are conducive to their vitality. Actions accordingly have bidirectional outcomes, for adjustments to environments in turn modify the lives of organisms. As Dewey summed up, "there is no such thing in a living creature as mere conformity to conditions."[50] Trevarthen echoes this position. Driving it home, he quotes Alfred North Whitehead, who offered views similar to Dewey when he observed that "there are two sides to the machinery involved in the development of nature. On the one side there is a given environment with organisms adapting themselves to it. The other side of the evolutionary machinery, the neglected side, is expressed by the word creativeness. The organisms can create their own environment."[51] Whitehead added that creativity nearly always necessitates a community and, with that, joint action. This is because isolated organisms are nearly helpless and thus ineffectual, creatively or otherwise—a position also defended by Dewey.

The observations of Dewey and Whitehead emphatically apply to children. This is not only because they are dependent; it is also because they act to change their social world, cementing bonds and cultivating and discovering the integration of their own bodies in the process. Infants and, indeed, neonates show self-synchrony, which is to say, parts of their body coordinate together spontaneously—something that is a precursor to coherent perception in Dewey's and Merleau-Ponty's views.[52] As seen, infants also engage in what Trevarthen calls "inter-synchrony."[53] This means that interpersonal coordinations align with the looks, movements, and vocalizations of adults. Over weeks and months, these sympathetic and synchronized negotiations between infants and adults develop "into a mastery of the rituals and symbols of a germinal culture, long before any words are learned."[54] These increasingly complicated rhythmic interactions suggest selective awareness, affective appraisals, and growing levels of purposeful consciousness. Much of this is archetypally cognitive insofar as it entails attentional processes, symbols, and rituals.

More broadly, the foregoing suggests that experience, from day one, is cultural, and becomes increasingly so as life progresses. A few summative comments from Trevarthen are worth quoting. He writes: "Cultures depend on a ceaseless, highly creative learning process, which does not just come from instructing the young."[55] Indeed, a "talent for companionship in experience, which is mediated by an intersubjective transfer of intentions, interests, and feeling in conversations of rhythmic motor activity" also instills culture.[56] Even before acquiring language, therefore, "the child is already becoming a thinker and actor in cultural ways. The motivation for this learned transformation in activities and experience appears to be a direct outgrowth of the integrated mobility."[57] Trevarthen goes on to say: "What may be called socio-ceptive regulation of actions, in relationships and communities, leads to development of collective ways of behaving that provide an environment of common understanding: a *habitus* for cooperative life."[58]

AESTHETICS, BIOLOGY, EXPERIENCE

Last chapter we argued that perception and cognition integrate with motor systems and introduced neurobiological findings to affirm this. To review a few points, we noted that imagining or performing tasks can equally activate motor areas. Executing actions and perceiving others carrying them out likewise mobilizes overlapping neural regions. This suggests broad synchronization in our psychic engagements. Synchronization, harmony, and the like also feature in aesthetic encounters, and a thesis advanced at the beginning of the chapter, which we will now address, was that social cohesion has aesthetic dimensions. The particular model of aesthetics we have in mind is Dewey's, along with consonant work from biology as well as experimental and developmental psychology.

To lay out Dewey's concept of "aesthetics," we begin by reviewing what he understood ordinary experience to be. Suppose, as an illustration, that a hiker flexes her legs into a trail, the trail pushes back, and the hiker's body undergoes motion. As this pattern repeats, the hiker's gait settles into rhythms of doing and undergoing. In one instance, she stumbles over root-cobbled ground. Her posture synchronizes to compensate: her hands and arms move forward to restore balance, then protectively down (along with her gaze), all accompanied by a sharp intake of breath and perhaps an expletive. When she happens upon icy portions of the trail—the sort where ice clings to heavily trodden paths when snow elsewhere has mostly thawed—she moves to the side to avoid them, pressing into the adjacent leaf-strewn ground that offers a thin layer of muck and frozen earth below. Here, too, her body undergoes synchronized adjustments to the terrain. Likewise, when something tugs her attention and ears, a pivot of legs, torso, head, and eyeballs occur along with other intricate coordinations. On broader levels, the hiker's movements—and therewith her cognitive mapping of the area—coordinate to affordances of the environment.

Such integrations remain even absent vigorous movement. If writing a book at a computer (in effect, thinking through the computer), the woman's eyes might rove the table. Her gaze dances over the books strewn over it, pausing on an empty plate nestled beside her computer and the pencil tucked beneath the plate's lip. Fragrant bananas pull her attention to a fruit bowl resting on the lace tablecloth. Faint stains on the cloth from her now finished lunch tug at her gaze, arresting its motion. She feels the strain in her eyes, hands, wrist tendons, and knotted back muscles as she types. In a frustrated moment, she rocks gently, pulling one leg up, hugging it, and resting her chin on her knee, her foot pressed against the seat of the chair, enacting a total coordination as she types. Her attention focuses on the screen, but strays between clicking keys, the TV in the background, and the running water in the kitchen where her colleague cleans dishes. For the sake of an example she is constructing, she ponders why various phases of the moon are only visible at particular times of day, and arranges a mug and pieces of gum to solve the problem. In these and other ways, her experience entails doing and undergoing, with mild periods of tension and repose being part of the process. It is a mundane symphony of coordinated affective, cognitive, motor, and perceptual activity.

While anything qualifying as experience—as opposed to sensory excitations—has a basic level of coherence and quasi-narrative structure, the two scenarios just described remain somewhat scattered. Neither is "an experience" (a phrase Dewey introduced to connote aesthetic experience).[59] For Dewey, an experience is one that dramatically integrates into fulfillment and stands out as a unified whole, as opposed to being a collection of fleeting episodes. In his account, aesthetic experience involves disruption of habits and convention, which is part of what lends drama. Simultaneously, drama requires some continuity, so the habit-base required for almost anything we do is necessary for aesthetic experience too. Dewey accordingly described aesthetics as a union of the unsettled and the assured.[60] Thus, while paintings, movies, music, and other arts have unexpected twists that challenge us, this combines with predictable patterns we can follow.

The basic point is reinforced by experimental research—not to mention common sense—suggesting that adults and children prefer moderate complexity,[61] where complexity is computationally equivalent to the uncertainty resolved upon the reception of information.[62] In everyday terms, simplicity and predictability stultify, while over-complexity aggravates, and we like to balance the two. Pragmatically speaking, a level of predictability is useful, but novelty is also beneficial, as when exploring new areas and potentially discovering untapped resources.[63] Converging further with this picture, the basal ganglia are innervated and contiguous with the amygdalae (a pair of almond shaped structures typically referred to in the singular "amygdala"). The basal ganglia are critical for habit incorporation;[64] the amygdalae responsive to discrepant or unexpected aesthetic progressions,[65] and long understood to play a role in emotional processing. This mirrors Dewey's scheme, where tensions form when the established and unsettled or predictable and unexpected combine, heightening dramatic structure—in other words, what we experience as the aesthetic.[66]

As an illustration of the aesthetic, imagine now that the woman mentioned earlier hikes on a morning when the sun dances on trees coated in freezing rain. Despite a recent thaw, the air hovers around freezing temperatures, refreshing her face. The usual squirrels scurry and birds sing, but more actively today. At one point she startles a group of deer who, as they flee, almost seem to float away. The trail has interesting ups and downs, twists and turns. Sometimes it tunnels beneath ice-encrusted trees, sometimes through meadows, at one point alongside a bubbling creek. It is rugged, with demanding climbs that push the heart and lungs and build lactic acid in the calves, and stunning views to reward its tough ascents. In later portions of the hike, the woman happens upon an old logging road previously hidden by snow blanketing the countryside. She presses through expectantly, tugged around bends by her eagerness to see what she will discover. She loses herself in the activity of hiking and in her surroundings.

Under normal circumstances, the woman's gait necessarily synchronizes with the contours of her surroundings. On this morning this occurs too, but the woman is more fully engrossed, her mind not wandering to credit card bills, impending deadlines, and other banalities. Compared to normal experiences, she is more completely directed outwards and interested in what she encounters. This means that she is more bodily engaged, since the environment holds the focus of her ears, eyes, and emotions, mobilizing the turn of her head, torso, and gaze and, along with them, her attention and awareness. Moreover, her heart rate and respiration coordinate not just to the demanding climbs, but also to the anticipation of exploring new territory and gentle thrill of coming upon the group of deer. The twists and turns, ups and downs introduce additional rhythms of tension and repose. The previously unexplored area breaks routine, adding further drama and thus increasing interested engagement and the melting of self into world. Scurrying squirrels and singing birds supply additional embellishments, as does glittering ice on trees. The breathtaking views she works so hard to reach offer climatic moments. Together, this pulls what might have been a softly assembled scattering of episodes into a unified whole that stands out as *an* experience. To quote Dewey, the woman's day is "demarcated in the general stream of experience from other experiences."[67] It integrates tightly and carries "its course to fulfillment,"[68] or rather a series of fulfillments, with particular highlights. As with focal points and climaxes in art, these fulfillments and highlights pull the experience into a unified whole that "carries with it its own individualizing quality and self-sufficiency. It is *an* experience."[69]

Dewey's version of aesthetics renders it an everyday occurrence as opposed to a phenomenon confined to theaters and exhibits, resonating with enactive accounts and affordance theory. The reason for this is plain: Dewey's account frames aesthetic experience as a mode of skilled exploration. In the hiking example this is evident enough as to require no further explanation. However, it holds equally when it comes, for example, to perceiving paintings. Dewey offered an illustration. Suppose we encounter a painting, he wrote,

[and our] attention is first caught by the objects in which masses point upward: the first impression is that of movement from below to above.

This statement does not mean that the spectator is explicitly conscious of vertically direct rhythms, but that, if he stops to analyze, he finds that the first and dominant impression is determined by patterns so constituted by rhythms. Meantime the eye is also moving across the picture though the interest remains in patterns that rise. Then there is a halt, an arrest, a punctuating pause as vision comes in the opposite lower corner upon a definite mass that instead of fitting into the vertical patterns transfers attention to the weight of horizontally disposed masses. . . . [T]he close of one phase of order gives a new set to expectancy and this is fulfilled as vision travels back, by a series of colored areas dominantly horizontal in character. Then, as that phase of perception completes itself, attention is drawn to the ordered variation in color characteristic of these masses. Then as attention is redirected to the vertical patterns—at the point from which we set out—we miss the design constituted by color variation and find attention directed toward spatial intervals determined by a series of receding and intertwined planes.[70]

This illustration, highlighting the exploratory aspects of aesthetic perception, suggests that our eyes, so to speak, get educated by their cyclic tours around the painting so that they become sensitive to features formerly missed. As Dewey summed up in an earlier writing, "old habits are deployed in new ways, ways in which they are adapted to a more completely integrated world so that they themselves achieve a new integration. Hence the liberating, expansive power of art."[71]

Notice that this experience, like the hike, has dramatic punctuations, climactic focal points, twists, and turns. On a perceptual-motor level, intermodal coordination occurs. Because of this, Dewey argued that the experience of a painting is never purely visual. Instead, a "tendency to turn the eyes and head is absorbed into a multitude of other impulses"; and while "the eye is primarily active,. . . the color quality is affected by qualities of other senses overtly active in earlier experience."[72] Put more broadly, the act of looking recruits established and skilled habits even as the painting interrupts customary routines. This parallels how artists draw upon trained convention when producing work, even while pushing beyond it. This also occurs on a conceptual level, such as when art points to uncomfortable facts about violence or sexuality. This breaks routine. At the same time, such disruptions are only possible insofar as routines are already there. Nonetheless, if the experience is aesthetic in Dewey's sense, it will pull affective, cognitive, motor, and perceptual capacities into unity, albeit partly by challenging entrenched habits—since "more of the same" usually leads only to partial engagement and coordination.[73]

We want to propose that group action and social intercourse, especially in early stages, mirror what Dewey says about aesthetics. In turn, his notions of both ordinary and aesthetic experience (especially his idea of aesthetics as the unity of the unsettled and assured) relate to biological adaptation and change. One adaptive process, albeit a debated one, is "allostasis," defined roughly as the stabilization of life through change. The concept, first introduced by Peter Sterling and Joseph Eyer,[74] stands as an addition to homeostasis. One might say that whereas homeostasis is the maintenance of equilibrium through small adjustments—for example, drinking to restore hydration levels—allostasis is larger scale, more coordinated, and entails anticipatory alterations, sometimes under stress. An example might be cortisol and adrenaline release or increased heart rate and blood pressure, accompanied by the neuronal alterations and hypervigilance that occur in expectation of challenges. Moreover, whereas homeostasis characterizes shared optimal thresholds, which apply in cases such as sodium balance, allostasis points to circumstances in which optimal conditions vary with situation.

Brown fat, to consider an illustration, is important for thermogenesis and possibly insulation.[75] Optimal levels of brown fat for people in cold winter climates who have limited ability to heat their environment—say, the Inuit around eight hundred years ago—accordingly differ from those for people in temperate regions. Optimal levels also vary with altered routine, as when travelling between climates. Moreover, light levels interplay with neurotransmitters and hormones that moderate circadian rhythms and moods.[76] This suggests that optimal states vary with circumstances, so that sleep and lethargy might have been beneficial for those dealing with winter conditions in the past. This is insofar as torpor preserves calories when hunting and food gathering opportunities are limited, meaning that what we call seasonal affective disorder may have been adaptive. Notice, moreover, that lethargy brought on by such conditions once again entails a multitude of synchronized coordinations of body and brain activity with environmental circumstances, albeit on a more temporally extended level than typically considered by sensorimotor theories.

Changes can run deep as organisms constantly adapt to circumstances. Neurogenesis appears to take place in the hippocampal regions,[77] a pair of coiled, tube-like, subcortical structures especially associated with memory. Studies further find dendrite atrophy in this same area in animals that are subjected to stress by daily immobilization. This is accompanied by dendritic branching and, thus, expansion within the amygdala.[78] Seasonal variations in entire brain regions likewise occur in bird species.[79] Notice, moreover, that while changes in laboratory studies are inflicted, like a punch in the face, under normal circumstances patterns are not simply impressed on the organism from without. As mentioned, a person could move to a more temperate climate, sometimes quite easily, such as by relocating from a higher altitude to a lower one. People can also engage in immediate undertakings that affect hormone and neurotransmitter levels, such as exercising or staying inside watching television. Actions likewise moderate learning, stress, and neurobiological outcomes. While the seasonal neuro-plasticity in birds appears to occur largely in anticipation of mating, it also corresponds to migration. Hence it, too, is partly undergone by virtue of what birds do (as would align with Dewey's views). The processes described here are therefore temporally extended, multisystem coordinations of doing and undergoing in which synchronizations between organism and environment occur.

The word "homeostasis" is somewhat redundant; both elements in the compound suggest invariance. Allostasis, by contrast, appears contradic-tory; it combines the Greek *allo* for "variable" with the root "stasis," which in English means "unchanging" and in Greek, "standing" or "stoppage." One of the term's progenitors characterizes it as "stability through change,"[80] though a better rendering might be "vitality" or "maintenance of life through change,"[81] as the example of winter climate indicates. Simultane-ously, the verbal formulation of "stability through change" hits close to what Dewey says about aesthetics. He emphasizes the union of the assured and the unsettled, which could also be described as a union of the stable and the changing. In Dewey's aesthetics, moreover, it is through change that dramatic structures and culminations arise, allowing what might be scat-tered and fleeting moments to gather into a structured experience that gains stability and endurance as a unity.[82] Orson Welles's 1958 film, *Touch of Evil*, for example, is not encountered as a collection of separate shots and sound. Rather, it changes and builds dramatically in time, acquiring structure, unity, and endurance and thus standing as an identifiable—in some sense, a stable—whole.

There are concrete connections between aesthetics and stabilizing life through change, and they reiterate the exploratory aspects of aesthetics. This is not to say that aesthetic preference correlates to increased exploration, though this is often the case. Rather, it is to suggest that aesthetic sensibility organizes exploratory capacities and partly arises out of them. The earlier illustration of winter torpor illustrates the point, for it turns out that lethargic moods correlate with aesthetic preferences for enclosed, protective spaces. This inclination might keep one inside during winter months, limiting exploration. Conversely, and not surprisingly, energetic moods correlate with preferences for open, explorable spaces.[83] Lending further credence to these exploratory conceptions of aesthetics is neuroimaging research that has detected activity in brain areas associated with emotion and motor-response when people view art or listen to music.[84] That aesthetic—and, by extension, emotional—perception relates to what we feel capable of doing receives additional support from experiments on affordance theory, which find that people judge distant grades steeper or further away when wearing heavy backpacks, fatigued, in poor health, or suffering from low blood sugar (see chapter 2). Similarly, perceived steepness and distance, which are often accompanied by emotional deflation, relate to the ease or difficulty of navigating one's body—which is to say, they relate to the situation one finds oneself in, characterized by both the local environs and what one can do in them. This affects how tempting or forbidding a space is likely to be. In other words, aesthetic preference relates to explorability. Taken together, these findings reinforce the claim that aesthetics can be characterized as the stabilization of vitality through change.

SOCIAL AND AESTHETIC EXPERIENCE

Early development in humans typifies the maintenance of vitality through change, and this simultaneously characterizes the aesthetic. This is not merely because vitality necessitates nutrition intake; the growth of muscles, heart, kidneys, and other vital organs; and rapid neurobiological alterations. It is additionally because exploratory processes that unify routine with the unexpected are critical. Children show interest in objects that violate expectations, spending time exploring them and engaging in hypothesis-testing experiments such as banging on things.[85] Outside the laboratory, such behavior occurs in playful interactions with adults and other children, which demand group coordination. The processes described by developmental researchers, insofar as they involve established patterns and playful variations of them, are loosely allostatic and also aesthetic and cultural in Dewey's senses of the terms.

At early stages, infants engage in games of joint attention centered on goals and outcomes in a shared world, their intentional, sympathetic collaboration increasing with age.[86] Trevarthen stresses the aesthetic dimensions of this, which he also connects to cultural learning—an account that resonates with Dewey's understanding of aesthetics and experience as culture. This is not surprising given Trevarthen's early training. Trevarthen spent some of his formative years in Jerome Bruner's laboratories. There, Trevarthen and colleagues examined whether young infants built social trust through shared games that developed into story making and eventually linguistic communication.[87] Games and stories have dramatic, narrative-like qualities. Aesthetic experience, in Dewey's account, holds together dramatically; the combination of settled patterns and the unexpected turn is part of what lends drama. In early experience, drama and narrative involve establishing rituals and violating them within bounds, as one regularly observes in games between parents and young children. As such, early growth and experience appear to revolve at least partly around the aesthetic.

Like Bruner, Trevarthen acknowledges an intellectual kinship to Lou Sander's systems theory that characterizes shared experience or co-consciousness as a mutual regulation of emotions and brain activities.[88] He also credits Edward Tronick,[89] who advances a dyadic theory of early consciousness that emphasizes the interlocking of infant's and caregiver's minds. According to this account,

> each individual is a self-organizing system that creates its own states of consciousness—states of brain organization—which can be expanded into more coherent and complex states in collaboration with another self-organizing system. When the collaboration of two brains is successful each fulfills the system principle of increasing their coherence and complexity. The states of consciousness of the infant and of the mother are more inclusive and coherent at the moment when they form a dyadic state. . . . [90]

Claims about synchronizing brains are admittedly speculative. However, they are given credence by mirror neuron research and other studies, cited earlier, that showed like regions mobilizing regardless of whether subjects performed actions or were indirectly engaged in them. Additional support is offered by research indicating the same regardless of whether people experience pain or witness loved ones suffering it. This is more so when combined with studies showing that infants have greater dislike for the smell of cigarettes and alcohol when around people who are using them to deal with dissatisfaction, suggesting empathetic synchronization.[91]

Tronick's and Sander's positions highlight another commonality between Dewey's aesthetics and the developmental research of Trevarthen and others. This is the co-emergence and hence intermingling of self with world—specifically a social world centered on caregivers in early stages. This, as intimated earlier, is not only characteristic of the aesthetic, but also a means to it. It involves outward interactions that lead to an interlocking of self and world—in this case a predominately social world.

Other connections between aesthetics and child development exist. Some researchers argue that fetal, neonate, infant, and child movements all exhibit the four-part structure of narrative; that is, introduction, development, climax, and resolution.[92] Protoconversation begins early, at about four weeks. At two months, dialogues "exhibit . . . rhythmic steps, affective melodies and narrative envelopes of energy cycles."[93] Play emerges around three months, when games are repeatedly enacted and remembered with pleasure; at about five months, games acquire such sophistication that caregivers begin good-natured teasing, and infants reciprocate provocatively, feigning disobedience for fun; by six months, clear-cut cases of narrative and song enter into exchanges.[94] Games and play arguably have roles in integrating the child's experience of self and world and in increasing coherence of both. Trevarthen suggests that "meaning grows by confirming and 'cultivating' innate rhythms and values." This adds dramatic structure such that meaning and communication are "fabricated with *aesthetic sensibility*."[95] While all of this entails rituals and established habits, part of the fun for caregivers and infants is introducing small deviations—play between the assured and unsettled.

In advancing an aesthetic account, Trevarthen often references Stephen Malloch's theory of communicative musicality, which suggests that rhythm, pulse, expressive sounds, and movement engender drama in such a way that

protoconversations gain melodic form.[96] Here, too, biological evidence reinforces the point. Specifically, it indicates overlapping neural origins for
music and language, especially in Broca's area, but other regions as well.[97]
This makes sense given that language and music both entail syntactic
form[98]—that is, systems of ordering, albeit often flexibly deployed. Moreover, premotor and motor areas, as well as the basal ganglia, activate in
response to music.[99] As discussed earlier, the basal ganglia have a role in
syntactic regulation besides more customarily cited motor-habitual programs. Motor areas in the cortex likewise respond to language, particularly verbs. Additionally, this and earlier chapters have already laid out
arguments for theoretical connections between motor activity, language,
and cognition. That music mobilizes these regions fits with the everyday
role of motor-activity in making and enjoying music.

Connections between language and music are further affirmed by the
plain fact that infant-directed vocalizations nearly universally employ
melodically and rhythmically exaggerated patterns called "motherese"
(although the label is misleading, since both males and females employ it).
This is done to soothe, be playful, and emotionally engage. In the same way
that syntax familiarity helps us anticipate possible paths of sentences, infant
directed vocalizations—like musical melodies—carry expectations of certain progressions and closures. Exaggerated pulse, tone, and emotional
expression may additionally serve as a kind of over-punctuation, increasing legibility of early communicative attempts and introducing affectively
charged meaning—a point reinforcing the intertwining of cognitive and
emotional capacities. Research using magnetoencephalography (an imaging technique measuring magnetic activity as an index of electrical patterns
in the brain) has also found that exposure to music increases neural
responses to temporal violations in both music and speech.[100]

Connections between music, language, and early learning reiterate group
aspects of experience and cognition. Language is unequivocally cultural,
and may be a means of brain-to-brain coordination.[101] Music is likewise cultural. Darwin postulated that song facilitates social bonding and specifically mating in both human and nonhuman worlds, the latter as in the case
of birds.[102] Hormones play a role in bonding too, with tentative links to
music. Among the two hormones better known for this function are the
peptides vasopressin and oxytocin, which are produced in the brain as well
as extraneural structures such as the placenta.[103] Vasopressin inhalation

mobilizes brain regions that may enhance recognition of beliefs and desires in others.[104] Oxytocin similarly relates to social engagement.[105] Human experiments find, for example, that oxytocin inhalation augments contact, recognition, and social memory,[106] and higher levels correlate with cooperation in trust-dependent games.[107] Though one does not want to oversell these results given the sparseness of research and the variety of findings, several studies show that oxytocin increases after either hearing or performing certain kinds of music.[108] One of these found elevations following participation in jazz improvisation,[109] which involves significant attunement to others. Elevations likewise appear to accompany affectionate parenting behavior, including the use of "motherese."[110]

On an overt level, music often involves group coordination, such as audiences swaying reciprocally with the patterns of doing and undergoing that performers enact on stage. Language can likewise mobilize group coordination, whether in back-and-forth banter, talking someone through a task, or synchronizing a complicated group endeavor. As affirmed by developmental research and everyday observation, similar coordinations dominate parent-infant interactions. Quasi-musical utterances are critical in these interactions, this arguably even in deaf children, who may still be guided through rhythmic movements, hand clapping, and the like. This is to affirm more broadly that early interactions overwhelmingly have an aesthetic character, which is accordingly important to the social development of mind.

RECONSTRUCTION IN AESTHETIC AND SOCIAL LIFE

"Reconstruction"—a term favored by Dewey—expresses the idea that things are not formed from scratch, but re-formed from what is already at hand. This stands as a reminder that we are always in a medial position, in the middle of things, already engaged in a world—a world, wrote Dewey, "that is human as well as physical, that includes the materials of tradition and institutions as well as local surroundings."[111] It is a reminder that experience arises from interacting with one's social and physical environment and, by extension, rearranging portions of it. Mutual adaptation means mutual change of what preexists interactions. It thus entails mutual reconstruction. Dewey claimed that this "is true . . . of every experience. The creature operating may be a thinker in his study and the environment with which he interacts may consist of ideas instead of stone."[112] This indicates that thinking, like art and experience in general, is also reconstructive, a point Dewey explicitly made when he wrote that "thinking is no different in kind from the use of natural materials and energies, say fire and tools, to refine, re-order, and shape other natural materials, say ore";[113] or when he claimed that "we cannot lay hold of the new" nor "even keep it before our minds, much less understand it, save by the use of ideas and knowledge we already possess."[114]

Historians use the term "reconstruction" to connote the activity of putting the past together. In interpreting a past event, however, historians have knowledge of what followed it; they also have access to conceptual resources—theories and so forth—that did not exist at the time of the event. This knowledge inevitably qualifies what the past event is understood to mean. Dewey made a similar point about means and ends in his famous 1896 reflex arc article and elsewhere, arguing that while means temporally precede ends, ends are often analytically prior to means.[115] In other words, means and, with them, meanings, are often grasped from the standpoint of ends finally reached. In the case of aesthetic experience and its narrative-like structure, culminations and fulfillments are end-standpoints that recast elements within works of art. When our eye falls upon a focal point in a painting (which is such only by virtue of what leads to it), we experience how other elements within the work become the means by which the focal point is accentuated and one's attention is directed towards it. When we reach the climax of a play, earlier portions acquire new significance and meaning.[116] In these and other ways, our engagements (which are exploratory in the enactivist and Gibsonian sense) reconstruct works of art from the standpoint of where they finally lead. We thereby come to experience individual elements within works as forwardings (i.e., means) that jointly contribute to an overall culmination, which is part of the reason why Dewey claimed that aesthetic experience builds into a unified and *mean*ingful whole.[117]

Leon de Bruin and Sanneke de Haan, while not focusing on aesthetics, have likewise observed that narrative practices are reconstructive.[118] While sympathetic to enactive approaches, they question whether sensorimotor coupling accounts for narrative.[119] Their criticism would indeed hold if sensorimotor processes were exclusively a coupling of stimuli excitations and actions. Yet, we have argued—and will continue to build this point—that human cognition and affect infuse action, and thus also the sensorimotor patternings that make perception. Dewey captured this in his illustration of the child and fire discussed last chapter.[120] There, he described organisms acting according to their own structures and those of their environments. These actions introduce changes to environments and the organisms undergo corresponding consequences, which lead in turn to additional actions. Dewey's specific example, found in his 1896 reflex arc critique and later works such as *Reconstruction in Philosophy*, recounts a busy infant putting its finger into a fire. The initial action is aimless and without reflection—what some have described as motor babbling.[121] In consequence, wrote Dewey, the "child undergoes heat" and "suffers pain."[122] He elaborated: "The doing and undergoing, the reaching and the burn, are connected. One comes to suggest and mean the other."[123] The sensorimotor coordination therefore comes not only to have perceptual character, but also cognitive and affective import.

Notice that the protogames developmental researchers describe have much the same form as that found in Dewey's illustration. That is, both have loose narrative structure: events culminate in a semiclimactic moment that recasts what came before and welds what might otherwise be individual episodes into a meaningful event. This is not to deny a level of abstraction, imagination, or environmental "decoupling," especially on the side of the caregiver. That said, abstraction relates to action, as Lakoff and Johnson's exhaustive surveys show.[124] They point out, for instance, that "balance" characterizes not only surefootedness, but also everything from justice to mathematical equations to emotional states. Decoupling, in any case, does not undermine the claim that embodied sensorimotor engagements (both linguistic and prelinguistic) are primary from early stages. Furthermore, if decoupling happens, recoupling simultaneously occurs insofar as narratives structure environmental interactions.[125] Narrative experience is not exclusively sensorimotor in the narrow sense (that is, limited to the meanings of the two words in the compound term), and human affairs are virtually never so restricted. This is insofar as action is cognitive and affective, which means that perception is too—a point some affirm to be true for all biological organisms.[126] Actions occur in the world, and without a world, cognition, emotion, and perception would have next to no meaning or value in life.

Aesthetic experience is, of course, more emphatically dramatic than the example with the burn; hence, it is more integrated. A movie, to repeat, coheres around a combination of predictable as well as unexpected events along with climactic moments that pull incidents together, such that we see it as *a* movie, a whole. From early stages, many infant-caregiver interactions satisfy similar conditions. Dewey asserts that emphatic cases of aesthetic experience involve a dissolving of self into world, as when getting lost in movies, paintings, or hiking.[127] Interestingly, this last aspect appears to be the default starting point of human experience. As Noë puts it, "children are not separate; they are not observers; they are regulated by their mothers' soothing or alerting tones, eye contact, gestures, and touch. A mother is literally one of the structures constituting a child's psychological landscape."[128] Simultaneously, infants appear to gain an increasing sense of themselves as individuals. By about six months, they acquire "a proud performer's personality," which they express by "showing off."[129] They correspondingly seem to grasp others as intentional agents, turning to look at the same object that draws their caregivers' attention, as opposed to merely mimicking their caregivers' movements.[130] They exhibit a grasp of manners and sensitivity to identity, displaying "teasing happiness in the company of familiar playmates, shyness with intrusive approach of a stranger, and shame when unable to sustain . . . a familiar performance with someone who does not play their part."[131]

In short, caregivers form loci around which behaviors and arguably experiences of infants coordinate. Infants do the same for those engaging with them. The integration of doing and undergoing between caregiver and child is tightly bound. These everyday interactions have a quasi- or overtly aesthetic character: they reconstructively mix routine with playful twists, and they build dramatically and narratively. In these early experiences, action, affect, cognition, and perception are fused, just as they are throughout all of human life, albeit here in ways that are more overt and easier to track.

ACTION, AFFECT, COGNITION, PERCEPTION

We conclude this chapter by returning to Dewey's theory of perception, which, along with the work of James, Merleau-Ponty, and others, forms the intellectual lineage that Gibson continues.[132] Gibson's affordance theory posits that we see the world as openings and closures for action. He hints that affordances can be read like faces, and that we might see and indeed conceive an expression or a rushing river as threatening, thus avoiding both.[133] More recent theorists direct Gibson's work to the social world.[134] We review this along with how aesthetic sensorimotor accounts apply on cultural levels, emphasizing the integrated nature of action, affect, cognition, and perception.

In a vein similar to Dewey's and building on Gibson, psychologists Rachel and Stephen Kaplan assert that perception is affective, arguing that some settings have an enticing quality that pulls us into them.[135] Specifically, their experiments indicate that we are tempted by open spaces partly occluded by foliage, or trails disappearing around bends. Though they do not put it this way, the openness *assures* by offering a degree of legibility and predictability. Hidden portions, on the other hand, offer novelty: the promise of discovery. Thus, aesthetic settings, like works of art, combine the assured and unsettled; that is, they offer predictability with challenging twists. This entices penetration, as in the common experience of being pulled around bending trails or city streets or through Japanese gardens that never disclose a full view from any one vantage point.[136] Such scenes have a sense of "mystery," defined as allure arising from things "partially hidden" that "tempts one to explore further."[137]

Stephen Kaplan is a computer scientist as well as a psychologist, and he and his wife Rachel connect their work to information theory.[138] Information is here equivalent to the number of yes/no questions—zeros and ones in computational language—needed to code a message, image, and so forth.[139] The key point is that more information equals a greater number of possibilities, and hence more uncertainty. Thus, the union of openness and occlusion that characterizes aesthetics tends towards moderate uncertainty, as Daniel Berlyne prominently argued.[140] In a pragmatic vein, moreover, the Kaplans stress the adaptive utility of aesthetic preferences. That is, they posit that there are evolutionarily instilled desires to explore novel areas in search of untapped resources, but that they occur within bounds of reasonable predictability.[141] Relating aesthetic and affective perception to action and cognition, the Kaplans write that aesthetic perception reflects "a very rapid (albeit unconscious) assessment of what it is possible to do in the setting,"[142] and conclude that it falls within what Gibson called affordances.[143]

The Kaplans' and Gibson's positions can be extended to the social because human expression and gesture structure worldly experience by introducing constraints and openings not present in the brute physical world. In developmental contexts, this can include tracking and responding to the intonation of adult frequencies and co-vocalizing while postures, expressions, gestures, and utterances organize interpersonal space.[144] People focus especially on faces of partners and reciprocal actions, which cultivates affective intimacy and communication. So far this adds little to what developmental psychologists, not to mention Dewey and Merleau-Ponty, have been arguing for decades. Nonetheless, the concept of social affordances introduces something important: the recognition that human expression, gesture, and social interaction structure our experienced world by simplifying choice.[145] Specifically, gestures and other forms of expressive engagement draw attention to and, indeed, establish social affordances within interpersonal space—affordances that both constrain and open avenues of interaction.

This emphasis on interactive, jointly constituted, and coregulated openings and constraints resonates with Deweyan, Gibsonian, and enactive views. At the same time, social affordances offer avenues not available in the brute physical world and not emphatically dealt with by Gibson. This is because, at any given time, possibilities in the brute world can be acted on or neglected according to interests, whereas social milieus have allowances and constraints that cannot be ignored in the same way.[146] In other words, social affordances constrict interpersonal space such that people, if attuned, feel that available options are fewer than they were moments ago.[147] Simultaneously, options may be more numerous. For instance, if someone extends a hand, an option for contact emerges that was heretofore unavailable. At the same time, it is difficult to forego this option, meaning that possibilities shrink. With these concurrently expanded and constrained options, there also comes a possibility of increased drama insofar as handshakes and welcoming smiles integrate group activity into connections of means-consequence, while also forming mini-climaxes that unify interactions into identifiable wholes. Interestingly, refused handshakes, awkward glances, and social ruptures can do the same, reinforcing Dewey's observation that aesthetic experience does not have to be beautiful, though it typically is.[148] In either case, gestures and body language scaffold "spatial arrangements that simplify choice by constraining or cueing social affordances" that "sculpt the attention" and "ease . . . epistemic burden."[149] Attention simultaneously involves cognitive and affective weighing along with action selection and perceptual filtering.[150] It accordingly follows that action, affect, cognition, and perception knit together. This is even more so when emotional, attention-directing, and group-synchronizing aspects of language are factored into the equation.

This account distinguishes social affordances from standard perceptual ones while acknowledging that they can at times be identical. It also connects the idea of experience as culture back to motor explanations. In Deweyan and Merleau-Pontian accounts—not to mention Gibsonian and enactive ones—we perceive things as we do because of actions afforded by both the structure of our bodies and things encountered, and also because the same structures exclude certain actions. Thus, whereas playmates can roll balls to one another, the same action and therefore experience is impossible with cinderblocks. We accordingly encounter balls and cinderblocks as affording different things and come to perceive and conceptualize them differently.[151]

In all this, bodily movement remains primary. Yet, our first world is also overwhelmingly social. Early interactions are largely dyadic; that is, between infant and caregiver. Over time, interactions become increasingly triadic, which is to say, organized around child, adult, and the world to which they jointly attend.[152] So although many embodied accounts start with examples of handling balls and whatnot and build their perceptual and cognitive theories from there, we argue that movement does not precede social engagement, but rather intertwines with it. Further, while the social world affords and constrains actions and therefore experiences in ways similar to the primarily physical environment, it also introduces constraints not present in the immediate brute world. Thus, the notion of experience as culture is more than a restrictive sensorimotor account, even though the two are related. While this is so, we have also argued that no experience whatsoever can be reduced to a mere combination of sensory and motor activity.

4

PRAGMATISM AND AFFECTIVE COGNITION

The historically received view among those recognizing links between affect and cognition has been that it leads reasoning astray. David Hume, for instance, allowed that cognition depends on emotionally driven mental habits, but cited this as cause to doubt our beliefs.[1] Friedrich Nietzsche suggested something comparable, although adding that reason stripped of emotion is life denying and pushes us even farther from reality.[2] Plato also derided emotions, and when he—or his dialogues—permitted them an important role, the suggestion was that they be contained to certain spheres of life and kept subservient to reason.[3]

Classical pragmatists similarly held that thinking, abstraction, and indeed experience knot with emotional, interested and aesthetic life. However, they broke with the bulk of Western tradition by arguing that the emotional, interested, and aesthetic sides of thought contribute to its rationality, bringing it into touch with what may colloquially be called "reality." This was especially so for James[4] and Dewey, who not only affirmed the rationality of affective cognition, but also its biological and evolutionary utility.[5] Even Peirce, while wary of what he saw as James's mushy emphasis on personal interests, saw a role for emotion in inquiry, particularly the irritation of not knowing that provokes investigative seeking.[6] Mead likewise believed that emotions drive cognitive processes, including language.[7]

This chapter focuses on James's affective account of cognition. It also attends to Dewey's integration of emotion, cognition, and action with support from Peirce and Mead. Specifically, it examines how classical pragmatism meshes with comparable pictures that have emerged in neurobiology in recent decades. Some major neurobiologists echo the pragmatic position (repeated in existential phenomenology) that little in the way of differentiated thought and reason occurs in the absence of emotions and interests.[8] Pragmatic and recent accounts also share an oversight: namely, a widespread failure to emphasize ties between emotions and interests despite obvious experiential, conceptual, and neurobiological overlap.[9] In some cases this perhaps follows from overly narrow conceptions of emotion and interest that draw too sharp a distinction between the two.[10] Together, these oversights degrade what might otherwise be a more expansive account of affective cognition.

Pragmatic views on affective life also resonate with and augment a number of cognitive scientific frameworks. This includes theories emphasizing cognitive and perceptual shortcuts—for example, Gerd Gigerenzer's notion of "ecological rationality"[11] and Andy Clark and Karl Friston's accounts of active sampling and inference.[12] It additionally includes research discussing satisfying closure and coherence in cognition and highlighting how emotional motivations push inquiry.[13] In what follows, we consider the affective, active, and searching sides of cognition (cognition's relationship to active perception will be more thoroughly addressed in the next chapter). The concept of "foraging" nicely captures these aspects of cognition.[14] It can be taken either as a literal manipulation and selective gathering that leads to knowledge, or as something achieved through selective attention and emotional weighing—a kind of cognitive grazing, as it were. Along these lines, thinking is often an appetitive, driven, and active search for what is cognitively sating. Regulation, coordination, and synchronization encompass a great deal of what we say insofar as emotions systematically direct attention and action, which circles back on affective life.

In light of this and what has been said in previous chapters, the aforementioned outlooks lead easily to an account of perception. We touch upon this in the current chapter, giving more emphasis to perception's affective or valuative dimensions than is typical while nonetheless stressing its cognitive and active sides. In the next chapter, we expand upon the emotionally active, cognitive sides of perception, knitting this with Gibson's affordance theory and related experimental work. We also integrate this with Gestalt psychology, phenomenology, and, of course, classical pragmatism, all of which are in the lineage leading to Gibson.

EMOTIONS AND INTERESTS

In his celebrated 1884 paper, "What Is an Emotion?," James formulated what came to be known as the James-Lange theory. He later elaborated on it in works such as *The Principles of Psychology*. Noting that many presume we suffer loss, experience sorrow, and weep, or see a bear, become frightened, and run, James proposed the reverse: that "we feel sorry because we cry" and "afraid because we tremble."[15] That is, we first perceive affairs, then undergo physiological changes such as tingling and engage in related actions, and only after that do we experience emotions such as grief and fear. James added that, without physiological alterations following perception, our experience "would be purely cognitive in form, pale, colorless, destitute of emotional warmth." In such a case, we might "see the bear, and judge it best to run, receive the insult and deem it right to strike, but we should not actually feel afraid or angry."[16]

Though James did not claim that this account unexceptionally captures emotion,[17] his position made a lasting contribution and continues to influence ranking scholars. This includes figures such as Prinz and Damasio,[18] both of whom cite James's theory of emotions as foundational. Merits of the theory aside, a good deal of what James got right is missed in emotion literature because of the narrow focus on the account introduced in his landmark 1884 paper. In other words, the popularity of the account is part of its failure. Thus, Damasio and others who emphasize emotional bases of rationality using the 1884 article miss key writings such as James's pointedly titled "The Sentiment of Rationality," which knits together interests and cognition. Rose McDermott's "The Feeling of Rationality," for example, advocates that "decision making . . . rests on an integrated notion of emotional rationality."[19] It references James's 1884 "What Is an Emotion?," but, despite sharing a nearly identical thesis and title, not his "Sentiment of Rationality" (or related pieces).

Part of the fault lies with James, who arguably focused overmuch on what he called "standard cases" of emotion.[20] He identified these with peripheral feedback, which indeed characterize archetypical instances, as when fluttering bellies signal a budding romance or mark the agitation of stumbling near a precipice. Yet, we can cherish a beloved grandmother or academic topic without intense bodily tinglings,[21] just as we can see a gloomy landscape when cheerful and thus find correspondingly dour visceral feelings to be absent.[22] Furthermore, while visceral feedback theories would predict diminished emotional capacity following spinal injuries, evidence for this is equivocal.[23] More critically, as handed down to us in language and culture, emotion refers to something that none of the contending scientific theories completely capture. The reason is that the word "emotion" antecedes and in fact makes scientific discussion of the phenomenon possible.[24] This is very much unlike terms such as "cerebral cortex" and "electromagnetic radiation," which are more exclusively products of scientific inquiry and about which everyday culture has comparatively less to say.

The argument is not that emotion lacks a visceral aspect. In archetypical cases it is there unequivocally, but emotion is more complicated than this. The same follows for interests, which James treated as categorically distinct from standard cases of emotion. As he put it in his 1884 emotion article, "there are feelings of pleasure and displeasure, of interest and excitement, bound up with mental operations, but having no obvious bodily expression."[25] In the visceral account defended there, this also meant no manifestation of anything archetypically emotional. As argued, however, the presence or absence of visceral feedback does not unequivocally demarcate either emotions or interests. Moreover, emotions and interests overlap and almost invariably appear together, regardless of whether there are corresponding visceral feelings, and they share functional roles in directing attention.

At the same time, James—perhaps because he wrote in a literary and nontechnical style or perhaps simply because his views shifted—occasionally acknowledged overlap. Though not his view *per se*, his 1884 emotion article cites a self-description from a patient who suffered emotional deprivation concurrently with loss of "lively interest."[26] James more explicitly stated a relation between emotions and interests a half-decade earlier in his 1879 "The Sentiment of Rationality." Having indicated emotion as a central theme in the article via its title, he went on to connect emotion to reasoning while also articulating the role of interests in cognitive attention and abstraction. For just these reasons, overlooking this piece and others expressing comparable insights is a significant lapse for those who cite James as an influence when connecting emotions to reasoning.

The oversight is compounded by glaring conceptual and experiential overlap between emotions and interests. Emotions and interests both direct attention, making certain features salient,[27] as when gazing at beautiful things or focusing on threats. Love obviously entails intense interest. Intense interest in someone—while not necessarily implying love—implies at least an emotional fascination. Emotional fascination sometimes occurs with intense visceral feelings, as in budding romance; yet it might go either way, as with fascination for an academic topic, which may be accompanied by intense visceral feelings, but more often without much. Emotions and interests, as understood in everyday language and culture, may accordingly have or not have visceral dimensions while appearing, most of the time, to be mutually implying.

The overlap between emotions and interests is further substantiated on neurobiological grounds insofar as many researchers in the field emphasize the interweaving of emotional and attentional capacities (the latter are treated as equivalent to interests in James's scheme). The amygdala, once considered a fear area and more generally an emotion center, is now believed to perform a diverse range of cognitive, perceptual, and emotional tasks in coalition with other brain regions. Not surprisingly, therefore, it also regulates behavior, and selective attention is central to all of this.[28] *Descartes' Error: Emotion, Reason, and the Human Brain*, the book that launched Damasio's popularity among lay audiences, advances the same point. There, Damasio observes emotional and attentional deficits accompanying damage to areas in and around the anterior cingulate cortex, which is interinnervated with the amygdala.[29] Elsewhere, Damasio elaborates upon physiological connections between emotion and attention.[30] He notes contiguous interconnected nuclei in the brain stem, which directly connects to the amygdala.[31] These brain stem areas appear to contribute to both affective and attentional processes, thus regulating focus, visceral activity, bodily feedback, and arousal, as well as many emotional feelings. Emotion obviously connects to the appropriate direction of attention. Among much else, emotion supplies a "signal about the organism's past experience with given objects," which provides "a basis for assigning or withholding attention relative to a given object."[32]

While most of the major scientists and philosophers arguing for emotional-cognitive integration would not contest such points, there is nonetheless one area of general disagreement. Specifically, some suggest that emotion and cognition are initially separate (neurobiologically and otherwise), but that they go on to contribute to one another.[33] Others maintain that they are integrated all along.[34] Neurobiological evidence—along with everyday experience—arguably weighs in favor of the latter view. Indeed, most major brain areas appear to have some role in emotion, making it difficult to disentangle affect from cognition (though the resolution of imaging technologies is limited and anatomical divisions can be pretty broad). Leaving this aside, most serious brain researchers have abandoned the classic view that emotion is predominantly buried in "irrational" subcortical areas such as the amygdala bulbs. The latter do, however, offer convenient examples because they are well studied and fairly differentiated, anatomically speaking. Among some of the cortical structures involved in emotional reasoning are the cingulate, insular, and frontal cortex. They innervate portions of the basal ganglia and amygdala, which, in turn, project into other subcortical regions such as the hypothalamus and brain stem.[35] Many of these cortical and subcortical areas appear connected to visceral processing and appraisal systems fundamental to the organization of emotional experience.[36]

Debates about whether cognition and emotion begin as integrated or instead contribute to one another do not make much practical difference for our purposes. The widely relevant and agreed-upon point is that emotion and attention biologically overlap. Damasio claims it is biologically expedient "that structures governing attention and structures processing emotion should be in the vicinity of one another" and that in some cases they be one and the same.[37] This is because "the consequences of having emotion and attention are entirely related."[38] Other research affirms that attention relates to visceral awareness, and that higher interoceptive sensitivity predicts superior attention and performances on certain tasks.[39] This is not to say that emotion—understood interoceptively or otherwise— always improves attention and performance, for there are cases in which this is clearly not so, such as when panic causes us to miss obvious solutions. However, this is not unique to emotion. Detached formal logic can also lead us astray, as Nietzsche was fond of pointing out and as Gigerenzer has more recently affirmed in the case of choosing romantic partners and making many other life decisions.[40]

SCULPTING EXPERIENCE, DIFFERENTIATING THOUGHT

As discussed in chapter 1, James's concept of selective interests has roots in Darwinism,[41] and his first two publications (both from his student years) dealt with it. The first reviewed a work by T. H. Huxley, the formidable and philosophically sophisticated biologist remembered as "Darwin's bulldog."[42] In this work, James mildly rebuked Huxley's radical materialism, yet credited him for proposing that humans are, like other species, subject to transmutation. James's second piece discussed Wallace's 1864 *The Origin of the Human Races*,[43] which advanced a position somewhat contrary to the first: that humans are not subject to natural selection in the same way as other species, for humans are social, sympathetic, and intellectually complex. Wallace reasoned that the complexity of the human mind enables division of labor.[44] For example, the swiftest may hunt and the less athletic may gather food or mold clay, while social arrangements foster sharing. Wallace accordingly speculated that physical weakness does not necessarily entail death. Thus, whereas other species survive environmental changes through physical advantages, which then propagate, Wallace proposed that human evolution is confined to intellectual adaptations, such as conceiving better tools or reciprocal social arrangements. For Wallace, this meant that "man's *body* will have remained generically, or even specifically, the same, while his *head* and *brain* alone will have undergone modification."[45] This was, of course, overreaching, since Wallace's account implies that human evolution, by not specifically prizing physical prowess, might fail to extinguish incompatible features, leading to a less athletic species.

It is not clear that James ever believed that selective pressures exclusively act on the human brain or mind. But crucially, James did retain the idea that "social affections" and intelligence alter the "survival formula" so that individuals may get by, even when poorly adapted to the raw "natural 'outer' environment."[46] These include the "story-teller, the musician, the theologian," and others who receive a livelihood in return for satisfying wants of their community—"wants," James urged, that "are pure social ideals, with nothing outward to correspond to them."[47]

The merits or weaknesses of Wallace's evolutionary position are not important for current purposes. What is crucial is the view of mind James derived, anticipating Damasio and others such as Gibson, as will be seen in the next chapter. As James maintained in a variety of mature works, consciousness is foremost a selecting agency that emphasizes certain things and suppresses all the rest, which it does according to interests. This incidentally also relates to the concept of free will—defined as the sustaining of or focus on one idea over another—that James appropriated from the French neo-Kantian Charles Renouvier during his student years.[48] James's general point was that mind reacts not only to environmental stimuli. It generates interests and future goals not immediately given in the brute environment. By this and functionally similar means, it selectively attends to the world. Interests and goals also engender environment-changing actions, in this way, too, affecting what shows up to cognition and perception.

James—to review and add to what was said near the beginning of this book—turned this outlook against theorists who reduced organisms and their experiences to passive clay shaped by environments. James had British empiricists and neo-Lamarckians in mind, although he obviously exaggerated their views. Had James lived long enough, behaviorists in the vein of B. F. Skinner would doubtless have also attracted his ire. Darwinian thinking, specifically the claim that variations arise independently of environmental pressures that reinforce or extinguish them, provides a foil to these views. Phylogenetically, this is because evolution often involves organisms acting in ways that exploit adaptive advantages and is accordingly not passive (see chapter 1, especially the discussion of Gould). Assimilating this perspective on an ontogenetic level, James mounted two interrelated explanations of how mind can fit environments without being passively shaped by them. First, James noted that the "excessively instable human brain" sometimes spits out "accidental out-births of spontaneous variation"[49]—novel ideas not directly elicited by the environment. Then, depending on whether or not the new ideas help us cope, zero in on what is important, recognize critical relations, or deal fruitfully with matters, the environment either preserves or eradicates them.[50] A case in point might be a novel conversation opener that people use to engage others who interest them romantically, with its continued use or abandonment contingent upon whether it leads to further engagement or recurring rebuffs. Second, and more critically, James asserted that the environment provides sensory variations, which we either notice or ignore or connect this way or that, depending on our interests.[51]

James offered theoretical and empirical reasons for insisting that interests dominate cognition. He noted that in both science and everyday life, raw observations would overwhelm if not for the narrowing effect of interests.[52] Thus, as pointed out earlier, scientific hypotheses foster biases in the nonpejorative sense of shaping not only what we look at, but also our research methods. This, in turn, limits what observations can show up. Something similar occurs when focusing on a conversation at a party to the exclusion of background hubbub, smudges on walls, the air conditioner hum, the smells of food, lint on our colleague's jacket, and grit on the floor. Moreover, even supposing we could absorb every phenomenon, it remains that things typically relate in myriad ways.[53] To repeat a telling example from chapter 1, this is even so in simple geometric figures, so that Necker images appear in various ways depending on how attention is focused, yet never all ways simultaneously (figure 4.1). James accordingly concluded that selective interests are necessary to keep experience from degrading into "utter chaos."[54]

The phenomenon of selective attention is well studied, and its importance is not seriously disputed. Yet, it has often been treated as a cognitive process rather than an affective one. By throwing interests into the mix,

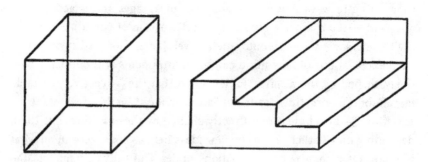

FIGURE 4.1 **More Necker illustrations**

In the image on the left, opposing planes appear as being front or back depending on how you focus your attention. The image to the right inverts depending on whether you focus on the upper or lower non-oblique plane. Adapted from R. Woodworth, *Psychology: A Study of Mental Life* (New York: Henry Holt, 1921), figs. 43 and 44.

James offered a prescient correction to this, illustrating how cognition and affect merge, and anticipating models that later began creeping into the psychological sciences. A prominent example is the Nobel laureate Herbert Simon's classic paper, "Motivational and Emotional Controls of Cognition," published in the *Psychological Review* (the journal having been a forum for more than one pragmatist decades earlier).[55] Simon simplistically reduced cognition to serial processing, the standard design principle for computers at that time and also for his model for AI. He also distinguished fairly sharply between motivating goals and emotions that redirect attention to urgent contingencies. However, Simon may be credited for listing a raft of affective drives involved in cognition. These ranged from forward-pushing aspirations to alternative-searching impatience to action-concluding discouragement or satisficing. These last two end a process either when it takes too long or has too many obstacles, or else when a good enough solution is reached.

This mirrors what James said in "The Sentiment of Rationality" when he stressed that irritation from inconsistencies and inordinate complexity push us forward, with satisfying relief and pleasure marking when we have moved adequately away. These hedonic qualities arguably do more, however, than just signal closure. Positive emotion, like aesthetics of a beguiling sunset or ravishing meal, make the happy outcome attention worthy, giving it weight and fortifying it in memory. So similarly with the negative emotions that motivate us away from unpleasant situations: they often secure events in memory such that we avoid irritation-inducing missteps in the future. This is at least the conclusion drawn by AI researcher David Moffat in his example of a terrified boy fleeing from a tiger, then joyfully escaping up a tree.[56] The fear motivates flight from the threat and makes it salient in memory, which precipitates increased caution in the future. The positive experience of the escape is similarly reinforcing: the boy celebrates the event and shares it fondly with fellow villagers, instilling it in his memory and that of his community.

Recall also that James articulated how sentiments underlie what we normally think of as purely logical, asserting that we experience feelings of "and," "if," "but," and "by."[57] While arguably not enough to fundamentally change the nature of logic, James's analysis suggests that everyday life and language outstrips that discipline. It does so because conventionally equivalent statements such as "P but Q" and "P and Q" have different affective

qualities, with the former expressing a shift in feeling between the two conjuncts and the latter being more consistent. For such reasons, they have
different connotations,[58] just as the double negation in the phrase "I am not
unsatisfied" has a different feel and therefore meaning than the phrase "I
am satisfied."[59] Gigerenzer's 2007 *Gut Feelings* partly elaborates on these
points, albeit strangely just barely focusing on feelings and other forms of
affect. For example, when discussing the sentences "Peggy and Paul married *and* Peggy became pregnant" versus "Peggy became pregnant *and*
Peggy and Paul married," Gigerenzer notes:

> We know intuitively that the two sentences convey different messages. The
> first suggests that pregnancy followed marriage, whereas the second
> implies that pregnancy came first and was possibly the reason for mar-
> riage. If our intuition worked logically and treated the English term *and*
> as the logical AND, we wouldn't notice the difference. *And* can refer to a
> chronological or causal relation, neither of which is commutative. Here
> are two more pairs:
>
> *Mark got angry* and *Mary left. Mary left* and *Mark got angry.*
>
> *Verona is in Italy* and *Valencia is in Spain. Valencia is in Spain* and
> *Verona is in Italy.*
>
> We understand in a blink that the first pair of sentences conveys oppo-
> site causal messages, whereas the second pair is identical in meaning.
> Only in the last pair is the *and* used in the sense of the logical AND.[60]

While insightful and mostly on mark, what Gigerenzer says is not quite
right, for the logical AND is present in every sentence, but does not capture enough. In the examples with Mark and Mary, if the statement is true,
then so are both the conjuncts. The same holds with the examples involving Peggy and Paul. This is according to formal logic, and there is nothing
incorrect here. It is just that everyday language expresses more than only
this, as Gigerenzer nicely conveys.

One reason that everyday language expresses more is because of its valuative or affective connotations. This fits with James's account of logical

connectors, but also with Gigerenzer's illustrations, insofar, for instance, as the straight-laced feel of "Peggy and Paul married *and* Peggy became pregnant" contrasts slightly with the accusatory tone of "Peggy became pregnant *and* Peggy and Paul married." These valuative or affective variations, of course, partly tie to cultural norms—the ecological contexts into which cognition extends. This connects the foregoing account to discussions in chapter 3 of group activity, experience, cognition, linguistic communication, and indeed emotionally expressive protomusicality during developmental periods. In a pair of articles from the *Psychological Bulletin*, Mead specifically located the origins of language in syncopated, emotional social acts.[61] While speculative, it should be recalled from previous chapters that Broca's area and the basal ganglia are involved in both music and language. Furthermore, it is not difficult to imagine someone expressing anger through vocalizations and gestures, and his or her companion in turn exhibiting placating behavior through identical means. This would supply a basis for protolanguage. Emotional tone obviously remains central in language. Thus, while intellectual sorts are wont to malign small talk, often language gets closer to a hug or handshake than relaying facts and reasoned positions, which is to say, it can be predominantly an emotional exchange.

In addition to this, and also connecting emotional valuation to reasoning and language, pragmatists and especially James tied affective life to more elementary functions such as concept formation. In doing so, James wed his pseudo-Darwinian account of selective interest to Peirce's pragmatic maxim, which held that ascertaining conceptual meaning required considering practical effects that objects of conception might have.[62] James thereby emphasized the affective or interested sides of conceptualization in ways that vexed Peirce. Whereas Peirce said that an object conceptualized as "hard" conceivably has the effect of scratching other things, James emphasized personally relevant interests and consequences, such as a "hard" and "heavy" object having the injurious effect of crushing any toes upon which it falls. The already cited but helpful illustration that James supplied was that of a mechanic conceiving oil primarily as a combustible or lubricant, and a carpenter as a darkener of wood.[63] In other words, conceptualizations vary according to values, interests, and ends.

Gigerenzer offers some anecdotal but still compelling evidence for James's position, especially as it pertains to memory (though again not emphasizing affect, in spite of his book being titled *Gut Feelings*).[64] He cites the neurophysiologist Alexander Luria's account of Solomon Shereshevsky, a man with a remarkable memory.[65] He could recall conversations as if recorded and repeat strings of thirty, fifty, or seventy words or numbers, the latter even years later if asked. What he could not do well, however, was remember the gist of things. He struggled to summarize stories that he could repeat word-for-word. He equally struggled to recognize faces, and this *because he remembered them in exact detail*, with all the fluctuations that come with alterations in expression, mood, lighting, and more. This meant that the same face was remembered differently under varying conditions to the point of unrecognizability. Gigerenzer quotes James to have said: "If we remembered everything, we should on most occasions be as ill off as if we remembered nothing."[66] James prefaced: "This peculiar mixture of forgetting with our remembering is but one instance of our mind's selective activity. Selection is the very keel on which our mental ship is built."[67] The Shereshevsky case illustrates problems that emerge when selective attention fails to weigh and accentuate features, and, as importantly, ignore what is less relevant.

Some neurobiological findings support what James, Luria, and Gigerenzer say about memory.[68] For example, amygdala structures innervate hippocampal ones. The former are important for emotional and attentional processes,[69] the latter classically regarded as memory areas. Evidence suggests that amygdalar activity modulates hippocampal activity and that this enhances memory of emotionally or hedonically weighted events.[70] Speaking more generally about cognition, Damasio asserts that having a piece of knowledge in awareness is possible only on the condition of being "able to draw on mechanisms of basic attention, which permit the maintenance of a mental image in consciousness to the relative exclusion of others."[71] This is unequivocally a Jamesian point, only James emphasized interests more than emotions,[72] and a good number of neurobiologists who focus on attention start their articles by citing James's ideas.

As a pointed illustration of cognitive impairments coinciding with affective deficits, Damasio talks extensively about a patient known as Eliot.[73] Eliot was a young man who underwent surgery for a brain tumor that resulted in damage to both his prefrontal cortices and the axons beneath them, the right more than the left. This area comprises a large percentage of the cortex and plays a role in emotion and many other processes, in addition to which it is heavily innervated by the amygdala bulbs. One unsurprising outcome, therefore, was that Eliot had severely diminished emotional experience. More unexpected—from the standpoint of long-held views of the emotions as irrational—was that this was accompanied by reduced decision-making ability. Stranger still was that Eliot remained capable of listing the pros and cons of options and still scored high on IQ tests. In fact, he appeared rational until asked to make a decision. Despite detailing advantages, disadvantages, and consequences, he struggled to state what he would do if faced with a choice. Eliot seemed to have little to guide him in weighing one option over another, analogous to being unable to select items from a menu because of lack of preference and therefore emotional pull. He also seemed to lack the sustained emotional interest to carry through a good many tasks. Outside the laboratory, his decisions were similarly personally and professionally disastrous, costing him his finances, job, and more than one marriage.

As of 1994, Damasio had a dozen other patients with comparable damage, all displaying similar impairments in emotion and decision making. One such patient had a stroke that damaged the dorsal and medial frontal lobe regions in both hemispheres. Lacking speech, movement, and expression, she appeared to suffer locked-in syndrome. Yet interviews after she recovered revealed to some extent that this was not so.[74] She reported having felt little. Unperturbed by her disabled state and consequently having nothing to express at the time, her passivity reflected the numbing of feeling she underwent. In this condition, no decisions were made or implemented, and normally differentiated thought appeared absent, reflecting not just James's views about reasoning, but also his theory of concept formation as interest-based abstraction.[75] Expanding on the difficulties these patients faced, Damasio echoed James's hypothesis. About Eliot, for example, he said that

I began to think that the cold-bloodedness of [his] reasoning prevented him from assigning different values to different options, and made his decision-making landscape hopelessly flat. It might also be that the same cold-bloodedness made his mental landscape too shifty and unsustained for the time required to make response selections.[76]

In James's terminology, lack of emotional engagement suppressed selective interests, leaving Eliot ill-equipped to assign values to different options and formulate aims accordingly. He consequently had limited bases for making decisions.

Throughout his career, James saw mind as a teleological mechanism and thinking as teleologically (that is, goal or interest) driven. Damasio has arrived at the same conclusion, remarking that "there appears to be a collection of systems in the human brain consistently dedicated to the goal-oriented thinking process we call reasoning, and to the response selection we call decision making."[77] These systems, he adds, also connect to emotional feelings, and thus to valuative weighing and differentiation that allow for reasoning. Other neurobiologists advance comparable conclusions.[78] Using amygdala structures to illustrate, Luiz Pessoa observes that damage in this area alters decision making in both humans and animals.[79] More specifically, these neuronal regions appear to help in "biasing the representation of value."[80] In short, affective life "appears to contribute to outcomes frequently linked with reasoning and cognition,"[81] perhaps to the degree that we cannot think in its complete absence. This is what James's classic views and recent neurobiology both suggest.

ANTICIPATION, ACTION, COGNITION, WORLD

Emotion motivates us, as with fleeing the earlier mentioned tiger, or with last chapter's illustration of bending trails pulling us aesthetically and physically into them with their mysterious allure. In these illustrations, emotions are not just reactions to events; they also involve perceptual coupling between organism and environment. In the case of the tiger, coordinated perceptual action circles back to modify emotion when the boy finally escapes up the tree. In the case of the bending trail, there is at least an adjustment of gaze if not exploration of the environment. Emotions also overwhelmingly anticipate future situations, shaping them by redirecting actions and therefore thought and perception too. This appears central to mysterious settings, which entice us specifically because not everything is open to view, yet they show signs of traversability, and these qualities together promise future discovery.[82] Though perhaps more trivial, the fear driving the boy from the tiger likewise reaches forward to his survival, as does his elation insofar as it instills a strategy that may work again.

These are just a few instances of emotion searching ahead, anticipating and predicting. By virtue of this, emotion becomes more than just that: it becomes cognitive. As an opening illustration, consider the sound of an out-of-view car approaching. Mild anticipation characterizes the experience. This is more so if we are expecting someone up our driveway, and here the emotional and cognitive aspects of "Who's that?" or "What's that?" or "Is she already here?" become entangled. Dewey offered examples of a blacksmith timing tasks by intently watching the color and texture of heated iron;[83] or a physician searching for symptoms pointing in some definite direction; or a scientist eyeing an ongoing experiment expectantly. These all entail coordinated coupling with the world, if not by overt action, then at least by direction of attention. Degrees of suspense also pervade the occurrences, knotted with a kind of thoughtfulness and sometimes inference-making, even if not explicit or fully formed.

A key point, then, is that the building tension of anticipation tends strongly and simultaneously towards the emotional and cognitive. Thus, somebody reciting "A, B, C, D, E, F" engenders an integrated emotional and cognitive expectation of hearing "G" for those familiar with the alphabet. Affairs are similar in melodic culminations, as with *The Sound of Music* song "Do-Re-Mi" hovering around the seventh note (*ti*) in the major scale and then resolving into the tonic center (*do*) near the end. Everyday utterances offer further examples, such as the phrase "in the event" anticipating the word "that" or "of" with high probability, but not "elephant."[84] Because the latter would be out of place, most would be apt to register it as a mild emotional and cognitive disruption (the two again experientially entangled), possibly accompanied by expressions of surprise, as when leaning forward with widened eyes. The same might follow if a colleague poured and handed us a mug of coffee that turned out to be cold, or if a sudden gust of wind tossed a frisbee away from our waiting hands.

By definition, thwarted expectations entail anticipation, and are accordingly cognitive in fairly standard senses because they entail grasping a future that did not occur. They can also be epistemically productive insofar as they maintain emotional and cognitive interest, since frisbee would be less engaging if the disc always glided perfectly to our hands. It is on such grounds that painters, filmmakers, musicians, writers, and some teachers endeavor to include unbalancing twists. Threading this into an emotional-aesthetic account of reflection and inquiry, Dewey elaborated:

The rhythm of loss of integration with environment and recovery of union not only persists in man but becomes conscious with him; its conditions are material out of which he forms purposes. Emotion is the conscious sign of a break, actual or impending. The discord is the occasion that induces reflection. Desire for restoration of the union converts mere emotion into interest in objects as conditions of realization of harmony. With the realization, material of reflection is incorporated into objects as their meaning.[85]

To put it in terms more comfortable to experimental psychologists, unforeseen interruptions can motivate anticipatory searching and active inquiry,[86] just as trails disappearing around bends invite expectant exploration in order to fill in missing details.[87] If acted on, exploration typically reduces uncertainty, which is mathematically equivalent to processing information (see also chapter 3).[88] The anticipatory drive to explore is accordingly knowledge seeking, and it balances the assured and unsettled, thereby moderating uncertainty into aesthetically optimal levels (see chapter 3).

The idea that anticipation is central to human psychology and living organisms is as old as it is obvious. In his *Rhetoric*, Aristotle described fear as the anticipation of destructive or painful future events.[89] Roughly two and a half millennia later, James wrote: *"The pursuance of future ends and the choice of means for their attainment, are . . . the* mark and *criterion of the presence of mentality* in a phenomenon."[90] More recently, Daniel Dennett has said that the "purpose of the brain is to produce future." According to Dennett, key to this "is the ability to track or even anticipate the important features of the environments, so all brains are, in essence, *anticipation* machines."[91] All of this implies action impulse, which is actualized much of the time. Most of the time, therefore, it also implies coordinated coupling with things of concern. This view has been given various mathematical formalizations by figures ranging from Hermann von Helmholtz, to the behavioral learning theorists Robert Rescorla and Allan Wagner, to cognitive scientists such as Clark and Friston.[92] What we want to do is reinforce the anticipatory, searching, and therefore cognitive aspects of emotion, which simultaneously entails identifying the emotional sides of the seeking and foraging that characterize cognitive activity.

The anticipatory side of emotion, which implies cognitive function, is in fact central to many of the perspectives already laid out. Damasio, for example, has explicitly framed it as such, and a research team working with him experimentally illustrates interplay between emotional anticipation, decision making, and conceptualization.[93] The experiments focused on gambling decisions in card games with sizable amounts of money staked, comparing people with ventromedial cortical damage (similar to what the patient Eliot had suffered) to those without such trauma. The first group evinced no increase in skin conductance—a sign of emotional arousal— prior to taking cards from risky piles, regardless of the number of trials. By contrast, after a certain number of trials, the second group unexception- ably came to exhibit "anticipatory skin conductance responses . . . when- ever they pondered a choice that turned out to be risky, before they knew explicitly that it was a risky choice."[94] At some time around or after the mid- point of the task, moreover, most participants in the second group entered what the research team calls the "conceptual period." This is to say, they developed overt knowledge about which card piles were risky, and even those participants who had not reached this stage still gravitated towards risk-averse decisions. On the other hand, only a minority of brain-damaged patients entered conceptual phases, and even these continued to anticipate disadvantageously, which is to say that they did not, in effect, anticipate at all.

James's account of emotional motivation towards the resolution of incon- sistency and inordinate complexity can likewise be framed in terms of anticipatory searching. Building somewhat on James's views and citing him, George Loewenstein suggests just this in his landmark article on curiosity in the *Psychological Bulletin*, which also published pieces by James and Mead.[95] As in the account of inquiry that James delivered in "The Senti- ment of Rationality," Loewenstein emphasizes both aversive and hedonic phases of curiosity. Specifically, he suggests that curiosity reflects a desire to close knowledge gaps, which are experienced as unpleasant, in contrast to the agreeable satisfaction that arises from filling in missing information and patching holes in understanding.[96] The appeal of Loewenstein's account is reinforced by common utterances. We speak of "piquing" or "provoking" curiosity, and the words equally mean to arouse, inspire, and vex or anger.[97] Though there is perhaps a risk of over-application, this points to emotional and more specifically aversive sides of curiosity. We also speak of sating curiosity just as we speak of quenching core drives such as hunger, thirst, or sexual desire,[98] thereby also highlighting hedonic dimensions.

The connection between anticipation and Loewenstein's account, along with that of James, is that curiosity or inquiry portends expectations of some result or closure, even if currently undefined and ultimately disappointing. Loewenstein's heavy citation of the psychologist and philosopher Daniel Berlyne reinforces the anticipatory sides of curiosity. Berlyne, it will be remembered from the previous chapter, introduced the view that we aesthetically prefer moderate uncertainty, as mathematically defined by information theory. Uncertainty coincides with anticipation, and in the final decade of Berlyne's life, the psychologists Rachel and Stephen Kaplan—also discussed last chapter—evolved the former's ideas into an aesthetic theory of epistemic exploration, this time in the literal sense of being pulled through environments in search of new discovery, as when foraging or just looking around—or what we have termed *coordinated coupling*. Such circumstances imply the promise or anticipation of obtaining currently unavailable information as well.

Dewey's work also captures much of what Loewenstein says, albeit accentuating the backward-looking aspects of curiosity and anticipation along with more obvious future-directed tendencies. He wrote:

Curiosity, inquiry, investigation, are directed quite as truly into what is going to happen next as into what has happened. An intelligent interest in the latter is an interest in getting evidence, indications, symptoms for inferring the former. Observation is diagnosis and diagnosis implies an interest in anticipation and preparation.[99]

Earlier on the same page, Dewey highlighted the cognitive side of all this:

Specific and wide observation of concrete fact always, then, corresponds not only with a sense of a problem or difficulty, but with some vague sense of the meaning of the difficulty, that is, of what it imports or signifies in subsequent experience. It is a kind of anticipation or prediction of what is coming. We speak, very truly, of impending trouble, and in observing the signs of what the trouble is, we are at the same time expecting, forecasting in short, framing an idea, becoming aware of meaning.[100]

In other works and especially his 1934 *Art as Experience*, Dewey related anticipation to habits and the aesthetic. Habits bring coordinated continuity to action and thought and hence to experience, while instilling future expectations and ways of engaging or coupling with things. They are inclinations to move towards, seek, want and chase after something; they are ways of handling. Future-oriented and, in this sense, anticipatory, habits embody expectations of things unfolding and being managed. At the same time, as "deep-seated habits or organic 'memories,' "[101] they look backwards and in fact inculcate and embody the past. All of this filters and focuses attention and activity, connects current episodes to past and present, and thereby supplies continuity.

Within the more formal cognitive-behaviorist learning theory of Rescorla and Wagner,[102] prediction similarly couples with expectations and their breakdown; or, in other words, prediction couples with the amount of surprise generated by the association of an outcome with a preceding event. This might also be framed as a breakdown of habits insofar as Rescorla and Wagner specifically examine cues with conditioned expectations. When these are violated or thwarted, new problem-solving and search principles emerge. This once again gets close to views that James conveyed in the late 1870s, which Peirce also advanced during the same period.[103]

Indeed, Peirce's views were sometimes remarkably similar to James's. In "The Fixation of Belief," for example, he asserts: "The feeling of believing" indicates "some habit which will determine our actions. Doubt never has such an effect." The article continues:

> Doubt is an uneasy and dissatisfied state from which we struggle to free ourselves and pass into the state of belief; while the latter is a calm and satisfactory state which we do not wish to avoid, or to change to a belief in anything else. . . .
>
> Belief does not make us act at once, but puts us into such a condition that we shall behave in some certain way, when the occasion arises. Doubt . . . stimulates us to inquiry until it is destroyed. This reminds us of the irritation of a nerve and the reflex action produced thereby; while for the analogue of belief, in the nervous system, we must look to what are called nervous associations—for example, to that habit of the nerves in consequence of which the smell of a peach will make the mouth water.[104]

Summing up, Peirce stated: "Irritation of doubt causes a struggle to attain a state of belief." He termed this "inquiry," a function that entails the consequent development of new approaches and solutions to problems.[105] In the same work, he emphasized tenacity, lauding it not as the most reliable standard, but nonetheless a way in which achievements and convictions are produced and maintained. This includes everything from the erection of architectural wonders to dogged pursuit of scientific inquiry in the face of repeated failures—though coerced labor has historically also played a role, as indentured workers and junior scholars both know too well.

Another piece from about ten years earlier drew even tighter connections between cognition and emotion. Indeed, while we advocate emotional-cognitive integration, Peirce was too cognitivist for our tastes, describing feelings as logically determined predicates and representations, and likewise characterizing emotion as "a sign and a predicate of the thing."[106] In all this, he was amazingly dismissive of emotions, even while going on to detail their utility, seemingly without recognition of the inconsistency. He argued that we deem "emotions more as affections of self than other cognitions" because they are "more dependent upon our accidental situation at the moment than other cognitions; but that is only to say that they are cognitions too narrow to be useful."[107] This last part was explained by his belief that hypotheses must be simpler than things explained, and that emotions, being among the narrowest of cognitions, must by definition be simpler than intellectual hypotheses. At the same time, he wrote that everything "in which we take the least interest creates in us its own particular emotion,"[108] which is relevant to attention and, in turn, critical to cognition. He also tacitly framed emotions as anticipatory forces in inquiry, observing that

> the emotions ... arise when our attention is strongly drawn to complex and inconceivable circumstances. Fear arises when we cannot predict our fate ... If there are some indications that something greatly for my interest, and which I have anticipated would happen, may not happen; and if, after weighing probabilities, and inventing safeguards, and straining for further information, I find myself unable to come to any fixed conclusion in reference to the future, in the place of that intellectual hypothetic inference which I seek, the feeling of anxiety arises. When something happens for which I cannot account, I wonder.[109]

Though it is not clear the extent to which Peirce recognized it in this article, this description of emotions once again highlights them as knowledge seeking insofar as they anticipate future possibilities, motivate inquiry, and direct attention.

These emotional processes also stimulate behavior, thus pointing to active foraging or what we called experimentalism in chapters 1 and 2, which in turn implies coordinated coupling. Dewey expressed the point in an early piece on emotion published in the *Psychological Review*.[110] Emphasizing cognitive and active dimensions, he wrote that "expressions of emotion are to be accounted for not by reference to emotion, but by reference to movements having some use, either as direct survivals or as disturbances of teleological co-ordination."[111] He added: "Emotion in its entirety is a mode of behavior which is purposive, or has intellectual content, and which reflects itself into feelings or Affects, as . . . subjective valuation."[112] Mead, in a pair of articles published in the *Psychological Bulletin*,[113] expanded this account of emotional-cognitive action, developing it into a theory of language.[114] As discussed in the previous section, it conceived protolanguage in terms of emotional, dialogical gesturing and posturing, almost of a musical variety if vocalizations are involved. This too entails anticipation, so that one person might express deference through intonation and posture to forestall an anticipated angry reaction, and another might gesture to indicate all is well.

Emotional cognitive action also has a coordinating and binding function on group activity, and not just in the sense that language does. This is because emotional expression can be an anticipatory strategy (often enacted unconsciously) that corrals the future behaviors of others, synchronizing them with an individual's mood.[115] (This reinforces outlooks we defended in chapter 3.) Thus, we might tiptoe around someone in an angry mood; and in response to sad expressions, we may go out of our way to do small or large things to ease the situation of the sufferer. Emotional expression thus shifts habits and alters our worlds, where worlds are constituted of doings and undergoings that organize around things and agents that concern us. Emotions may also have a self-coordinating effect on those who express them. In line with this, some studies suggest that dampening expression with Botox injections or altering them by having participants hold pens in their mouths decreases or modifies emotional experience,[116] though these results are debated.[117]

In chapter 2, we argued extensively that coordinated actions—not brains, nervous systems, or sensory excitations alone—generate perception. We offered considerable empirical support for the position. In chapter 1, we made similar claims about cognition. Here we suggest the same about emotion, adding that emotion has a critical role in actions that coalesce into cognitive experience. This idea is central to classical pragmatism, especially James's position that beliefs are action-producing emotional stances that are grounded in habitual attitudes (Dewey gives even more emphasis to the last part). James specifically argued that one's willingness to act measures one's strength of belief, such that one is not staunchly against poverty if one never does anything to alleviate it. Also critical—and likely arising from his scientific background—was James's argument that acting on beliefs generates consequences that supply evidence after the fact.[118] This occurs in certain scientific experiments, as when the thing tested is not really known beforehand. It also occurs in everyday life, as when discovering that we can do something only after acting on the belief that it is possible.

Strong belief, conviction, or faith, moreover, often has emotional undertones in addition to obvious cognitive ones, which explains why scientists can be as passionate in defending views as religious individuals are.[119] Though it sometimes leads us astray (as in the case of climate change skeptics), passion supplies the necessary tenacity not only to keep us believing in the face of obstacles, but to keep us acting as well. So it was with the painstaking observations and roughly twenty years of mathematical work that yielded Johannes Kepler's laws of planetary motion, a task motivated by his search for God's perfection in the cosmos.[120] Although all this demonstrates an especially Jamesian standpoint, it also comes close to views framed by Peirce from the late 1870s onwards. He, too, regarded beliefs as habits and therefore as action-producing and emotionally motivated impetuses. He sometimes made the latter explicit by emphasizing the role of irritation and tenacity in inquiry. He also tacitly emphasized beliefs as habits, depicting them as not only behaviorally and cognitively anticipatory, but also engendering of emotional expectations. By suggesting that thwarted expectations generate irritation while nudging habits, he offered an account of how anticipatory behaviors develop and evolve, leading to different kinds of searching (or what we have called foraging).

Though some of the key figures considered in this book have exhibited discomfort with talking about beliefs, this pragmatic account generally aligns with phenomenological, Gibsonian, and 4E views. It does so by positing active and searching minds that forage the world for information and understanding: this is not merely in the sense of mentally attending, anticipating, and seeking, but also of actively doing and coordinating, as illustrated in our discussions of habits and the experimentalism introduced in the first two chapters. Experimentalism, again, is the idea that we uncover properties and make sense of the world by actively and somewhat systematically messing with it.[121] When in the midst of figuring out what things are, we thump, shake, rattle, and roll them or otherwise alter conditions under which we observe them (such as by looking through lenses).[122] Much of this occurs as a matter of course when we work or deal with practical matters in our surroundings. In these ways, things manifest according to what we can do to them, and them to us.

Moreover, nonhuman organisms (setting aside the issue of their cognitive sophistication) find themselves in comparable situations. Thus, a cat actualizes sinewy toughness of a wood post when digging its claws in, whereas we realize lacquered smoothness when caressing its polished surface, as argued in earlier chapters. For certain species, dung and rotting flesh are food; for us, these things tend to be perceptually and cognitively repulsive and potentially dangerous, though manure may be fertilizer, fuel, or even building materials for dwellings in some cases. Dung and rotting flesh are certainly not ingestible from our standpoint, barring fetish or perhaps desperate hunger in the case of spoiled meat. Here, once again, things become what they can do to organisms, and organisms to them and with them, though this is neither to impute nor deny the organism's own awareness of this in cases such as an insect devouring carrion. As with perception, then, which is limited by what we can do and consequently how actions and sensory input couple with one another and the world (see chapter 2), cognition is similarly restrained.

ANTICIPATION, COORDINATING, COUPLING

While we are agnostic about debates over hardwiring, it remains the case that our biological needs and availability of possible fulfillments and satisfactions bind with motivations and anticipatory emotional drives, and thus with actions, cognitions, and perceptions too. Anticipation also knots with coordinated patterns of doing and undergoing. As Dewey often reminded his readers, organisms engage in approach and avoidance behaviors at a primordial level, moving towards nourishment and away from noxious substances or predators; they take in and expel, whether through eating and digestion, breathing, or cellular respiration. This is conveyed in the Greek root of the word "psychology," from *psukhē* for breath, life, or soul; and Aristotle framed cognition, perception, and ingestion as equally psychic activities that incorporate and assimilate the external world.[123]

Without defending Aristotle's view, what more recent psychologists characterize as core drives—thirst, hunger, and the like—infuse everyday ways of speaking about cognition. We say a proposal smells bad or fishy or leaves a bad taste in our mouths.[124] "Repulsive," "repellent," and "revolting" are other words for bad tastes and odors, and we revolt and retreat from rotten ideas and schemes.[125] Conversely, inadequate information implies deprivation, and a consequent hunger to fill gaps. We thirst, ache, and find ourselves starved for knowledge, analogous to how we do similarly for sustenance and sexual contact.[126] This, in turn, anticipates approach. Indeed, all these ways of talking suggest, yet again, anticipatory dimensions. We seek alternatives to a fishy plan because we anticipate problems. Hunger for information motivates expectant searching for what is missing. Discouraged by endless wrong turns and anticipating insurmountable obstacles, we lose appetite for further pursuit. This too can be epistemically productive, pushing us away from dead ends and towards more fruitful avenues.[127]

Anticipatory behavior occurs on all levels, answering immediate and more global conditions. Even plants, for example,

> process the information from a changing environment to develop and reproduce. Communication between cells and tissues is essential for plant fitness, which involves an integrated signaling system that includes long-distance electrical signals, vesicle-mediated transport of IAA and production of chemicals known to be neuronal in animals.[128]

Serotonin is one such signaling agent common to plants and animals. In the former, it moderates physiological and developmental functions, shoot formation, flowering, and defense mechanisms.[129] Responses can be surprising, with indications of plants releasing airborne chemical messengers when under attack by herbivores, leading others nearby to increase noxious compounds such as tannins in their leaves; or sending signals that attract predators that eat the insects gnawing on them; or more simply, plants appearing to alter root courses to avoid toxins.[130] These behaviors seem anticipatory because they activate before threats are fully realized, while keeping in reserve those defensive measures with potential costs elsewhere (such as tannin elevation, which has been shown to reduce leaf production).[131] Herbivores, in their turn, sometimes appear to anticipate too, either going upwind or moving to foraging areas far enough away for the plants there not to have received the tannin-eliciting signal.[132]

Aristotle attributed souls or psyche—*psukhē* in Greek—to all living things. While this sounds odd, it is less so when one remembers that the Greek word can literally be translated as "life." That said, Aristotle's point was deeper than this, for he proposed parallels between plants taking nutrients in and other organisms absorbing the world through perception and cognition. Perhaps Aristotle was also sensitive to organism-environment coupling that characterizes cellular life in the plant, animal, or any other taxonomic kingdom, as Herbert Spencer unequivocally recognized millennia later.[133] The aforementioned cases of plant anticipation serve as an illustration, and there are many other examples of coupled coordinations, some of them more global and long term. Just such a case is the process of flowering, which synchronizes with times of year and consequently with clement weather, pollen release from other plants, and the presence of insects to carry the pollen. Blossoming patterns, moreover, occur on all sorts of time scales, sometimes ranging up to years and occasionally in synchrony with lunar cycles.[134] Without suggesting the presence of emotions or consciousness, much of this might be loosely conceived as a plant analogue to arousal systems in animals, which have comparable anticipatory outcomes.

Corticotropin-releasing hormone, or CRH, to offer an illustration, ties to twenty-four-hour anticipatory circadian rhythms, declining during sleep and elevating during wakefulness.[135] Evidence suggests the presence of CRH receptors in the retinas of some species,[136] which makes sense given the connection of circadian rhythms to ambient light. Research further indicates that these may project into nuclei within the hypothalamus, which is thought to have a coordinating role in circadian rhythms.[137] During wakeful periods, moreover, CRH elevates when dealing with threats, and, unsurprisingly, lowers during foraging, courting, mating, and other behaviors not relevant to immediate survival.[138] CRH is involved in energy metabolism, unsurprisingly given increased demands during threats. Levels vary in nonthreatening conditions, where elevations are associated with anticipation of food or glucose reward, especially in the presence of correlates of past rewards.[139] All this, again, relates to energy metabolism. Variations in wakefulness, sleep, arousal, threat exposure, response to danger, ambient light, and foraging and feeding all obviously knit with drives, mood, and activity.[140] Moreover, these oscillations are nearly identical to attention patterns, varying with wakefulness, threat exposure, and opportunities to eat or to pursue mates. Unequivocally tied to attention and doings and undergoings, these synchronized shifts are—according to arguments offered—replete with cognition, not to mention perception.

Emotions and moods contribute to longer-term regulatory and anticipatory functions than this, again entwining with cognitive and perceptual activity. Posttraumatic stress disorder, or PTSD, may be used as an illustration. This condition can follow from any number of unfortunate events, ranging from car accidents to assault to sexual abuse to wartime experiences. There are neurobiological differences between PTSD sufferers and nonafflicted individuals, although it is difficult to track the extent to which these follow from trauma or antecedent variations increasing vulnerability. The sample sizes in studies are small. While many focus on a particular cohort—for example, the survivors of the 1995 Tokyo nerve gas attack—circumstances run the gamut, and measurements are taken at widely varying intervals after the events associated with PTSD onset. Symptoms include numbing or detachment, avoidance, sleep problems, concentration difficulties, impaired inhibitory control, hypervigilance, and anxiety.[141] Much of this, incidentally, is associated with elevated CRH.[142]

In everyday contexts, some of these symptoms might be interpreted as paranoia. Many might label as paranoid a man "furious because thousands of people he has never met are trying to kill him," as the book sleeve of Joseph Heller's 1961 *Catch-22* says of the protagonist Yossarian. The novel goes on to add that "Hitler, Mussolini and Tojo . . . were all out to kill him" too.[143] In point of fact, this would not be paranoia, but a fairly accurate assessment for a WWII bombardier. This is not to deny that many symptoms of PSTD are products of wear and tear on the system. It is instead to assert that in some circumstances, the heightened arousal that goes with the condition can be life preserving, even if it becomes maladaptive upon return to safer life. This might apply to hypervigilance and anticipatory sensitivity associated with angst, in turn cultivating wakefulness and expectant hair-trigger reactions to sirens warning of attacks. Under such circumstances, anticipatory arousal—which here amounts to total body coordination to long term conditions—is simultaneously cognitive: it involves emotional worry and sensitivity to future threats woven into thoughts—or "representations"—about dangers not immediately realized. Indeed, even when these worries become unwarranted, as when back in civilian life, they remain unequivocally cognitive-emotional.

A last example of long-term anticipatory emotional-cognitive coordinated coupling with the environment was touched upon in the previous chapter. This is seasonal affective disorder, a term referring to annual cycles of increased sleep, fatigue, and lowered mood, usually in winter months when light is less and weather cooler.[144] Lowered mood obviously affects how we see and think about things. This is not just in the sense of heightened pessimism or lowered self-worth, but also in less obvious ways such as preferring more enclosed spaces when we are depressed[145] and hills appearing steeper when we are sad.[146] This makes proximate sense, as discussed in chapter 2, insofar as lowered mood correlates with lethargy. Lowered energy, in turn, makes strenuous activity more difficult, so that hills appear as greater obstacles and one's inclination to explore declines.

While perceptual, such consequences are also cognitive, not to mention anticipatory, more so if findings hold outside of the laboratory. This is because preferring a protected space when outside means inclining towards something out of sight and, in this sense, held in thought. Simultaneously, it implies emotional desire to be somewhere else. Moreover, although emotional inclination and mood, like any human faculty, can be maladaptive, and while seasonal affective disorder is overwhelmingly so in contemporary contexts, annual changes in mood and arousal were arguably not always so. Thus those inhabiting Iceland during the Middle Ages would have had relatively little to do outside in the winter. Increased sleep would have preserved scarce calories, as would have lethargy and the accompanying inclination to stay indoors. Without any pressure to do serious work, this arguably would not have been experienced as depression, because part of what defines seasonal affective disorder and makes it troubling is the difficulty it causes in finishing tasks. Likewise, part of what increases irritability and the like is not being able to get desired sleep, which, again, would have been less of an issue, since leisure time arguably increased during winter months. So, once again, actions circle back to affect mood and thought. Affective life, cognition, and behavior are mutually defining.

Taken together, PTSD and seasonal affective disorder point to broad regulatory functions of emotions, mood, and arousal on cognition, as well as on daily activity, perception, and life in general. The cases also imply and connect to fairly immediate ways in which affective life knits with cognition, perception, and activity, while emphasizing the coupling of emotions with environmental conditions both immediate and temporally extended. These rhythms, moreover, are anticipatory, and expectant responses to events are again fundamental in our cognitive capabilities. The biological clocks mentioned earlier, for example, motivate preparation and expectation of action, even if it is an action as simple as searching for coffee in the morning. They accordingly precipitate forethought [147] and highlight the organism—neural and nonneural—as an active searcher that looks to what is to come, reacts to challenges, marshals responses, and forages the world for resources. Behavior, cognition, and anticipatory patterns of emotion, mood, and arousal are mutually coordinating and, indeed, co-constituting.

FACT, VALUE, ACTION

Emotional motivation, arousal, mood, and interests figure into the abstraction of concepts, weighing of options, and decisions made. Much of this is grounded in selective attention, but is mutually coordinating with action. Thus the carpenter's knowledge of soft or hard wood follows from uses to which such materials can be put—that is, from possible actions.[148] This, in turn, implies certain ends and therefore emotionally motivated interests, which circle back to entail certain uses (uses = environmentally embedded bodily actions).

In many ways, this resonates with Gigerenzer's notion of "ecological rationality."[149] It argues that cognition rarely employs universal decision mechanisms, relying instead on heuristics matched to the contours of particular situations. Gigerenzer more specifically maintains that successful strategies in everything from catching a ball to investing money to selecting a life partner neglect the bulk of available options and eschew as much computation as possible. This is because very little of either is relevant most of the time. In other words, and in line with James, he suggests that partial or bounded ignorance often yields more fruit than all-encompassing knowledge. For example, he provides evidence demonstrating that partly ignorant investors who do not recognize many companies and choose equal distributions of stocks based on name familiarity typically outperform experts, indeed vastly so. But Gigerenzer ends up almost entirely missing the affective dimensions of decision making and action. He fails to consider that attraction to familiarity might occur because its emotional comfort exceeds complete unknowns. More critically and generally, he fails to recognize that interests and emotions forge selective attention. By this means, we take probing cognitive samples while excluding most data, especially irrelevant aspects. In short, Gigerenzer's account does not detail how emotions are cognitive and more specifically heuristical—which is to say, emotions are rough and ready problem-solving tools that are part of our adaptive armament.[150]

Though we argue that emotions are overwhelmingly adaptive and rationally productive, we also recognize that they can be successful or unsuccessful, helpful or misleading. However, this is just like any other form of problem solving, which can fail, especially in extreme cases. Accordingly, obsessive logic—or what might be thought of as the "absurdly rational"—can be obstructive to everyday human relationships and life,[151] a point Gigerenzer also defends on conceptual and empirical grounds.[152] Similarly, many political narratives—for example, those prompting xenophobia—are overwhelmingly harmful and grounded in emotionally charged misinformation.[153] Mass addiction to social media and the emotionally fueled choices that it encourages are a related example.[154] At the same time, when it comes to immediate doings and choices dominating everyday life, we mostly do all right. Thus, while many savor cuisine that is bad in excess, most have an emotional interest and hence inclination towards nutrient-rich foods, and are emotionally repulsed by rotting meat and other pathogen-infected fare. In this case, emotional attitudes knit with biologically substantiated concerns for health and are rationally grounded in meaningful senses of that term.[155] So it is with James's ideas about concept formation: the fact that dry cleaner operators may be *emotionally* interested in removing stains, and may hence conceive of and indeed value oil as a fabric tarnisher, does not strip oil of this quality nor undermine its importance in the laundry business.[156]

This account suggests that what we call facts only show up by virtue of values. The oil illustration is an example, and there are others, scientific activity among them, which many nonetheless take as a value-free arbiter of facts. Values, however, enter in at many levels,[157] and not as contaminants, but as heuristical facilitators of knowledge production. Theories are always underdetermined by evidence, and sometimes competitors mesh equally with data, in which case scientists choose the more elegant (aesthetically pleasing) and economical (easier to use) of otherwise equally compelling positions.[158] Scientists, moreover, adopt metaphysical standpoints. Common among these are that all reality is causal, material, or mechanistic—all of which are fruitful assumptions at times. This tends to relegate alleged evidence for alternative realities to the fallacious, so that scientists maintain metaphysical positions as starting points not because of observations, but irrespective of them.[159] These starting points are accordingly not empirically verified, and they impose limits on what investigators will look at ahead of time, what observations count as genuine versus confused, and the methods that will be employed. Values enter scientific inquiry in other ways as well. Theories that fit with other accepted accounts are more comfortable and, all else being equal, preferred. This is not to suggest that they are irrational, but rather that evidence and formal logic are not the sole criteria of scientific inquiry, and further that values are always involved. In effect, these strategies are valuative heuristics, and ones that have proved enormously successful. Taken together with evidential standards, they also prove very rigorous and difficult to achieve, as scientists know well and as James recognized.[160]

A "just the facts" attitude is accordingly impoverished and false. Someone might object, saying, "Here is a hard fact: rain is falling." Yet our conceptions of water and rain tie to their value in various spheres of life, and this is regardless of whether one is a scientist viewing it as droplets of H_2O, a farmer grateful that it is increasing crop yield, or a thirsty dog appreciating accumulated pools. Appraisal systems—whether those of science, philosophy, or organisms coping with environments—are accordingly laden with value, and pragmatists were particularly sensitive to this. As Hilary Putnam says:

> The classical pragmatists, Peirce, James, Dewey and Mead, all held that value and normativity permeate all of experience. In the philosophy of science, what this point of view implied is that normative judgments are essential to the practice of science itself. These pragmatist philosophers did not refer only to the kind of normative judgments that we call "moral" or "ethical"; judgments of "coherence," "plausibility," "reasonableness," "simplicity" and of what Dirac famously called the beauty of a hypothesis, are all normative judgments in Charles Peirce's sense, judgments of "what ought to be" in the case of reasoning.[161]

As James additionally pointed out, even the dictum to reserve decisions until there is full evidence—which is never achieved—is an emotional decision; that is, based on inclination and hence valuation, not data.[162] Outside of the sciences and other formal disciplines, much the same holds. We value the potential of a spouse, house, fruit, and other things, here defined in the context of our interests, and do likewise in considering the dangers or aesthetics of a space.[163]

In the next chapter, we will examine how perception is rife with value in pragmatic, phenomenological, Gestalt, and Gibsonian accounts, and how this fits with the biology of perceiving organisms. By undercutting the fact-value distinction here and later on, however, we do not mean to repudiate the warnings that come with it. In particular, we agree that just because something is natural or pervasive does not mean it is good or an exemplar of how things ought to be. But the problem here is not one of drawing normative claims out of supposedly value-free facts. Instead, it is one of introducing values that are different from ones already there, and not appreciating, moreover, that human beings can judge certain things to be undesirable or morally problematic regardless of whether they are inherent in biological nature or otherwise pervasively there.

The main point we have tried to advance in this chapter is the knotted character of cognition and affect and thus also valuation. Cognition straddles diverse appetitive and consummatory experiences.[164] "Appetitive" reflects the active search mode and selective attention, and "consummatory" indicates diverse forms of satisfaction involved in cognition, not to mention the narrative-aesthetic dimensions detailed in the last chapter. Replete with goals and expectations, cognitive activity such as conceiving, decision making, and memory processes are grounded in appraisals—affective valuations—that permeate, chisel, and give form and meaning to our experiences. Cognition weaves with emotion and interests, which, in turn, are bound to arousal systems and bodily regulation, all of which is embedded in activity in environments. Affective dimensions thus play a part in bringing about coherent experience of both self and world, and are accordingly a precondition of rationality too.

5

PERCEPTION, AFFECT, WORLD

In his treatment of interests, James not only anticipated neuroscientific developments connecting emotion to reason. He and other pragmatists—most notably, Dewey—additionally anticipated a great deal of what Gibson would go on to articulate in his theory of affordances. Specifically, pragmatists foresaw Gibson's position that we perceive the world in terms of possibilities of action, lessening the divide between subject and object. Embodied phenomenologists advanced similar positions and were also among Gibson's influences.

Affordance theory holds that organisms perceive according to bodily capacities in environments. Insofar as bodies and environmental resources are involved, this means in terms of what is biologically needful and hazardous. For this reason, affordance theory resonates with valuative frameworks that suggest that perceptual systems are affectively geared towards avenues of satisfaction and avoidance.[1] These are the main points we will address in this chapter. Throughout the chapter, we examine standard cases typically considered under the auspices of embodied philosophy, 4E cognitive science, and affordance theory—cases that entail gross bodily movements such as walking, reaching, and grabbing. We also examine less-considered processes, specifically, how perception organizes around vital functions, such as gustation and digestion, that link to core drives and thus affective life.[2] Because nourishment is essential for all cellular life, these functions run from elementary levels of sensorimotor organization on up to the most complicated forms, in turn connecting to overt movement in the world. In the same way that we earlier followed Dewey in arguing that unicellular life has preconditions of experience, we suggest the same holds for affect and values.[3]

Gibson cited affective and valuative renderings of perception offered by Gestalt psychologists, and his work was also influenced by pragmatists and phenomenologists.[4] In this spirit, we draw on these schools, along with experimental psychology, to argue that what researchers in Damasio's vein say about cognition also applies to perception—in short, that emotional and interested dimensions shape perception. Understood as aspects of perception, emotion and interests are cognitively endowed, information-rich ways of being in touch with the world. This position fits within a sensorimotor framework insofar as perception organizes around emotional-interested shifts in attention and action, the latter circling back to affect what we notice and how we feel about it.

Building on the last chapter, this one introduces empirical research emphasizing the appetitive sides of perception, or what might be called gustatory affordances,[5] a concept hinted at in Dewey's work.[6] As with cognition, gustation entails appetitive satisfaction, active searching, reaching, and manipulating. In the previous chapter, we characterized this as "foraging," understood in a knitted cognitive-behavioral sense. An additional claim we make is that the body itself selectively filters environmental stimuli. This is not just in the sense that brains and desires have evolved according to physical needs and resource availability; irrespective of biologically based psychology, bodies sift environmental information, their physical structures automatically functioning as mediators. Thus, for example, eardrums, bones, hair cells, and other parts of the auditory system only vibrate within certain frequency ranges, just as certain manipulations fall outside the scope of what hands can do with things. Such restraints are prevaluative insofar as they set limits on what falls within perceptual interest. As such, they are additional coherence-inducing mechanisms, and for this reason are important to perception and cognition.

As we push standard conceptions of perception in these ways, we continue to develop a situational metaphysics and epistemology: the view that interrelated systems are where things with determinable and hence perceivable and knowable properties first show up. Once again, this outlook undercuts skeptical frameworks suggesting that we are forever one step removed from reality, trapped in worlds of representation. It does so by emphasizing that few if any determinable properties—including primary ones—are expressible outside relationships; it points to the arbitrariness of elevating observations achieved through scientific tools while diminishing those engendered through human interactions; and it suggests that insofar as representations are at play, their job is not primarily to mirror the world, but to aid action in it. It indicates, in short, that perceptual apparatuses are tools for negotiating environments. A tool such as a shovel is not judged according to how well it imitates or mirrors the soil it digs, but on its ability to perform an action and reach different ends in varying situations.[7] So it should also be for perception. These points are in line with what we have called the experimentalism of pragmatism (see chapters 1 and 2), and are reinforced by fairly recent realizations in AI, all of which we consider in the context of emotional life in this chapter.

An additional and related position we defend was also touched upon in chapters 1 and 2, and this is that emotional qualities are not solely inside organisms, but are instead qualities of interactions in which both organic and extra-organic things partake.[8] That is to propose that perceived emotional qualities (for instance, a somber landscape) arise in the same general way that other perceptual qualities do. This is to affirm, once more, that emotions are information-rich ways of being in touch with our surroundings. It is to further suggest that emotions are nearly meaningless and unidentifiable outside of situational contexts in which cognition, motor activity, and perception simultaneously arise.

ACTION, PERCEPTION, COGNITION AND VALUES

The last chapter cited James's claim that recollecting everything amounts to memory problems. This was borne out in the case of Shereshevsky, the man with a nearly perfect memory that left him incapable of getting the gist of things and recognizing faces.[9] The problem for him was that details obscured generalities. When it came to faces, for instance, memories included variations in expression, lighting, and more; lacking selective filtering, he failed to abstract defining features. Shereshevsky suffered impaired memory therefore precisely because he remembered too much (see chapter 4). James's repeatedly cited if somewhat artificial example of the carpenter versus mechanic illustrates a comparable point, where the two are interested in different effects of oil and accordingly abstract different concepts of it.[10] The basic precept can be extended to perception so that, for example, the carpenter not only values and hence conceives oil as a wood darkener, but *sees* it this way as well.[11] A hiker likewise might perceive a river as cooling, freezing, drinkable, navigable, or obstructive depending on what she is interested in doing. This injects values and action into perception, and also into cognition insofar as the process entails identifying substances and environmental contours according to needs.

While values vary between the mechanic and carpenter, James observed that some interests remain reasonably persistent, thus rendering widely shared conceptualizations.[12] The same applies to perceiving. Hence nearly everyone perceives flooding rivers as threatening even if they are enjoying a view of them from safe distances—something that can be attributed to the relatively stable interest in avoiding deadly risk.[13] Other interests vary more widely, and the Gestalt psychologist Kurt Koffka, quoting Kurt Lewin, observed along Jamesian lines that "a person's world undergoes a fundamental change when his fundamental aims are changed."[14] Though lacking sustained discussion of interests, recent scientific work supports this view. For example, studies find that cigarettes appear longer to deprived smokers, glasses of water taller to the thirsty, and tools such as shovels larger if emotionally inclined towards tasks such as gardening.[15] This relates squarely to selective attention, more so if increased size is a measure of heightened salience, as seems likely. Neurobiological findings, moreover, suggest that emotion and attention have comparable roles in cognition and perception: that of filtering and organizing the world along with its options and choices.[16] In fact, a good amount of neurobiological and experimental work on selective attention, which is regarded as archetypically cognitive, revolves around perceptual sorting tasks.

A second line of defense for valuative, affectively charged perception comes from Gibson's 1979 book on ecological psychology, which references James in its opening pages. Gibson's insight, again, was that we perceive in terms of the ease or difficulty of bodily actions we might take in the world. This obviously resonates with pragmatism and was influenced by it, along with Gestalt psychology and phenomenology. Though it does not emphasize interests and it shies from any talk of subjective experience, Gibson's work is rife with discussions of perceptual value. Accordingly, his account approaches the pragmatic idea that interests delineate our worlds and that things show up in terms of possible use-values (though this is not how Gibson typically put it). Thus, to adapt some of his examples, filthiness makes water salient as a solvent and hence a cleanser; thirst emphasizes its drinkability, just as a rock may show up as a hammer, missile, paperweight, or pendulum bob, depending on one's goals.[17]

Here one might also consider the emotional pull towards familiar faces in crowds; or consider serene or angry cloudscapes, dull classroom settings, or, again, a cozy Christmas scene with a tree, lights, and decorations, with a cat warming itself by a fire. Gibson nicely condensed much of this by summarizing the Gestalt suggestion—developed from the work of their contemporary Heinz Werner[18]—that environments have "physiognomic qualities."[19] This suggests that they convey emotions almost as readily as faces. Facial expressions, in turn, can be invitations to approach, avoid, tread lightly, and more. On this Gestalt view, as Gibson elaborated, "the meaning and the value of a thing seems to be perceived just as immediately as its color." Gibson went on to remark, quoting from Koffka's 1935 *Principles of Gestalt Psychology*:

> "Each thing says what it is. . . . a fruit says 'Eat me'; water says 'Drink me'; thunder says 'Fear me' [. . .]." These values are vivid and essential features of the experience itself. The postbox "invites: the mailing of a letter, the handle "wants to be grasped," and things "tell us what to do with them." Hence, they have what Koffka called "demand character."[20]

At first glance, this seems to invert James's account insofar as it speaks of the world projecting interests and emotions onto us rather than the reverse. Notice, however, that this is only possible in the presence of an organism with a body. Bodies have biological requirements. In the case of humans, this is manifested in desires and interests, thereby squaring the position with James's.

Gibson recognized these points, noting that demand characters show up and disappear depending on current interests, whereas affordances—like physiognomic qualities—remain regardless of temporary concerns.[21] Thus, the same person might be lovable or not to different individuals or even to the same individual at different times. By contrast, an angry, flooded river remains an obstacle and threatening regardless of intentions to approach or avoid it. Similarly, we can register pained expressions regardless of whether we feel sympathy or an inclination to give succor. A lesson to be drawn is that wakeful humans find the perceptual world affectively qualified. In other words, we experience emotions not just as internal feelings, but as characters of environments, comparable to expressions on faces or in body language.[22]

This view or ones approaching it are widespread in pragmatic, phenomenological, and Gestalt traditions, and also in outgrowths of these schools. The late psychologist Nico Frijda built on phenomenological and Gibsonian scholarship when he wrote that "emotional experience is primarily a perception: a mode of appearance of the situation,"[23] which echoed Heidegger, Merleau-Ponty, Jean-Paul Sartre, and likeminded thinkers.[24] Frijda went on to add that emotional experience is "objective" insofar as "it grasps and asserts objects with given properties" that are "out there," such that emotional experience is "perception of horrible objects, insupportable people, oppressive events" and the like. Though it cuts against popular opinion to point this out, we regularly speak this way. Koffka observed that we describe a book as "proud" or a young birch as "shy," and that poets speak truly of "the glee of the daffodils."[25] A year before him, Dewey noted that we say: "Situations are depressing, threatening, intolerable, triumphant." According to Dewey, we talk as if emotions are "*to* or *from* or *about* something objective."[26] Merleau-Ponty likewise argued that a purely introspective examination of affective life reduces it to nothing more than pangs, tremors, and heartthrobs that cannot be differentiated into specific emotions.[27]

In short, it is difficult to characterize emotions without worldly reference: we love this person or are happy when our favorite team wins; we get angry at a bad driver and sad because our grandmother died. Dewey excepted instances of breakdown from this, as when people are severely depressed independently of what is occurring in the world.[28] Yet he added that, even here, emotions demand objects beyond themselves, such that depressed people tend to see doom and gloom in nearly everything.[29] However, it is worth adding that experimental work on affordance theory and emotion suggests that this may amount to appropriately perceiving the world according to biological conditions. This might be so if negative views incline individuals to seek shelter in times when depressed energy leaves them ill-equipped to deal with strenuous circumstances (see chapters 2 to 4).

Decades before Dewey, Koffka, and Merleau-Ponty, Peirce likewise emphasized the world-directed cognitive import of emotions, writing that

> every emotion has a subject. If a man is angry, he is saying to himself that this or that is vile and outrageous. If he is in joy, he is saying "this is delicious." If he is wondering, he is saying "this is strange." In short, whenever a man feels, he is thinking of *something*. Even those passions which have no definite object—as melancholy—only come to consciousness through tinging *the objects of thought*.[30]

Indeed, while we say emotions are in us, we equally speak of finding ourselves in emotional moods,[31] which is to say, in the midst of public emotional atmospheres or affective situations. This is so even when nobody is with us, as when gazing upon a gloomy landscape or an emotionally cozy setting with a fir, garlands, and wood fire. The perceptual character of all this is reinforced by the fact that we often sense emotions we do not feel, as in Koffka's example about seeing a gloomy landscape when cheerful; or the common experience of sensing the merriment of a party even when emotionally down; or again of seeing faces expressing emotions we do not feel. Taken together, this suggests that emotional characters in settings and things are more than aggregates of what we presume individuals are experiencing, or indeed what we are feeling.[32]

A number of recent commentators capture some, but not all, of this. Nico Frijda is clearly among them.[33] So too are Rudolf Müllan and Jan Slaby, who detail the public character of some emotions in their introduction to Herman Schmitz's "Emotions Outside the Box."[34] Shame, grief, and anger "seem to possess a room-filling authority that regularly affects or 'grips' even total strangers." This makes them "radically different from other kinds of corporeally moving impulses or stirrings such as hunger and thirst or vigour and languidness, which are felt only by an individual person and by no one else."[35] In another piece, Slaby notes that "situation and feeling are inextricable" and that "gestalt features of the environment and our embodied experience fuse into one another."[36] He connects this to 4E and antecedent movements by adding that all of this is "inextricable from our readiness and willingness to act, including our sense of ability, of strength and control."[37] Yet notice once more that, similarly to seeing color, we can perceive emotional qualities without becoming or being gripped by them.[38] We do this when sensing merriment while sad or when beholding the fury of a social setting or cloudscape while cheerful. The same occurs with art. Thus, while Prinz expresses awe at how minor musical harmonies send listeners into an emotional abyss,[39] more remarkable is the fact that we can sense anguish in Pergolesi's Stabat Mater when happy and hear exultation in Beethoven's "Ode to Joy" when blue.[40] This shows that we can sense emotions in the world without having them, and that we regularly do when perceiving facial expressions and postures.

Dewey's ideas about perception, particularly of art, help elaborate this outlook and specifically why it is misguided to dismiss emotional characters of situations as psychological projections.[41] As discussed in chapter 1 and elsewhere, Dewey urged that subject and object are completed in the same interaction,[42] anticipating Gibson's assertion that affordances are equally about the environment and behavior.[43] Such is exemplified when smoothness and roundness are undergone as consequences of fingertips handling bottles. Dewey asserted that the same holds for art. To reiterate what was said in chapter 1, Dewey asserted that it is misguided to say a picture of a landscape—painted or photographed—causes aesthetic effects in us.[44] This is because the picture emerges as a total effect of an interaction between human capacities and things encountered, so that certain organisms would not see a painting or photograph as a picture. The picture accordingly only appears to entities equipped to see it as such. This further means that its aesthetic qualities, as parts of the total effect, belong to the picture as much as the rest of its properties do. Consider a painted or photographed landscape with melancholic mountains, bashful birches, and angry clouds. In Dewey's analysis, these aesthetic-emotional characters, being also outcomes of overall interactions, are as objective an outcome as the picture itself. The same holds if encountering such characters in melancholic mountains, bashful birches, or angry clouds in a setting such as the Rocky Mountains.[45]

Similar extensions of emotion beyond the skin of organisms that take biological conditions into account are presented in the studies on affordance theory mentioned earlier. These experiments, to reiterate, indicate that people perceive graded inclines as steeper or lengthier to traverse when suffering from fatigue, poor health, low blood sugar, or the burden of heavy backpacks. Lowered energy is common to these situations, as is emotional deflation upon perceiving steepness or distance when suffering fatigue. Decreased energy also characterizes sadness, and not surprisingly, an increase in perceived steepness accompanies this emotion.[46] As also discussed earlier, depressed and therefore lethargic moods similarly correspond to desire for closed and hence action-limiting spaces. Conversely, happy and hence energetic moods correspond to desire for open and thus explorable spaces. All these emotional tones, again, relate to the difficulty or ease of navigating one's body. Consequently, they are not purely in the head, but are ways of grasping bodily capabilities relative to environments—or, in other words, affordances. At the same time, these shifts in outlook depend on biological conditions since they vary depending on whether our energy levels are deflated, neutral, or elevated. Yet this is just a variation of the Gibsonian view that we perceive according to bodily capacities relative to environments.

Our thoughts and perception about the world also depend on phylogenetic histories of doing and undergoing in environments to satisfy biological needs. Wintertime fatigue is one example linking mood, cognition, perception, and action for reasons outlined in the last two chapters. Another case in point, albeit one not explicitly tied to affective life, is the fact that the human range of sensitivity to visible light centers on the peak output of electromagnetic radiation from the sun. This can be compared to cave-dwelling fish that have lost visual sensitivity and, in fact, their eyes as well. As also argued in earlier chapters, it is arbitrary to suggest that any of this makes our contact with the world merely subjective. Instead, it makes our contact relational, and this holds broadly, as when one particle hitting another engenders subatomic observations, or when primary properties such as length and mass vary with relative velocity (something similar holds for color—see chapters 1 and 2).

Notice that in these scientific cases, relationships and the instruments altering them become a kind of system that limits what can show up. The human emotional-perceptual-cognitive system is a variation of this. Like scientific instruments, it introduces changes into relational schemes through which the universe shows up. More specifically, it might be seen as a screen that filters relevant information within the biological limits of the particular organic system, which includes the active creature and its surroundings.

Interests and emotions do this by shifting attention and action. Actions, in turn, focus our attention on the contours of tasks at hand, thus circling back and changing our emotional-interested stance.

VALUES, LIFE, AND STRUCTURE

Emotions entail "action tendencies."[47] In this regard, emotional-interested impulses—which fall within the domain of what we have called valuations— parallel systems in embodied and extended AI platforms that bring structure to action, therewith producing the analogues of perception and cognition. Though this book is not on AI, and though AI is far from achieving anything approaching emotional life, insights from this field are informative, particularly the growing realization that actual worlds are better models than any internal representational system. This position implies a less controversial, pragmatic thesis: that perception and cognition require worlds for their completion. What has been said so far in this chapter suggests the same for emotion.

Brooks is the best-known advocate of embodied and extended approaches in robotics and AI. Among his key insights (which apply to humans as well) is that intelligence is not a sole product of central processing, whether in brains or CPUs. Specifically, his work casts doubt on the imperative of constructing world-models inside processing systems by symbolic means such as predicate logic. While classic symbolic modeling, sometimes called "good old-fashioned AI," has achieved enormous success in some areas, it has overwhelmingly failed to handle complexity involved in everyday practical tasks, including menial ones. Appreciating this, and perhaps overstating his case somewhat, Brooks has asserted that "the power of intelligence" arises "from the coupling of perception and actuation systems,"[48] and urged that under such arrangements, internal models of reality become less important. In this chapter and previous ones, we have advanced the same position with regards to emotion, cognition, and perception, arguing that they integrate in worldly action.

A central reason that worlds need not always be built inside central processing systems is that they are always already available. "The world is its own best model," as Brooks famously and repeatedly declares,[49] adding that it is better than anything that can be constructed. "It is always exactly up to date. It always contains every detail there is to be known. The trick is to sense"—or sample—"it appropriately and often enough."[50] The "world" referred to here encompasses physical surroundings and what embodied beings can do in them. Brooks's claim means that it is the world itself, as opposed to a centrally constructed model of it as described by Dawkins or Hoffman (see chapter 2), that is the guiding principle of action. He writes:

> To build a system based on the physical grounding hypothesis it is necessary to connect it to the world via a set of sensors and actuators. Typed input and output are no longer of interest. They are not physically grounded.
>
> Accepting the physical grounding hypothesis as a basis for research entails building systems in a bottom up manner. High level abstractions have to be made concrete. The constructed system eventually has to express all its goals and desires as physical action, and must extract all its knowledge from physical sensors.[51]

This does not imply that no commands are programmed, but rather that they are simple, and that the world is left to fill them out. This means that a sensorimotor system of some sort is necessary. This might be in combination with a program leading to random wandering, other programs, causing the robot to halt when it encounters obstacles, and still others leading it to approach distant objects. In the case of Long's attempts to robotically model living systems (discussed in chapter 2), this includes light sources that stand in for food.[52] When additional programing layers like this are in place—which, again, do not represent the world, but drive action in it—complicated and intelligent behaviors emerge.

Emotional impulses or action tendencies are in many ways akin to the simple command structures just described, especially because they cease to be simple once enacted in the world. With the world supplying sophisticated grounding, emotions likewise become complicated perceptual-cognitive-motor devices. In fact, absent the world, emotions are arguably unidentifiable and meaningless, little more than pangs, tingles, tremors, and flushes, with virtually nothing to distinguish between love, anger, or hate.[53] In addition to reinforcing evolutionary and biological arguments that stress the importance of action in the world, this also fits with the pragmatic notion that psychological functions (whether emotional, perceptual, or cognitive) are incomplete until discharged in acts (see chapter 1). And acts, of course, occur and therefore are grounded in the world. From this perspective, then, purely internal and fully formed emotions do not exist in the first place, since they are only realized and identifiable as such in the context of the world. Put otherwise, love, anger, hate, and the like are world-directed, and even moods such as general malaise that do not point to specific objects are nonetheless characterized by changes introduced to preferred actions, and therefore to perception and cognition too.

This account goes some way towards dissolving a central debate between internalists and externalists when it comes to emotional experience. Two such individuals are Jesse Prinz and Charles Nussbaum. Prinz rejects situational accounts (at least in his older work), arguing that emotion perception involves sensing internal bodily changes, which, in turn, points to relations between us and the world.[54] "There are important differences," he writes,

> between fear experiences and color experiences. The conscious feelings associated with mental states that represent colors are projected out into the world. When we experience redness, we experience it as if it were out there on the surfaces of objects. Not so with emotions. The feeling of rage, for example, is not projected onto the object of rage; it is experienced as a state within us.[55]

Nussbaum responds to the above quotation by stating: "Like Frijda and like Sartre before him, I think this description of the phenomenology of emotional experience is just mistaken. Rage is not experienced as a state within us, but is projected onto the offending object, which is perceived as hateful."[56] Yet, while experienced in the offending object, this does not mean that rage cannot simultaneously be felt within us.[57] It clearly is, but the feelings, visceral stirrings, and impulses we undergo only make sense and properly become rage when grounded in environing circumstances, without which there is little to distinguish what the emotion is.[58]

Taken together, this suggests that the selective interests of James and the demand characters of Gestalt psychologists and the situational emotions of Dewey are of the environment as much as they are from the organism, such that the emotional "to me" or "for me" dissolves into worldly properties, as Frijda wrote.[59] This is meant experientially, but it also holds literally. It echoes Dewey's general view that "perception and *its* object are built up and completed in one and the same continuing operation,"[60] and that perceptual qualities are consequently "qualities of interactions in which both extra-organic things and organisms partake."[61] Thus, when fingers glide over a glassy surface that does not bite flesh, smoothness is realized. When hands adjust to a bottle, roundness manifests as an outcome of what we can do and what our surroundings afford, to use Gibson's terminology. As Gibson maintained, therefore (essentially repeating Dewey),

> an affordance is neither an objective property nor a subjective property; or it is both if you like. An affordance cuts across the dichotomy of subjective-objective and helps us to understand its inadequacy. It is equally a fact of the environment and a fact of behavior. It is both physical and psychological, yet neither. An affordance points both ways, to the environment and to the observer.[62]

Although Gibson further insisted that affordances remain whether or not organisms are attentive to them, the theory implies interested capacities. This is the case insofar as affordances emerge from organism-environment relations. By extension they are grounded in bodily life, which relates to biological needs and hence emotions, interests, and attention. A wooden post affords scratchability for cats, less so for humans. This being so, cats are emotionally interested in scratchability in ways humans are not.

A common neuroscientific view up to at least the early 1990s goes against the position defended here. This view suggested that emotional and cognitive processing are separate even if they sometimes end up contributing to one another. Out of this, there arose a distinction between "cognitive computations" and "affective computations," each largely carried out at an unconscious level.[63] Joseph LeDoux,[64] for example (whose views have since changed), suggested defining "cognitive computations" as those yielding information about stimuli and their relations to other aspects of the world.[65] He then proposed a second class of cognitive processes that he called "affective computations," which, according to him, interpret information about how a stimulus relates to the well-being and goals of the organism itself. In LeDoux's scheme, cognitive computations tend to lead to additional analysis and further elaboration, whereas affective computations tend to lead to behavioral and physiological action. In our account—as well as those of thinkers such as James, Dewey, Merleau-Ponty, and Gibson—these distinctions do not make sense. Every cognition and perception is grounded in bodily action in the world, meaning that all necessarily relate to the biology and needs of the particular organism. However, for reasons discussed throughout this book, this does not make them merely subjective, since they come to full realization and become what they are through bodily action in the world.

In spite of vacillating a little on the affective character of affordances, Gibson made comparable assertions—for example, declaring that a cliff edge affords falling and accordingly looks dangerous and in fact is so.[66] This again relates to both the environment and the organism since precipices pose little peril to species that can fly or with low mass. Similarly, frigid Arctic water offers different risks and opportunities for polar bears and humans. In no small measure, therefore, the emotional character of perception follows from what bodies can do in environments and what environments can do to them.[67] This goes some way towards explaining why, as Dewey put it, the world starts as "kind and hateful, bland and morose, irritating and comforting, long before she is mathematically qualified or even congeries of 'secondary' qualities."[68]

Dewey beautifully reiterated these points. To express the encompassing nature of emotional perception and cognition and their connection to action, he characterized the thickening of space and time that accompanies the experience of constricted movement:

> Space is room, *Raum,* and room is roominess, a chance to be, live and move. The very word "breathing-space" suggests the choking, the oppression that results when things are constricted. Anger appears to be a reaction in protest against fixed limitation of movement. Lack of room is denial of life, and openness of space is affirmation of its potentiality. Overcrowding, even when it does not impede life, is irritating. What is true of space is true of time. We need a "space of time" in which to accomplish anything significant. Undue haste forced upon us by pressure of circumstances is hateful.[69]

This rendering of the emotionality of space and time fits with Dewey's notion that experience and worldly properties are outcomes of actions taken and effects undergone in consequence. Dewey pointed out that the "old saying that the gods were born of fear" engenders misconceptions, that it insinuates that individuals are first endowed "in isolation with an instinct of fear," only afterwards "irrationally ejecting that fear into the environment."[70] Yet, Dewey insisted, "fear, whether an instinct or an acquisition, is a function of the environment."[71] By this, he meant what Gibson did when the latter pointed out that cliff edges look perilous because they are dangerous.[72] As Dewey summed up, "man fears because"—or at times when—"he exists in a fearful, an awful world."[73]

For Dewey, this means that organisms do not project emotions into perception; rather, perception is inherently emotional. But the claim entails more than just this. As Dewey went on to say,

> even such words as long and short, solid and hollow, still carry to all, but those who are intellectually specialized, a moral and emotional connotation. The dictionary will inform any one who consults it that the early use of words like sweet and bitter was not to denote qualities of sense as such but to discriminate things as favorable and hostile.[74]

Approaching and avoiding—for example, reaching for nutrients and withdrawing from noxious substances—are fundamental to living processes, and this is conveyed in everyday language. It goes almost without saying that emotional, interested, and aesthetic attractions and aversions are central to action; by extension, they are critical to sensorimotor activity and hence perception too.[75] This stands as a more primordial iteration of what Gibson said when he characterized environmental interactions as a "process of perceiving . . . value-rich ecological object[s]. Any substance, any surface, any layout has some affordance for benefit or injury to someone."[76]

GUSTATORY AFFORDANCES AND THE VISCERA

The foregoing suggests that our attitude is nearly always one of wanting to do, get, or avoid something and therefore one of concerned interest. Most of the time, food and water are among the most immediate needs. Though Gibsonian and 4E traditions do not emphasize the imperatives of digestive systems much, the fact that organisms spend so much of their lives struggling to satisfy it suggests that many affordances express relationships between the viscera and environment, in concert with the rest of the organism.

This brings to attention, once again, the idea that embodied information-processing systems—both perceptual and cognitive—extend beyond the brain. Visceral systems depend on the brain to function as a coordinating and processing vehicle. Yet these systems also have reciprocal relations with the brain as well as with the surrounding environment and actions in it, which means that visceral systems also modulate cognition and perception.[77]

The action of eating a noxious substance, for instance, might lead to a synchronized gastrointestinal and nervous system revolt. This is likely to be registered as illness. Here, peripheral organs such as the heart, vascular system, and skin contribute to that registration, perhaps by increasing pulse, blood pressure, and perspiration. Illness, in turn, lowers energy, and as we have seen, studies show that this shifts cognitive-perceptual preferences. Some of these may relate to a desire for protected spaces or a disinclination to explore as environments manifest as being more strenuous.[78] Shifts may also be more specific, such as when an aversion to an illness-inducing substance closes it off as an affordance for further consumption. Other foods might be registered more favorably and therefore remain attractive within systems of perceptual searching. This suggests that gustatory aversions and hedonics, which are affective, combine with other visceral functions such as heart activity to enter the human affective-cognitive-motor-perceptual problem-solving arsenal (here conceived within the affordance framework). Such processes appear to be widespread, with even unicellular life activities being shaped by parallel occurrences.[79] While the latter cannot plausibly be described as emotional or conscious, a great deal of single-celled activity is unequivocally sensorimotor. It is also cognitive insofar as surprisingly complex problem solving occurs. The behaviors of such organisms might further be described as at least prevaluative. This is insofar as single cell creatures exploit affordances with biological use-value and avoid what is destructive to their continued existence.

The single-celled organisms that comprise the plasmodia of *Physarum polycephalum*, a slime mold, offer a compelling illustration. Their sensorimotor activity is complex, indeed approximating the robotics models described by Brooks, albeit with greater sophistication, and some biologists make this exact link.[80] In this case, their behaviors are layered not through programing *per se*, but through chemosensory mechanisms and dynamical relations between themselves and their environment. Thus, just as some of Brooks's robots have random wandering programs, *P. polycephalum* increases allocation toward exploratory growth when nourishment is depleted.[81] Brooks's robots also have approach-avoid programs, and something similar shows up in *P. polycephalum*. Accordingly,

> when the slime mold senses attractants, such as food, via specific binding to receptor molecules presented on the outer membrane surface, the oscillation frequency in the area closest to the food increases, causing cytoplasm to flow toward the attractant. Additionally, binding of attractant molecules to sections of the surface membrane reduces the tension at that section, leading to a difference in internal hydrostatic pressure, such that cytoplasm flows toward the source of attractants. When repellents such as salts and light are detected, oscillation frequency decreases and membrane tension increases[, leading to withdrawal].[82]

This behavior is in fact collective, meaning that *P. polycephalum* exhibits protosocial activity, in line with arguments from chapter 3. Collectivity manifests because "each unit oscillates at a frequency dependent upon both the local environment and its interactions with neighboring oscillators. . . . The collective behavior of the oscillators, each passing on information to entrain its neighbors, drives the organism's locomotion."[83] Moreover, when suffering from desiccation, *P. polycephalum* forms hardened, protective walls that are resistant to temperature, light, and dryness, and can remain dormant for years, waiting to be revitalized when conditions more favorable to life arise.[84]

In addition to all this, *P. polycephalum* shows an ability to anticipate periodic timing of nonfavorable conditions.[85] These creatures also display remarkable foraging capacities, preferentially migrating towards optimal combinations of proteins and carbohydrates when faced with multiple choices.[86] Just as impressively, they navigate labyrinthine mazes and solve other shortest-path problems.[87] One specific mechanism they employ to achieve this are secretions of slime that they avoid in future explorations until other options are exhausted. By recording past movements externally,[88] they are effectively laying down openings and closures for movement—affordances.[89] Comparable strategies are common in nature, though the specific implementations vary. A variety of insects leave pheromone trails to mark areas already foraged or explored.[90] In the case of egg laying (parasitic or otherwise), some insects do similarly to signal that a host or an area has already been exploited.[91] A variety of mammalian species likewise mark depleted food caches with urine.[92]

Taken together, this suggests variations of Andy Clark and David Chalmers's extended-mind thesis, in which memory is externalized to devices such as notebooks,[93] once again highlighting that cognition is not an exclusively internal affair. This also fits with Gibson's thesis that openings and closures in the environment—as opposed to representations in the head—dictate perceptual or sensorimotor behavior: it resonates with the pragmatic insistence that sensing the world is a matter of actions performed in it and consequences undergone by the organism through changes introduced to its environment (see chapters 1 and 2). All this also supplies a good illustration of Brooks's injunction that the world is its own best model. Besides these implications for memory and the extended-mind thesis, all of the aforementioned behaviors of slime mold, insects, and mammals stand as preconditions for what the Kaplans' research[94]—along with other clusters of work that were at their height in the 1980s and 1990s—characterize as aesthetic experience (see chapter 3).[95] Of course, we do not go so far as to say that *P. polycephalum* enjoys such experience. Yet it does have preconditions of aesthetics because, in the Kaplans' account, which is grounded in affordance theory, aesthetic attraction in environments is about sensing markers that indicate navigability, combined with the promise of coming upon something novel and new. In other words, their concept of aesthetics revolves around environmental exploration in search of resources; and while we typically scope out places just for fun in today's contexts, this does not abrogate the evolutionary origins of the pastime.

For other organisms, exploration predominantly occurs in search of food, sometimes mates, and sometimes other vital resources such as shelter against predators or the elements. This is just as it was in our evolutionary past, and in a way remains so, whether, for instance, one is exploring new pubs for food, potential romantic partners, or to escape rain or a harrassing creep on the street. In the case of at least cephalic organisms, moreover, such explorations connect to core drives, and for this reason have affective timbre. This is not to reduce exploratory problem solving to rumbling stomachs and other visceral events; yet neither is exploratory problem solving reducible to eyes, ears, or even brains. It instead depends on the organism as a whole acting in its surroundings. Understood as part of a synergic whole, viscera such as the stomach and intestines are part of a total information-processing system. They interact with surroundings as well as events in them such as volume expansion and increased acidity. The nervous apparatuses that detect these variations convey vital information, whether in the form of satiety or the presence of gas-producing pathogens.

As discussed last chapter, this relates to everyday concepts of cognition, which are often characterized in terms of the digestive system, and not just ingestion, but also food preparation and expulsion. Johnson elaborates:

> Derek has a real *thirst* for knowledge, a *huge appetite* for learning, and an *insatiable* curiosity. You don't expect me to *swallow* that *garbage*, do you? I'll have to *chew* on that for a bit. The whole idea *smells fishy*. He's known for putting out a bunch of *raw facts*, *half-baked* ideas, and *warmed-over theories*. Let's *put that on the back burner* for a while and *let it simmer*. What've you *cooked up* for me now? What a *rotten idea—disgusting, unsavory*, and enough to *make you puke*, if you have any intelligence at all. That's pure *bullshit*. You're not *shittin'* me, are you? We're gonna have to *sugar-coat* it to make it *palatable* to her, or even *force it down her throat*. Do you ever feel like you have to *spoon-feed* your students? His scheme left a *bad taste in my mouth*. There's too much here for me to *digest*. Where's the *beef* in your theory? The really *meaty* issue is sustainability. Let's just *chew the fat*. Finally, something you can really *sink your teeth into*! Now that's *food for thought*! Our philosophy teacher just wants us to *regurgitate* what she gives us in lecture—*just spit it back* to her.[96]

In his many accounts, Johnson shows that metaphors for conceptual thinking relate to perception, the body, and action in the world. A telling point to note, and one that Johnson offers evidence for but arguably does not adequately explore, is that in many cases, cognitive characterizations are not only grounded in perception, but are identical to it. Thus, Johnson quotes a woman's description of a high school crush, in which she recalled: "I wanted to eat him alive! He was yummy!"[97] In this case, the cognition, approach impulse, emotional desire, and perception all knot together, reiterating a central thesis in this book.

Given the pervasive importance of foraging and gustation, these interconnections are not surprising, and they are in fact embedded in the physical structure of the body. As one research team explains,[98] the nucleus of the solitary tract, an area in the brain stem, intercommunicates with diverse visceral sites. There is broad head-to-tail visceral organization here, with neurons receiving input from different organs such as the stomach, heart, and kidneys. Neurons involved in this lie in close proximity, suggesting "early integration of viscerosensory information across modalities that are linked through patterned responses to functional goals."[99] Correspondingly, the aforesaid area in the brain stem projects into subcortical regions such as the hypothalamus and amygdala, along with other areas in the brain stem that together "contribute to coordinated autonomic, hormonal, and even immune outputs."[100] These outcomes occur through synchronizations of visceral activity and reciprocal communication that are conveyed, for example, along the vagus nerve between the stomach, brain stem, and subcortical structures.[101] These structures, in turn, reciprocally communicate with other neural regions, and the brain reciprocally with the rest of the body.[102]

All these events and activity coordinating in synchrony impacts our perception of the world, and while neurons in the cranium are obviously critical, they only constitute a part of the overall affective-cognitive-motor-perceptual process. Recent 4E theorists emphasize that psychic life pushes beyond the head, yet this is already true of the brain itself, which threads into a diverse range of peripheral systems, and hence into physiological functions and actions in the world. Without question, therefore, problem solving extends beyond the cortex and, indeed, the cranium. Thirst, to consider an example, often entails conscious problem-solving orientations that are largely driven by regions outside the cortex.[103] It involves a range of subcortical and visceral appraisal systems that evaluate when water is needed, times it might appear, and where it might be located, either geographically or in foods. To a significant extent, these processes are moderated affectively.

As in the case of emotion, appraisal does not make a lot of sense absent action in the world, which suggests embedded, enacted, and extended components in addition to embodied ones. Along similar lines, a wide variety of species connect foods they have consumed with illnesses that have followed. These associations, which entail gustatory characteristics and gastrointestinal discomfort, can remain for long periods after.[104] Humans experience this too, developing revulsions after food poisoning incidents.

In such cases, the food may become perceptually-emotionally repulsive even before the tasting of it, thereby closing avenues for gustatory action. Cognition also plays a part insofar as the revulsion entails anticipating, identifying, and rejecting or avoiding what has caused past discomfort. Such learning is once again thoroughly embodied insofar as visceral information plays a critical role in bodily organization of action. For similar reasons, learning is also embedded, enacted, and extended.

The amygdala is important for integrating visceral information: gastrointestinal, gustatory, olfactory, and cardiovascular information are all routed through the amygdala.[105] In animal studies, lesions across the alimentary-gustatory axis impede the acquisition of aversions to combinations of odors and tastes that signal danger.[106] In humans, strongly aversive odorants correspondingly activate the amygdala as well as the orbitofrontal cortex,[107] the latter of which is involved in emotion- and reward-based decisions.[108] This reiterates the affective or valuative elements of perception, in this case of a gustatory or olfactory sort.

Visceral nervous structures also house autonomic and neuroendocrine systems,[109] where the latter is involved with both neurotransmitters and hormones, which in some cases are composed of the same chemicals. Chemical messengers such as CRH—discussed last chapter—convey information about bodily conditions that integrates into the organization of behavioral responses.[110] CRH is not only found in the brain, but throughout a variety of peripheral organs, including the gastrointestinal tract, and plays a role in modulating immune and stress responses and activities such as getting up in the morning.[111]

Angiotensin—a peptide hormone—regulates vasoconstriction and therefore blood pressure. It is also in the sequence that leads to water retention in visceral organs, and when synthesized in the brain as a neuropeptide, angiotensin mobilizes activity that elicits thirst and sodium hunger, which is essential for fluid homeostasis.[112] These chemical messengers can operate rapidly. Indeed, drinking sometimes occurs seconds after angiotensin infuses in the brain, whereupon multiple appraisal systems detect and signal need for water, focusing attention on likely locations of it (for example, foods rich with it). This indicates nearly immediate changes in perceptual search patterns and affordance arrays—shifts that reorient and synchronize a wide range of bodily activity in environments. The same peptide that acts in the peripheral systems to regulate fluid balance also shows up as a neuropeptide in the brain. This, in turn, has regulatory affects on behavior and physiological activity critical to the maintenance of fluid stability.

Throughout his 1890 *Principles of Psychology*, James depicted peripheral organs as epistemically active extensions of the brain. This is to say, he held that perceptual and cognitive sensitivity spreads across the whole body and is not confined to the head. His slightly younger contemporary, Ivan Petrovitch Pavlov,[113] famous for positing classical conditioning but also a reputed physiologist and neurobiologist, held a comparable position. His research on the gastrointestinal tract, for which he won a Nobel Prize in 1905, arguably led him to appreciate that psychic life was not exclusively a cortical affair. Saliva secretion, to give an illustration, is mediated by the brain, but is also tied to anticipation of occurrences in the peripheral system and the world. The release of insulin in anticipation of food when hungry likewise involves synchrony between brain, peripheral systems, and worldly actions aimed at obtaining food.[114] Anticipatory mechanisms, moreover, relate to time of day and the history of expectations.[115] This once again offers biological grounding for the thesis, especially defended by Dewey and Merleau-Ponty, that the body is not a collection of adjacent organs. Rather, it is a synergistic system that coordinates not just through internal means, but by acting in the world as well. Such total interactions are, as we have repeatedly posited, preconditions of coherent affective-cognitive-perceptual experience. The views offered in this section are also consistent with James's assertion that interests alter how the world shows up, for avenues for gustatory satisfaction vary depending on biological need. This view, in turn, connects to affordance theory, though Gibson insisted that affordances remain regardless of whether or not they are noticed.[116] The latter is arguably true, but it does not repudiate the relation between affordances and biological need for reasons already discussed.

All this together weighs against the thesis that there are pure "cognitive computations" (again, defined as neural processes dealing with worldly information unrelated to the organism's well-being). Salt detection may be considered for a final illustration. Notice, to begin with, that certain cells in the solitary nucleus (located in the brain stem) fire at a rate proportional to the saltiness of a stimulus—all else being equal.[117] However, things are rarely equal, and the firing rate adjusts not only with the concentration of sodium ions on the tongue, but also with the current balance in the animal. Saltiness accordingly does not seem to be an unadulterated sensory "given."[118] Brain areas involved include hypothalamic, amygdalar, and several circumventricular regions, for example, the subfornical organ. It also involves the peripheral nervous system. Together, these organic structures monitor the concentration of sodium ions in the animal's blood and respond to hormones related to the need for salt. These areas also project into the brain stem, triggering or inhibiting what are known as "salt-best" neurons. All this not only influences sodium-foraging behavior, but arguably how compounds containing salt taste.[119] Indeed, even activity in the chorda tympani—a cluster of nerves going from the taste buds into the brain—has been observed to vary depending on sodium need, where firing rates decrease in deprived animals.[120] Diminished firing rates plausibly make substances taste less salty, and behavior responses suggest that deprived animals find sodium more pleasant.[121] Under most circumstances, these outcomes would increase salt foraging and ingestion and, in the language of Gibson, make gustatory affordances for salt salient.[122] This suggests, again, that no pure sensory excitations are registered. Instead, sensation knots with biological needs and drives, and hence also with behavior and problem solving, reiterating the interweaving of action, affect, cognition, and perception.

AFFORDANCES, BIOLOGY, AND VALUATIVE LIFE

In this chapter, we have looked at valuative conceptions in pragmatism, phenomenology, and Gestalt psychology and their imprint on Gibson's thinking. By doing so, and by integrating all of these frameworks with biological work and consonant 4E research, we have endeavored to extend the scope of affordance theory. Expanding on the last chapter, we argued that gustatory and other visceral dynamics make affordances show up as they do. We also defended a thesis central to this book: that action, affect, cognition, and perception are integrated.

Under the influence of pragmatists, phenomenologists, and Gestalt theorists, Gibson's last book advances just this thesis, though the point sometimes gets obscured by Gibson's terminology.[123] In particular, Gibson avoided saying much about "cognition." This was arguably because he was uneasy with schematic models suggesting, for example, that objects stimulate retinas so that nerve impulses convert into images in the brain (whereupon neural operations then abstract meaning and engage in conceptual reasoning). Nonetheless, Gibson emphasized the knotted nature of action, cognition, and perception by publishing in collections such as *Perceiving, Acting, and Knowing*,[124] or by remarking:

> Knowledge of the environment, surely, develops as perception develops, extends as the observers travel, gets finer as they learn to scrutinize, gets longer as they apprehend more events, gets fuller as they see more objects, and gets richer as they notice more affordances. Knowledge of this sort does not "come from" anywhere; it is got by looking, along with listening, feeling, smelling, and tasting.[125]

Passages quoted earlier about the dangerous precipice or value-rich ecological contexts indicate that Gibson also saw affect as knitted with cognition, motility, and perception.

The knotted nature of action, affect, cognition, and perception predictably comes out in the next generation of psychologists working under Gibson's sway. For example, the Kaplans—already discussed at length—assert that

> an underlying criterion in making a preference judgment is an evaluation of the scene in terms of presumed possibilities for action, as well as potential limitations. Even in the very rapidly made rating, an important consideration involves an assessment of the scene in terms of what it makes possible, what it permits one to do. Gibson's concept of affordance is similar: An affordance is what an environment offers the perceiver, or, in other words, what the perceiver would be able to do in the setting.[126]

They go on to suggest that perception entails "the assessment of potential actions applied to scenes. People are evaluating an entire setting in terms of potential actions."[127] This again indicates that perception involves cognition because it is a judgment or evaluation; that it involves affect because it is an assessment of aesthetic preference; and further that it involves motor potentialities because it entails sensing out possibilities of action. It also involves expectations because the perceiver is working out potential but as yet unrealized movements. Motor-perceptual expectations are simultaneously affective and cognitive for the reasons detailed in chapter 4.

Intriguingly, the exploratory movements of slime mold and insect achieve similar outcomes, and while they cannot plausibly be said to be making aesthetic judgments, they are processing information. Furthermore, their explorations—carried out with the aid of slime and pheromone trails—might be said to be prevaluative and perhaps pre-aesthetic. This parallels organic life generally insofar as environmental information is always already filtered by virtue of what bodies can do and how they can respond. Thus, as intimated at the outset of this chapter, human ears and eyes are only responsive to certain ranges of sound and light frequency, and only what falls within them can elicit the total responses that generate perceptual experience. This further means that the range of what we can detect, even if aided by technological instruments, limits what can be valued. Hands, muscles, and other parts of the body similarly restrain how things show up for reasons discussed earlier (see chapters 1 and 2). The situation is comparable in the cases of slime mold, insects, and other creatures. They are, in effect, blind to certain chemicals and environmental protrusions while sensitive to others. Interactions within these spheres of ecological sensitivity—or *umwelten*—elicit total coordinations.

Whereas the Kaplans—incidentally still caught in the stimulus-response model that Gibson, Merleau-Ponty, and Dewey sought to avoid—argue that environments elicit certain emotional and therefore aesthetic reactions, experiments suggest that the reverse simultaneously occurs. Such is the case in the studies cited earlier that show anxious moods correlating with preferences for enclosed spaces and happy, energetic emotions correlating with preferences for open spaces. Notice, once more, that enclosed spaces are protected, something desirable when emotionally threatened, and open spaces are subject to greater exploration, which we incline towards when happy and full of energy. This fits with affordance theory, which holds that the perceptual world is a total outcome of what we can do—or it might be added, feel able to do—in environments. Dennis Proffitt's pioneering work focuses more directly on affordance theory than the Kaplans' does.[128] As we have discussed, Proffitt reports that people in poor health or who are fatigued or burdened with heavy backpacks judge grades to be steeper or objects to be further away. Perceived steepness or distance often comes with emotional deflation if we wish to climb a hill or reach a destination. The premise that interests are emotion-like (which we substantiated in the last chapter), when integrated with Proffitt's research and that of Dewey, Merleau-Ponty, and

Heidegger (which all stress the affective dimensions of lived time and space), considerably expands the thesis that emotions undergird perception and cognition. This is because there is little we encounter or do in a disinterested way.

In this and previous chapters we have offered neurobiological affirmations of the position we are advancing, but a brief summary and expansion is worthwhile. To begin with, Gibson's account and those of his pragmatic, Gestalt, and phenomenological predecessors all stand as forerunners to recent sensorimotor views, and were called "sensorimotor" by thinkers such as Dewey. Evidence from various quarters, in turn, indicates integration of sensory and motor functions on a neurobiological level (see chapter 2). An instructive go-to example for many defending embodied views on neurobiological grounds is the *Caenorhabditis elegans*.[129] This organism is a roundworm about one millimeter in length that has no respiratory or circulatory system, and has a rudimentary nervous structure of approximately three hundred neurons. The *C. elegans* appears to have neurons that perform both motor and sensory operations,[130] which fits with sensorimotor accounts. It has others that are simultaneously sensitive to mechanical disturbance, fluid balance, odorants, and damage to the organism. Yet other neurons are sensitive to something approaching taste, and also to odorants and damage.[131]

While the neural architecture of a roundworm does not necessarily track to human beings, the findings are nonetheless consistent with the integrated views we have been advancing. Research on sensory substitution devices, discussed in chapter 2, provides yet another line of evidence. Reinforcing the overall picture are numerous neuroimaging studies that detect simultaneous activity in brain areas associated with emotion and motor-response when people view art,[132] which one would expect based on the Kaplans' research. This also fits the claim—advanced by both Gibsonians and pragmatists— that perception and action are valuative. In everyday life, we experience this nearly constantly, whether with the temptation to go down a winding trail; hesitancy to approach a dark alley; the emotional tug that pulls us to both familiar faces and satisfactory answers; or, again, the selective interests that allow us to focus on one conversation to the exclusion of others.

Integrating Gibson's views with neuroscience is not without challenges, but not because evidence is lacking (as the discussion just offered indicates). It is because neuroscientists ubiquitously adopt a metaphysical standpoint contrary to that of Gibson—and metaphysical disputes are not resolved by experiments, but, according to pragmatists, by the amount of work they allow us to do. As Michael Anderson[133] and others such as Matthieu de Wit and colleagues[134] especially emphasize, ecological psychology has historically neglected findings from neuroscience. Neuroscience has done the reverse, either by ignoring Gibson's ideas or perverting them by adopting standard representational models that assume the same inner-outer divide as early modern thinkers did. The reasons for this disconnect are therefore obvious: the mechanistic, stimulus-response models that ecological psychologists jettison overwhelmingly guide neuroscience.

Pragmatists, especially Dewey (along with Merleau-Ponty), also rejected these models, but they embraced the nascent neuroscience of their day. Their views therefore stand in contrast to standard cognitive neuroscience. The latter operates on the assumption that events in the brain and mind are built up from almost discrete units causally interacting with one another. Dewey's and Merleau-Ponty's work, by contrast, start with situations, which include the totality of everything in the body and environing circumstances, both social and physical. Gibson and Heidegger did likewise without emphasizing biology much. Gestalt psychologists, whom Merleau-Ponty and Gibson regularly drew upon, fundamentally advanced the same idea, though once again not focusing on biology and getting less into the metaphysics of the position than Dewey, Heidegger, Merleau-Ponty, and Gibson did.

As with Dewey, who insisted that subject and object are built up in one and the same continuing operation, Gibson maintained that affordances are neither objective nor subjective, or perhaps that they are both. To give a Deweyan example already used, the smoothness of a wine glass is not in the object alone, yet neither is it "in" the organism. Instead, it is a consequence undergone when the subject actively meets and handles the object in question. Gibson similarly held that affordances are equally about the environment and behavior. They are physical and psychological, yet also neither, as he added cryptically,[135] and they reveal the inadequacy of subject-object divides. Gibson further observed that meaning is reserved for ecological contexts and that the brute physical world is bereft of it. This is another way of saying that human-environment interactions are always suffused with value.

Although it is not clear that neurobiological data unambiguously affirm this position, there is no shortage of findings at least consistent with it. For instance, activity in Broca's area is associated not only with language capacities, but, in different neural coalitions, also with movement preparation and action recognition along with variety of other functions, including aesthetic perception (see chapters 2 and 3). At the same time, insofar as this region is inherently connected to language and thus semantics, it is consistent with the thesis that perception (understood in terms of sensitivity to possibilities of action) is inherently meaningful and replete with cognition. A number of investigators have likewise suggested that the basal ganglia, in addition to the frontal cortex, simultaneously underlie motor capacities, grammatical rule processing, and the emotional weighing of options (see chapters 2 and 3).

Along with Dewey and Merleau-Ponty, one of Gibson's central goals was to do away with mental representations, which have long tyrannized accounts of mind. Overcoming such models is not merely a matter of empirical discovery. It is also a matter of philosophical standpoint. It is worth noting that while mental representations help explain some things—for example, the process of planning a dinner party,[136] or working out the phases of the moon by picturing their arrangements—a tremendous amount can be said without recourse to them. Moreover, even these representational forms of reasoning need not be (and often are not) carried out exclusively in the head, and indeed typically employ external aids such as notebooks or arrangements of objects standing in for the earth, sun, and moon.

It is a slogan among enactivists that the world is a better model than any representation of it, a dictum appropriated from Brooks. Of course, saying it is so does not make it so. Neither does claiming the contrary make it the case. We believe that there is obviously room for both positions. However, when judged from the standpoint of practical results, a great deal of progress has been made by embracing Brooks's precept. He and other roboticists such as Long[137] have largely avoided systems that construct full-fledged representations of the world through internal, symbolic amalgams that precede actions based upon them. They have instead developed perception-to-action arrangements in which much of the "computation" occurs through machine bodies sensing and exploiting available affordances in environments. In so doing, they have managed to succeed where older, symbolically driven machines have failed. In sum, we have tried representational models in various instantiations for a half millennium. On their own, they have not worked that well. Pragmatists would say it is time to add something else to the mix.

6

BROADENING ECOLOGIES

I n this book we have argued that it is impossible to trace mind to any one structure because it arises out of a totality of neural and extraneural capacities working in concert with environments, both physical and social. We have also emphasized that action, affect, cognition, and perception knit together.

These assertions cut against the history of philosophical, psychological, and neurobiological scholarship. These fields have tended to locate mind inside the head despite objections from classic and recent embodied thinkers. They have also inclined towards isolating behavioral, cognitive, motor, and perceptual life as discrete processes. Some of this may have been done for ease of communication. We obviously do not deny the helpfulness of the terms since we, too, use them, nor do we deny that there are archetypical instances of each. Yet all too many discourses crystallize these distinctions, framing them as real rather than nominal despite compelling objections, again, from embodied thinkers, especially Dewey and Merleau-Ponty.

Importantly, our main claim has not been that action, cognition, emotion, and perception "contribute" to one another. It has instead been that they are essential to one another: they make one another what they are. We have further argued that these endowments only function in global contexts, which include bodies, viscera, and a great deal more that synchronize through interactions with the world, and that all this together constitutes psychic life. Put otherwise, we have advanced an ecological account of mind.

Most understand the word "ecology" to pertain to the environment and systems in it. What many do not recognize is that we have ecologies within us. This is so in the literal sense that we host vast populations of microorganisms that modify bodily functioning and, indeed, psychic life—a point this book has only lightly grazed so far. The body and its organs and capacities synchronize to bring forth psychic life. This is further structured by interactions in the world that change both the living creature and its surroundings (see chapter 2).

In this last and concluding chapter, we aim to ravel together the view that cognition, emotion and perception, along with brain, body, viscera, and world, conceptually and biologically imply one another. To our minds, such a position goes without saying, but views to the contrary are in fact longstanding. It is only recently, for instance, that researchers are recognizing that the gut and microscopic organisms in it add significantly to brain functioning and psychic life. This line of research, which we explore in this chapter, adds to 4E accounts and their pragmatic and phenomenological antecedents. We also elaborate upon visceral factors in psychic life, continuing a discussion that we began in earnest in chapter 4. Specifically, we turn to a range of research that suggests that the microbiome inside the body and especially the gut profoundly shape mood and action, and therefore also cognition and perception of the world. It is our aim to integrate all of this in a way that meshes noncontentiously with available neuroscience. As exciting as explosive controversy may be, we aim for the opposite. Broadly speaking, we hope to frame our ecological views in ways that opposing experts and educated lay readers would equally see as uncontroversial.

Though not our main thrust, we would note that our position has epistemic ramifications. Many mainstream accounts see emotions as internal, private states and cognition as representations telling us something about the world. The position we have defended, by contrast, regards both as outcomes of interactions between the organism and extra-organic things. Standard accounts from the modern era onward advance an epistemological dualism. This is the view that perception and knowledge of properties are representations and therefore one step removed from reality at best. Our position insists that subject and object integrate and build up in one and the same operation. This means that the quality of smoothness, for instance, is in neither the subject nor object alone (or it is in both, if one wants to put it that way). More accurately, it is a product of the two interacting. Thus, fingers brushing over a glass tabletop might realize smoothness such that it is a quality of an interaction, as opposed to a property of the surface that is then represented in the mind.

Notice once more that this way of thinking implies integration in psychic life. Complex and simple motor engagements between organisms and environments mobilize attentional and therefore affective and valuative faculties for reasons that we have repeatedly outlined. Even when not moving, the perceptual array coheres around motor potentialities and selective attention. Furthermore, selective attention is simultaneously cognitive and perceptual, and it is also affective and hence valuative (see chapters 4 and 5). Taken together, the arguments in this book suggest that affective, cognitive, motor and perceptual life are ecologically bound together from the start, which, again, is not to deny the usefulness of the individual terms or the existence of archetypal examples of each. In short, we argue that cognition is emotional or affective, and emotion cognitive. Perception is likewise cognitive and emotional, and emotion perceptual and cognitive insofar as emotion helps us see and understand what is going on in the world. Inasmuch as cognition also achieves this end, it, too, serves perceptual functions. All these faculties are embodied, which means that they are knotted with action and hence the world.

FROM GUT TO BRAIN AND BACK AGAIN

One form of ecological integration that we have defended revolves around interplay between visceral conditions, brain activity, cognition, emotion, motility, and perception. A final expansion is worthwhile. We will focus on gut-brain connections and especially the role of animal-hosted microbes—organisms sometimes called "psychobiotics" because of their impact on affective, cognitive, motor, and perceptual life. Even here, we are obviously just scratching the surface of this topic, and we have limited ourselves in the same way throughout this book by, for example, giving only cursory attention to the role of hormones. Our agenda, however—which requires some technical discussion in this and later sections of the chapter—is not to deliver definitive statements on the subject, but rather to emphasize the enormously complex ecological character of mind and to suggest future research directions.

The organization of action and bodily sensibility are, as we have said, replete with cognitive-emotional-perceptual processing.[1] This helps account for why a considerable amount of data and many classic philosophical arguments suggest that much of our life unfolds unconsciously or at least pre-reflectively (see chapter 2). Conventional understandings often attribute this outcome to operations in the recesses of vast neural complexes. We do not reject this as a partial explanation, but have added that a great deal of what constitutes the "mental" happens outside the brain. The gut-brain axis is our culminating example of how this occurs.

Nobody seriously doubts that gut dysfunction produces illness, and this has obvious consequences on our moods, which are tied in turn to how we act, perceive, and think in the world (see chapters 3 to 5). However, pathways between gut activity and psychic life are more than just responses to discomfort. While healthy guts exhibit some permeability, allowing very small particles to pass into other parts of the body, overly porous intestines link to a range of disorders. These run the gamut from diabetes, asthma, and inflammatory bowel disease to psychiatric conditions such as autism, depression, and anxiety.[2] Schizophrenia is also associated with gut problems.[3] It is worth stressing, on the one hand, that correlations by themselves do not imply causation, and it is inadvisable to reduce psychiatric conditions to gastrointestinal disorders. That said, hormones, neurotransmitters, and immunological bodies released in the gut communicate with central and peripheral nervous apparatus.[4] Inflammation of the gastrointestinal tract, for example, leads to the release of cytokines as an immune response. Combined with elevated intestinal porosity, this increases the permeability of the blood-brain barrier and the brain's susceptibility to harmful molecules. This appears to contribute to heightened anxiety, depression, and memory loss,[5] among other outcomes. For this reason—in addition to others discussed—gastrointestinal conditions relate to psychic life.

Though it is difficult to pin down exact numbers of bacteria in the gut, it is at least in the trillions,[6] with upwards of five hundred different species.[7] Most bacteria are symbiotic: the human body relies on them to digest food, keep unwanted pathogens at bay, synthesize certain vitamins and amino acids, and carry out numerous other functions.[8] Common sense already suggests that gut dysfunction—which often entails lower numbers of desirable bacteria and too much of undesirable kinds—produce illness, which impacts mood. However, pathways between bacteria and psychic life are more than just a response to repose or discomfort. Animal studies show, for example, that the oral introduction of the probiotic *Bifidobacterium infantis* increases tryptophan, a precursor to the neurotransmitters GABA and serotonin.[9] GABA is involved in a range of inhibitory responses, including lowering the heart rate and blood pressure or suppressing inflammatory reactions that occur for immunological reasons. Though precise mechanisms are not well understood, decreased serotonin activity and inflammation link to negative affective states, most famously depression, as well as impaired cognitive functions in areas such as memory.[10]

This raises the possibility of treating chronic negative moods with probiotics, as Megan Clapp and colleagues suggest in their excellent review article, following the lead of others.[11] Researchers have explored this strategy and found tentative affirming evidence for the prospect. Male mice with depleted microbiomes exhibit increased stress responses, which can be ameliorated through the introduction of *Bifidobacteria* species. *Bifidobacteria* appears to alter communication of genetic information in GABA receptors and decrease cortisol levels in blood (cortisol is a molecule of energy consumption, but is also associated with stress). These changes, however, disappear in mice with severed vagus nerves, the primary signaling pathway between the viscera and brain.[12] This suggests once more that psychic events involve global synchrony throughout the body with things it encounters—in this case, bacteria.

Other studies demonstrate comparable points. Microbial imbalances in mice, to offer one case in point, correlate with inhibited maturation of B and T lymphocytes—varieties of white blood cells that carry out immune responses.[13] Such mice simultaneously suffer irregular interactions between the hypothalamus and pituitary and the adrenal glands[14] (the first two structures are located in the brain, and the last just above the kidneys). When treated with probiotic *Lactobacillus* species, the microbiomes in deficient mice improve; they show decreased intestinal permeability along with boosted functionality in the hypothalamic-pituitary-adrenal axis. Experiments on animals in stress conditions, moreover, suggest probiotics as a treatment for impeded hypothalamic-pituitary-adrenal function and irregular neuronal firing. This work further indicates that this intervention may promote neurogenesis and synaptic plasticity in the hippocampus.

Treated animals tend to exhibit lower levels of hormones associated with stress along with healthier intestinal permeability.[15]

Human studies obtain comparable results, with improvement among those suffering depression and anxiety occurring after three to six weeks of ingesting *Bifidobacteria*, a probiotic.[16] Meta-analyses suggest that this outcome depends on age, with treatments being less effective on those over sixty-five years old.[17] Encouragingly, however, sufferers scoring at the bottom of psychological diagnostic scales show the most improvement along affective parameters such as being agreeable versus angry, clearheaded versus muddled, composed versus anxious, confident versus unsure, elated versus depressed, and energetic versus tired.[18] Furthermore, studies on healthy individuals comparing the effects of probiotics and anti-anxiety medication show that those treated with the former have

> reduced cortisol levels and improved self-reported psychological effects to a similar degree as participants administered Diazepam, a commonly used anti-anxiety medication. Analogous studies found that probiotic therapy reduced depressive symptoms and improved [hypothalamic-pituitary-adrenal axis] functionality as well as Citalopram and Diazepam.[19]

Probiotics in mice have likewise achieved results similar to the antidepressant escitalopram, reducing anxiety and proving more effective than psychopharmaceuticals in maintaining healthy metabolism and body weight.[20] They also tend to have no adverse side effects.

A related line of research deals with gut abnormalities commonly shared by people with autism, which, in at least some cases, appear related to microbial imbalances. Francesco Strati and colleagues uncovered significant increases in *Firmicutes* relative to *Bacteroidetes* in autism sufferers.[21] They further found relative overabundances of the genera *Collinsella*, *Corynebacterium*, *Dorea*, and *Lactobacillus*. This is combined with relative decreases of the genera *Alistipes*, *Bilophila*, *Dialister*, *Parabacteroides*, and *Veillonella*. More than one study has also found increases of *Candida*—an already pervasive yeast fungus—within the guts of autistic individuals.[22] Some specific causal mechanisms have been proposed for this relation. For example, it turns out that interactions between ammonia and propionic acid released by yeast cells are associated with autism.[23] As Clapp and colleagues summarize,[24] this results in excessive amounts of β-Alanine, which has a similar structure to GABA and has been implicated as a contributor to autism, though this point is still debated. An additional explanation is that elevated levels of the organisms exacerbate autistic symptoms by impairing mineral and carbohydrate absorption, while simultaneously building up toxins.[25]

Toxoplasma gondii supplies a particularly grim illustration of how microorganisms moderate psychic life. Although it generally infects the guts and brains of warm-blooded animals, this parasite only reproduces in felines in a life cycle that is disturbing from the standpoint of host organisms. Joanne Webster and colleagues brought this to widespread attention some decades ago.[26] Specifically, her research teams have found—and this has been well replicated—that prey animals such as mice and rats hosting *T. gondii* behave in ways that make them more susceptible to predation. In mice, this includes increased movement and inclination to exposed areas,[27] behaviors that make them easier prey. Mice also show impeded learning and memory in tests involving maze navigation,[28] arguably degrading their ability to flee. As strikingly, they lose aversions to cat smells such as that of feline urine.[29] This is in spite of the fact that mice without the parasite have a natural fear of cats even when they have never been exposed to them. Something similar occurs in *T. gondii*–infected rats, though their learning and memory is less impaired, perhaps because the infections tend to be less acute.[30] Critically, however, rats have impaired fear of novelty, which under normal circumstances helps them avoid both predators and live traps.[31] Infected rats also show degraded aversion to cat odors, if not heightened interest in them.[32] Together, these changes make prey animals more vulnerable to attack. Once an infected animal is eaten, *T. gondii* reproduces in the cat, is redistributed through the cat's feces to prey animals, and the cycle begins anew.

T. gondii infections are common in humans as well, and though the condition usually appears asymptomatic and harmless, there can be a range of associated effects, some troubling and others less so. To begin with, infected men tend to find cat urine more pleasant than those who are not, with the reverse holding for women.[33] More worryingly, several studies find car accidents more prevalent among the infected.[34] One research team proposes that the parasite impairs perceptual-motor capacity,[35] citing a study that indicates degraded reaction time among the infected.[36] Also, most participants in these studies are men, and *T. gondii* correlates with elevated testosterone in males.[37] There are, in turn, widely accepted links between elevated testosterone and aggression, and recent work in fact suggests that increased aggression comes with *T. gondii* infections.[38] Even more troublingly, infection correlates with increased risk of Alzheimer's and Parkinson's disease, along with autism and schizophrenia.[39] Contrastingly, however, one set of mice experiments suggest that *T. gondii* has immunosuppressive effects in the form of anti-inflammatory mediators that reduce cognitive impairments associated with Alzheimer's symptoms.[40] The broader point, in any case, is that microbes within organisms alter psychic life, sometimes in profound ways. A few studies even indicate increases in suicide attempts,[41] though this may follow from factors that go with ill health.

While we do not want to overstate these findings—after all, infection in humans is common, and most of us do not suffer the dire consequences listed—the results do, in company with the earlier discussed research on microbia, indicate a relationship between the gut, brain, and mind. Everyday language conveys latent awareness of this. Giulia Enders,[42] a popular science author focusing on the gastrointestinal system and providing an accessible if mostly unreferenced review of effects discussed in this section, arrives at such a conclusion, writing that

> we humans have known since time immemorial something that science is only now discovering: our gut feeling is responsible in no small measure for how we feel. We are "scared shitless" or we can be "shitting ourselves" with fear. If we don't manage to complete a job, we can't get our "ass in gear." We "swallow" our disappointment and need time to "digest" a defeat. A nasty comment leaves a "bad taste in our mouth." When we fall in love, we get "butterflies in our stomach."[43]

This repeats accounts offered in previous chapters, especially portions discussing Lakoff and Johnson's ideas (see chapters 2, 4, and 5).

Leaving the microbiome aside, there are other ways in which the gut and brain interact. Visceral systems intercommunicate with numerous brain regions. A first interactive junction is the solitary nucleus in the brain stem, which receives information from not only the viscera but also the heart, lungs, ears, tongue, and other areas.[44] Other brain stem areas are involved too, including parabrachial structures such as the locus coeruleus. The parabrachial region appears generally involved with blood, salt, sugar, water, cardiovascular, and temperature regulation, along with control of hedonic responses.[45] The locus coeruleus handles a variety of emotional and homeostatic responses.[46] The brain stem communicates reciprocally with the amygdala bulbs (as discussed in earlier chapters), and together they handle a range of cognitive-emotional-motor-perceptual tasks such as selective attention and approach-avoid behaviors. The amygdala bulbs, in turn, are reciprocally connected to cortical sites and nearby subcortical structures such as the hypothalamic and thalamic areas.[47] A pair of date-shaped protrusions, the thalamic regions process sensory and motor signals, and are important in the regulation of consciousness, sleep, and alertness, as well as learning and memory functions.[48] They also communicate with the cortex, the hypothalamus, and other brain structures.[49] Proximately tucked beneath the cleft of the thalamic bulbs, the hypothalamus similarly plays a variety of regulatory roles related to temperature maintenance, food intake, thirst, circadian rhythms, emotion, and sexual behavior.[50]

While this brief account is but a crude survey of the array and interconnectedness of functions carried out by the brain and viscera, it nevertheless highlights the synergic nature of biological processes and psychic life. It specifically lends additional weight to the claim that affective, cognitive, motor, and perceptual capacities are integrated throughout the organic system, including not just the body, but also the surrounding world. This is because many of the regulatory processes described involve both the internal milieu as well as environmental openings and proddings, the latter including, for example, cycles of night and day, threat avoidance, and salt, food, water, and mate availability. These regulatory processes entail affective, cognitive, motor and perceptual synchronies of body and environment such as those characterizing core drives and their satisfaction. These synchronies simultaneously involve selective attention, identification, approaching, avoiding, and emotional pulling towards or pushing away from substances, things, and opportunities. The visceral-neural axis is accordingly environmentally situated, and enables both bodily sensibility and diverse forms of problem solving. The parabrachial region in the brain stem, to give an example, links to sodium hunger, taste aversion, and learning.[51] It has subnuclei that underlie cardiovascular, gastrointestinal, and gustatory processing,[52] which occur as both internal and external coordinations. All this together implicates these brain structures and visceral organs in the body's internal balance and action in the world.

Something similar holds with microbe-gut-brain interactions. This is more so in light of recent research, which suggests that bacteria differentially produce chemical signals that activate regions such as the hypothalamus to produce rewards depending on what and how much we eat.[53] This means that organisms within us—internal living ecologies—are interacting with our surrounding environment in complicated ways, getting us to do the same through total body coordinations with things in the world. These processes again mobilize global responses related to arousal, satiety, and food preference, which is to say, once more, that they mobilize affective-cognitive-motor-perceptual activity. In short, ecologies within us are literal, and they point to synergic processes of our bodies, both neural and extraneural, that achieve coordination through engagement with the world.

BODY, BRAIN, WORLD

Ecological orientations tend to go against modularity. Modularity is the thesis that the brain is comprised of distinct functional units dedicated to highly specific tasks, and that psychological faculties are similarly divided. Like many researchers in the neurosciences these days, we reject modularity. That said, we do not dismiss the thesis as patently stupid, and we agree that some areas in the brain are indeed fairly specialized. For example, Broca's area, while not operating alone, seems dedicated to speech comprehension and production, as discussed in earlier chapters. The fusiform gyrus and other parts of the temporal cortex likewise appear involved in facial recognition (lesions in this area are associated with prosopagnosia, that is, impaired facial recognition).[54] Damage to a range of highly local brain sites can likewise inflict an inability to see, remember, and imagine color.[55] These are just a few among a great number of cases that appear to support modularity, so it is not hard to see why the modularity thesis was tempting, especially since it suggested a fairly straightforward map for the brain and psychic faculties.

At the same time, nothing in these examples demonstrates that any specific region is alone responsible for the aforesaid capacities. Damage to them could instead indicate broken connections between normally integrated regions, as most neuroscientists readily acknowledge. In addition to this, neuroscientific research methods arguably overstate the degree of specialization. More than one scholar highlights this by drawing parallels between recent imaging tools and phrenology.[56] Franz Gall, a prominent figure in that dead science, argued that mental dispositions and talents correlate with enlargements and indentations on the skull. Though put to racist purposes and ultimately unjustified by evidence, the hypothesis as it was originally formulated was testable and not completely unreasonable. Contemporary cognitive neuroscientists have done something similar, mapping the brain such that they assume they can infer, for example, language mentation from fMRI measurements of local blood oxygenation. Among other problems, fMRI and other imaging techniques have limited resolution. As previously asserted, moreover, it is often difficult to ascertain the extent to which localized activity is solely responsible for a function or whether a given brain region instead handles a task in conjunction with others. In most cases—and we have discussed this at length—evidence weighs heavily in favor of the latter.

In addition to this, researchers are increasingly and consistently finding that given structures handle multiple tasks,[57] points we have also repeatedly detailed. Anderson emphasizes just this in his discussions of neural reuse.[58] Large and small brain regions, including those examined in this and other chapters, deploy to enable multiple functions. Varieties of examples testify to this assertion. One affirming case is neural reorganization that can occur after limb amputation such that, for instance, stimulation from the face can activate brain areas formally associated with the hand.[59] Broca's area—which is classically associated with language, but also involved in movement and music perception—is another illustration that we and others have leaned on, perhaps overmuch since it is a fairly large area in the cortex.[60] However, there are also highly local instances of reuse. This is so in neurons of the earlier mentioned C. elegans, which are capable of performing both motor and sensory tasks, among much else (see chapter 5). Studies on mammals similarly show multipurpose integrative neurons in the superior colliculus (see chapter 2). At a less local but still compelling level, reading braille has been shown to activate brain areas associated with vision in those suffering blindness and more tellingly in those made to wear blindfolds for a relatively short period (see chapter 2). This state of affairs meshes generally with the thesis that traditionally separated psychic faculties in fact integrate locally and globally in the brain, which is consistent with pragmatic insights and conclusions advanced by Merleau-Ponty. In short, that some regions exhibit a degree of specialization does not indicate that they perform only one task. Indeed, evidence suggests a fair amount of flexibility. In addition to this, we have shown that most (and almost assuredly all) brain regions work in concert with others and indeed with the rest of the body, in ways directed towards tasks in the world.

All of this fits the broadening ecology we seek, and it dovetails nicely with embodied standpoints. To elaborate a little, the human body has specialized organs and appendages—for example, hands, eyes, neck, and so on. However, when the hands are deployed, say, in typing, a total coordination takes place that affects not only the brain, but also the gaze and posture and therefore the neck, back, and more. Without this coordination, the activity (hands typing) is not possible. Evidence unequivocally indicates something similar as the norm for the brain. That is, it suggests that cognitive, emotional, motor, and perceptual life involves total coordinations of neural activity, and that no single region is ever activated alone. Additionally, to consider again the example of the hand, its specific synchronies with other anatomical regions depend on the given task, and reuse is possible here as well. Thus, while hands are normally used for active manipulations, they can be trained for comprehension and expression of language such as signing and reading braille.

Pragmatists, Merleau-Pontian phenomenologists, and 4E cognitive scientists have further observed that a great deal of processing occurs through body-environment interactions, as when the knee solves complicated problems of physics. Moreover, while nearly everyone rightly accepts that emotion has neurobiological components, evidence suggests that expressions and postures also play a role. Darwin in fact argued that bodily expression softens or dampens emotions.[61] More recent experiments—whether with participants holding pens in their mouths that encourage certain facial expressions or conditions limiting them such as Botox injections—testify to this point, though there are ongoing debates about the meaning and replicability of some of these studies (see chapter 4). James, Dewey, and Merleau-Ponty advanced similar arguments, and taken together this fosters a compelling case—in line with enactive views—that even emotional experience involves a sensorimotor loop. This is again reinforced by a variety of neurobiological findings. To name one, the cerebellum, which is historically regarded as a structure responsible for motor coordination, is also implicated in emotional experience, associative learning, and other functions.[62]

Numerous other neurobiological findings discussed throughout this book illustrate comparable points. These findings stand as a reminder of something nearly everybody accepts: that the brain is always involved. Theorists speaking about human minds know this, and 4E proponents specifically recognize that the brain coordinates motor and sensory activity in humans (this observation supplies the foundation for much 4E work). But too many in the 4E camp pass over detailed discussions of the nervous system. As suggested earlier, they may do so in the interest of illustrating—as Noë puts it in the subtitle of one of his books—that "you are not your brain."[63] Indeed, it appears that Noë gets some of the brain science wrong, for example, when he claims that sensory substitution devices do not activate the visual cortex (see chapter 2). While a case where only tactile neural systems are engaged might be more consistent with some readings of enactivism, activation of the visual cortex does not refute them. Indeed, it can reinforce enactivism, as argued in chapter 2.

Other findings are similarly supportive, as in the case of premotor and motor areas of the frontal cortex that make functions, bodily expression, and skilled actions such as hammering possible,[64] which is to say, the brain sciences go some way towards verifying embodied accounts. Another reinforcing example is that of phantom limb cases, specifically patients experiencing sensations of fingernails painfully digging into a palm that is no longer there. Researchers have speculated that this arises partly from outgoing signals to clench the missing hand, combined with the lack of feedback indicating that the deed is complete. Under such circumstances, sufferers may experience the sensation of unceasingly tightened hands and the discomfort that would normally accompany this.[65] This supplies neurobiological support for Merleau-Ponty's account of the phenomenon as a failure to adapt bodily-habitual schemata to current conditions.[66] It more generally fits pragmatic, phenomenological, and 4E explanations that regard perception as outcomes of sensorimotor loops, while also being consistent with a fairly uncontroversial understanding of the brain.

So although we have steadfastly defended the claim that we are not our brains, human sensorimotor perception, cognition, and emotion obviously require a nervous system. This is so uncontroversial that it hardly bears mentioning. However, getting beyond correlations between the brain and various psychic processes to an actual understanding of how the nervous system contributes to them is still a giant leap. Accordingly, more work is needed to integrate the brain sciences with emerging 4E views.

BROADENING ECOLOGIES

In a well-known article published at the turn of the twentieth century, James asserted that consciousness does not exist as an entity, but is instead a function.[67] In advancing this thesis, he understood himself to be countering the Kantian idea of the transcendental ego. This is the logical postulate that a pure self or consciousness must underlie all human experiences, even though it is not empirically observed, which means that nothing more can be said about it. For James, this position entailed dualism, which he strove to circumnavigate in his later career. James's knowledge of the history of philosophy and its impact on thinkers of his time was not nearly as developed as that of Peirce and Dewey. Kantians are unlikely to find his attack devastating, and we accordingly do not wish to dwell on it. What is worthy of expansion in the context of the ecological thesis of this book—and what James completely missed—was the resonance between ancient Greek views and his proposal that consciousness is a function.

The Greek root of "psychology" is *psukhē*, and as discussed in chapter 4, the word could mean "breath," "soul," or "life." For Aristotle, *psukhē* was precisely that which differentiates living from nonliving things. This accordingly makes the term "psychic life" redundant, though we have opted to use it quite a bit. In *Peri Psukhēs*, better known by its Latin title, *De Anima*, Aristotle defined *psukhē* as "an actuality of a natural organized body."[68] He repeatedly equated this to the functioning of various powers of living bodies, such as perception or nutrient absorption, through which the environment is assimilated into the organism. He excluded only reasoning from this set of powers because he could not identify a bodily function dedicated to it (the brain was understood by some at that time to be an organ for dissipating heat rather than as an exclusive site for psychic activity).

The gist of Aristotle's account of psychic processes is in fact conveyed in the term "human being," where the latter half of the expression is historically associated with the form, essence, or the "what is" of an entity. For Aristotle, what an entity is connects squarely to function, especially in the case of living things. By way of example, Aristotle argued that an eye removed from an animal ceases to be a real eye, even though its matter and structure remain the same.[69] In other words, he saw a real eye as a living or functioning organ, which is to say, one engaged in seeing. An eye that is removed no longer has the potential to see, which means, obviously, that it cannot actually see. The word "actually" is fitting because it is almost synonymous with "real" and hence the "what is" of the thing, and further because it contains the root "act." It thus emphasizes the active nature of functioning organisms. On this account, if the eye were an animal, sight would be its soul or *psukhē*, which is to say, its activity or function. So similarly with organisms in Aristotle's scheme: *psukhē* is the activity or actuality of the functioning body. It is actuality because, in acting, the body actualizes what it is, just as the eye, in seeing, actualizes what that organ is.

It is accordingly plain to see why Aristotle associated soul with actuality, and further why he had no problem unifying it with body: for him, the soul was the activity or function of the body. This also provides a sense of why form and function (*telos*) are so closely linked in Aristotle's scheme. When a natural being is alive, its matter is organized; it has a form, and all of its parts are working together to maintain its form. Put another way, everything contributes to maintaining the formal integrity of the organism. This activity or function of maintaining the form is the soul according to Aristotle, or what we have called "psychic life" (in all its etymological redundancy). The connection between soul and form is also made loosely in English, where the soul of something can be the "core," "heart," or "essence." As Aristotle reasoned more broadly, soul—understood as the activity of the body—is the principle, actualization, or act of life.

Though pragmatists did not lean on ancient terminology, they advanced comparable views and founded a school known as functionalist psychology, which used core ideas from Aristotle's account. It embraced a broadly ecological standpoint. In this view, which aligned with the views of embodied phenomenologists, Gibsonian psychologists, and 4E theorists, eyes and other organs do not function in isolation of the rest of the body. Instead, what we experience shows up as a matter of what our bodies can do, and this entails total coordinations of sensory organs, neural structures, viscera, and more. As we have argued, moreover, human experiences entail mobilization of cognitive, emotional, and motor capacities too. Experiences also implicate an environment, which not only offers subject matter, but pulls and integrates organs and faculties into a synergic system critical to having a coherent grasp of things. Seen accordingly, the activity of seeing is not isolated in the eyeballs and retinas. Nor is it in the combination of ocular and neural structures alone. It instead entails an overall mobilization of organs and capacities, as we have repeatedly argued, following Dewey and Merleau-Ponty and substantiating the case with experimental and biological data. The same generally holds for psychic life, which emerges from the body as a whole (including the organisms living in it) in combination with actions taken in environments—actions that pull doings and undergoings into synchrony.

Proponents of classical pragmatism recognized their debts to the past. This was why James subtitled his pragmatist manifesto "A New Name for Some Old Ways of Thinking." This was also why Dewey repeatedly insisted that the key to his cutting-edge concept of experience was found in ancient ideas. Without question, every generation resurrects something important from the past. In recent decades—and indeed during the period in which classical pragmatism emerged—one such idea was that mind is in body and body in mind. Chemero, Clark, Damasio, Gallagher, Hutto and Myin, Lakoff and Johnson, Menary, Rowlands, and Thompson are among some of the more prominent researchers who adopt this position today. One way in which neuroscientists such as Damasio have done this is by identifying how visceral emotional responses help us engage in problem solving. This understanding of problem solving suggests that visceral emotional responses perform cognitive functions too, which is to say that visceral emotional responses are coping mechanisms that lead to decisions, perceptual choices, and actions in the world (see chapter 4). In this, there is something of a revolt against reducing emotions to raw feelings, though too many—including Damasio—tend towards this, as we argued in chapter 4. While not focusing on neurobiology, Gibson has likewise pointed to the emotional and valuative sides of perception, again grounded in action in the world. This is all in line with what pragmatists, Merleau-Ponty, and Gestalt psychologists have argued, all of whom contribute to the lineage leading to Gibson and today's 4E proponents.

Speaking of emotion, James correctly—if somewhat tritely—argued that a hypothetical purely disembodied instance of emotion would amount to a non-entity.[70] But James, as we suggested in chapter 4, also identified emotion too narrowly with visceral feelings. Nonetheless, he at least recognized that pangs, tinglings, and other visceral stirrings do not become identifiable as specific emotions without worldly grounding, a point we flesh out in earlier chapters via Dewey and Merleau-Ponty. Emotion can involve bodily attitudes towards people, things, and events, which can occur through expressive behavior, attention shifts, and accompanying readiness of motor equipment, altered energy levels, or approach and avoid behaviors. All this integrates with cognitive and perceptual perspectives adopted in settings in the world, which circle back on our experience, coordinating and organizing it (see chapters 1 to 5). This in fact holds with anything falling within the domain of human psychic life. It means that all experience is simultaneously embodied, embedded, enacted, and extended, and also cognitive, motile, perceptual, and valuative or affective, in the broad senses we have used these terms. Some may take this as something of an empty truism, but historically this has not been understood so, and it challenges in-the-head approaches and the notion that the brain alone primarily generates experience.

Thus, to elaborate, romantic love is not identical with sexual desire or brain activity, for it also entails objects of love. Although love can encompass a variety of bodily feelings, it simultaneously involves worlds that organize habitually around the emotion and people towards which it is directed. As such, it moderates cognitive and motor activity as well as perception, such that reducing love to bodily feelings or brain states diminishes a rich concept that is central to a great deal of human experience.[71] While emphatically embodied, the emotion—if it is to show up specifically as love—is embedded, enacted, and extended. We see this in even relatively simple moments in which the bodily sensibility of the emotion might manifest in an active and enthusiastic embrace of a person in a place with special meaning. Typically, it enters into structures of attachment and appraisal, and consequently influences patterns of approach, contact, protective behaviors, and more besides.

Like numerous other emotions (not to mention logic, comportment, and a great deal else comprising psychic life), love, too, manifests as a normative goal. It is normally something desired, and many see better and worse ways of being in love, forming attachments, and valuing others.[72] This is just as there are norms for reasoning and ways of moving that are esteemed as more beautiful and healthful. Love, like other emotions, is an orientation towards the world; that is, love is a way of dealing with people and things in it, as Dewey and a good many others have insisted.[73] Few doubt that love has roots in the biology of attachment that anchors us, fosters social cohesion, and thereby enables our functioning in the world, helping us make sense of it and thrive. Yet love is not reducible to biology, much less to hormones such as oxytocin. Rich and abiding cultural factors have shaped what it is and how it is enacted, as is well known by those who have deeply engaged with the vast body of literature, music, painting, sculpture, cinema, and other arts attempting to elucidate and grapple with it, or sometimes cheapen it. So while we have advanced a biologically and neuroscientifically substantiated story of psychic life, we warn once more against overselling such accounts.

Gustatory processing is similar in that it involves more than just the interface between peripheral organs and the brain. Here, too, culture moderates what we find palatable and what sort of food and drink we desire and deem appropriate in different situations, as with a meal after a workout compared to a wedding feast carried out according to certain religious practices. In some cultural moments we may find ourselves suddenly enjoying what we formally disliked, as when eating a particularly pungent cheese in France or experiencing a flaming pudding with close friends and family at Christmas. In such ways—and this is speaking from personal experience—the span of a few moments in a cultural setting may permanently change our culinary preferences.

But while not reducible to biology, ingestion is an unequivocally biological function that is illuminated by that field of study. Ascending from the viscera to the central nervous system, one finds a variety of problem-solving mechanisms involved in nutrient and water balance. These link to semantic networks in the brain that help give this information relevance,[74] helping us decide on appropriate behaviors, for example, when at fast-food restaurants versus the snack table at funeral receptions. Visceral input at neural sites—whether in the brain stem, amygdala and olfactory bulbs, or the retinas—all operate within the information-processing, gustatory, decision-making network. Among much else, forebrain visceral regions are critical in regulating the internal demands amid external circumstances.[75] So too is the microbiome within us, as discussed earlier in this chapter. As Dewey seemed to recognize generally,[76] bodily sensibility and visceral and muscular-skeletal control—both self-directed and pointed at the world—are fundamental. This is to repeat an earlier claim: that cognizing, emoting, and perceiving, at their core, entail avoidance and approach, warding off and attaching. Alteration in motivational drives for water, food, safety, and more accordingly extend not just across the visceral system and brain, but also into the world, and they call continually upon integrated cognitive-motor-perceptual-valuative capacities.

Emotions are involved in this, and they, too, appear to be partly driven by semantic networks that traverse more recently evolved cortical areas,[77] albeit with subcortical regions also in play.[78] Fear and emotions involved in attachment are widely shared by a variety of animals. They are perhaps primarily orchestrated in older regions such as the amygdala bulbs, the hypothalamus, and the bed nucleus of the stria terminalis in the brain stem.[79] However, they also appear regulated by cortical sites.[80] Emotions such as envy, shame, and embarrassment are highly conceptual, if not cultural. Not surprisingly, therefore, these especially link to brain areas dealing with semantics.[81] This once again highlights the integrated nature of emotion and cognition. Insofar as emotion and cognition regulate attention and action in the world, neither operates independently of motility and perception.[82]

So while promoting beyond-the-brain approaches, which emphatically hold in noncephalic organisms such as slime mold (see chapter 5), we take it as another truism that neural structures ought not to be ignored in organisms that possess them. Electrophysiological studies on rats, for example, have recorded activity in the parabrachial region in the brain stem upon stimulation of the oral cavity, suggesting that neural structures in this area play a role in gustatory processing.[83] This is reinforced by the fact that ablating the aforementioned region alters feeding patterns. Rats normally exhibit rapidly formed and long-lasting aversions to substances causing illness, but after a recovery period, most animals that have undergone ablation do not display this behavior. Specifically, after ingesting a nausea-inducing chemical infused in sucrose and salt preparations, they fail to avoid these substances.[84] Other studies affirm that the parabrachial region is critical for learned associations between ingestion of substances and visceral discomfort that follows. This research further suggests that the same neural region is important in appraising the significance of salt in sodium-hungry animals.[85] Brain stem structures are in fact generally regarded as vital for core regulation of the internal milieu and helping to establish connections between visceral disruption and the organization of action to improve and reestablish bodily stability.[86]

The brain stem, in combination with a range of cortical and subcortical sites and peripheral systems—which is more or less to say, the body in its entirety—contributes to the processing of visceral information. In cephalic organisms, capacities to act, cognize, emote, and perceive collectively infuse information with meaning. Detection of discrepant changes is often a feature of consciousness or what James equated to selective attention, and this enables synchronized adjusting to environmental conditions. Other prevalent adjustments relate to restoration of homeostasis, in which selective attention often plays a role. Recognition of regulatory demands also facilitates behavioral options. Bodily sensibility in peripheral systems helps inform the central nervous system of critical considerations such as whether to feed or drink. The reverse also holds, with the brain moderating peripheral activity according to various needs. This only occurs in the context of an environment and accordingly contributes to patterns of action.

Pragmatists and later Merleau-Pontian phenomenologists have long argued that it is impossible to trace mind to any one structure because it in fact arises out of a totality of sensitivities and capacities working together in concert with the world. The account offered in this book reinforces this, adding the weight of recent biological, experimental, cognitive scientific, and philosophical research. From all this, it obviously follows that the reticular formation does not equal psychic life, nor do the amygdala bulbs, nor do information molecules such as hormones and neurotransmitters, nor still the brain and nervous system in their entirety. Bodily contact, practiced patterns, and habits are enabled by regions of the brain working in coalition and also in human action in the world, not to mention the microbiome. All of this and more constitute the organization of action, affect, perception, and thought, which are integrated processes. Bodies are inhabited by minds, and minds need bodies. Feelings and emotions are not detached from cognitive and perceptual systems in the brain, and they are not separate from bodily conditions and actions either.

CAUTIONS AND PROJECTIONS

Though many of the general claims we have advanced may seem obvious, there is longstanding resistance to them. Thus it was that over a hundred years ago, Peirce observed that scholars had never quite dispensed with the Cartesian notion of mind as something that resided in the pineal gland.[87] Nearly everybody laughs at this, he added, yet they nevertheless continue thinking of mind as something separate within the person that supplies a correlative to the world. A hundred years on, the same is carried in the notion that the brain is the seat of human subjectivity. This view is not patently foolish, and focus on the brain *per se* is not the problem. The difficulty instead emerges when scholars take the brain or nervous system to be the key to solving the mystery of psychic life. To be sure, we now understand that certain parts of the brain and nervous system are associated with particular functions. In spite of this, we are little closer than a century ago to understanding how we get from nerves firing to the fullness of human experience; we cannot currently explain this in the step-by-step ways that we use to describe the workings of the internal combustion engine or the processes leading to a supernova.

The outlooks we have defended do not settle the enigma. However, we believe that they have more explanatory power than neurocentric approaches, and that they highlight their inadequacies on biologically justifiable grounds. We also think they explain more about psychic life (here taken in a more inclusive sense than just consciousness). Extended navigational processes such as mountains funneling monarch butterflies to final destinations (see chapter 2) or groups of brainless *P. polycephalum* exploring their surroundings by avoiding secreted residue that marks where they have already been (see chapter 5) are just a few in a long list of testifying cases. Such processes track to human beings. So it is that roads and pathways form external memory markers guiding where we go—a view that resonates with the work of Dewey, Merleau-Ponty, Gibson, Brooks, and 4E proponents such as Clark.

Though elements of embodied, embedded, enacted, and extended ecological approaches trace to ancient sources,[88] conditions for an upsurge of something approaching them occurred in the eighteenth and nineteenth centuries. This was especially in the context of myriad evolutionary and, by extension, ecological theories put forth during this time. In this period, mind became increasingly understood in terms of biological adaptation framed as an ecological phenomenon (see chapter 1). Darwin, who insisted on looking at things "in ever-increasing circles of complexity,"[89] was one case in point, and there were many others such as Buffon's work from the mid-1700s and Lamarck's from the beginning of the following century. Though most scholars had thought of mind as something directed towards the world, evolutionary conceptions highlighted that it could not be understood in isolation of the body and the environments with which the body dealt. In short, evolutionary outlooks emphasized that the body, rather than being an appendage to information gathering, was instead at the heart of it. Versions of this idea appear in James and Dewey's ecological frameworks, which assimilated the biological and evolutionary turns of their day. Phenomenologists and Gestalt psychologists likewise delivered ecological orientations. Gibson did the same. More recently, 4E theorists have followed suit, incorporating ideas from all these earlier schools by way of Gibson. Some of them—for example, Chemero, Gallagher, Menary, and Thompson—draw directly on one or more of these now classic schools.

At the same time, biological developments from at least the nineteenth century onward have arguably led to some strides backwards, and the reason is not hard to see. Tibor Solymosi, who coined the term "neuropragmatism," has interestingly and rightly pointed out that there is an overabundance of "neuro-hype"[90] and "neurophiles" these days.[91] This overselling of neuroscience risks obscuring much that is important to human psychic life, reducing the latter to the language of the neuron,[92] and this is problematic for reasons we have outlined at length. It would accordingly be a good idea to preserve explanations at the level of lived experience. That is to say, it is advisable to consider the worlds in which we act and consequences we undergo as results of our behaviors, as Dewey, Merleau-Ponty, and others in pragmatic and phenomenological lineages (not to mention Wittgenstein in his later work) repeatedly urged. This means pondering practices carried out in broader social arrangements, and it means not relinquishing descriptions of experience in accounts of mind, contra what eliminative theories argue. In short, we ought not to lose sight of the transactional, participatory, exploratory, and cultural experiences that emerge through approaching, enjoying, avoiding, and generally coping with the worlds we encounter. However, in an age when the word "scientific" is colloquially used as a gold seal of epistemic legitimacy, and when philosophy departments are closing, too many attempt to vindicate their worth by claiming to do neurophilosophy, neurophenomenology, neuroethics, neuroaesthetics, and indeed neuropragmatism. Intended or not, the impression becomes one of science and specifically neuroscience acting as a last-ditch resuscitation of a fading discipline, the former increasingly put forth as the only recourse for understanding mind.[93] In such a context, a central if understated goal of pragmatism and, indeed, what Solymosi calls "neuropragmatism" is in danger of getting obscured: that of offering first aid to science, broadening and enriching its outlooks on matters of importance, as we have attempted to do in our synthesis of pragmatism with biological, psychological, and other philosophical research.

In addition to approaches that put the brain on a pedestal, a separation of cognitive, emotional, perceptual, and motor life regularly appears. Gigerenzer's work, which was discussed in earlier chapters, is a case in point. Gigerenzer advances a concept of "ecological rationality," which holds that rationality does not operate through universal decision mechanisms. Rather, it occurs through cognitive heuristics (including bodily activity) matched to specific contours of given situations. He argues, moreover, that successful strategies typically neglect most information and eschew as much computation as possible, since very little of either is relevant most of the time. These are worthy points. Yet, Gigerenzer and many other proponents of ecological rationality pay little attention to emotions. This is even though emotions are a large part of what leads us to rationally ignore most of the world and thereby focus our attention. Darwin in fact set the stage for understanding emotions as biologically adaptive and therefore as not opposed to rationality.[94] A rational choice, as we have argued (building on James, Dewey, Damasio, and others), is emotional. The reverse also holds, though we recognize that what is traditionally called "emotion"—and for that matter, "rationality"—can lead us astray at times.

So while we have maintained that emotions are typically rational, they can sometimes be the opposite. The United States, for example, is rife with hostility towards Latinx migrants who are in fact less likely to commit crimes than native-born residents.[95] There is also an extended history of prominent figures irrationally inflicting harm on others and themselves because of unchecked sexual desire or greed. And we all know what it is to have anger get the better of us. As argued, however, emotional tendencies more typically integrate fruitfully with action, cognition, and perception so that we largely incline towards what is healthful and display wariness towards what is not. Thus even anger can be a rational self-defense mechanism, whether when dealing with someone trying to cheat or otherwise hurt us, or responding to social injustice. We likewise experience anxiety when perceiving bad air in overcrowded subways, and this weighs into decision making and actions. As discussed earlier, we additionally undergo cognitive-emotional-motor-perceptual attraction to nourishing foods and prospective mates. This is even if evolutionary time lag leads to destructive sexual excess or makes us overly disposed towards substances rich in sugar, fat, and salt, which in past conditions of deprivation would have given us an adaptive advantage. While reason similarly serves us well much of the time, it can be problematic—or in other words, cease to be reasonable—when stripped of emotional sensitivity. Thus, someone who attempts to resolve a dispute with a romantic partner by constructing a logically succinct email with an opening thesis, supporting points, and a conclusion will likely not achieve the hoped-for end of settling the disagreement. More formal examples come from Damasio's case studies, in which brain damage that reduces emotional-visceral response (in other words, brain damage that impairs bodily sensibility) augurs disastrous decision making and actions. This is even though sufferers remain capable of weighing pros and cons. James, too, gave a variety of phenomenologically rich examples of how emotions and interests enter into conceptual abstraction, reasoning, and linguistic-logical interpretation, as well as into action (see chapter 4).

In the same way that cognition is often treated as separate from emotion, it is still very much a part of the neuroscientific literature to divide the brain along such terms. One place this manifests is in the ubiquitous juxtapositions of lower versus higher functions, in which the cortex is depicted as interpreting undifferentiated bodily agitations that first enter the brain through allegedly more primitive subcortical structures. Stanley Schachter and Jerome Singer's view of emotion, for instance, remains a dominant framework for conceptualizing relationships between peripheral fluctuations and cognitive processes. This account conceives the brain stem as emotional, and holds that it passes undifferentiated excitatory or inhibitory signals to the cortex, which then imposes form and coherence.[96] In philosophical terms, this is something like Kant's categories of understanding structuring the amorphous sensory manifold. In the case of Schachter and Singer, this is wrongheaded because brain stem regions, among much else, regulate peripheral systems and bodily behavior, and for reasons discussed at length, this brings structure to cognition and perception.[97] It is for similar reasons that pragmatists challenged the Kantian machinery of the *a priori*, even while granting that bodily systems, whether in the form of interests and emotions or overt actions, structure and indeed limit cognition and perception.[98] This is not to elevate subcortical regions above cortical areas. Nor is it to raise emotional-interested life or bodily action above cognition. It is rather to emphasize that all operate reciprocally not only to self-synchronize, but to organize actions, and that this in no small measure achieved by coordinating around structures in the world. At the same time, it is to assert that information processing—an affective-cognitive-perceptual operation—is much more than just a cortical affair. It is further to highlight the critical role of motor and affective life in such processes. In short, it is to maintain that action, affect, cognition, and perception integrate reciprocally with brain, body, and physical and social surroundings, and that all are mutually defining.

In ways paralleling the division between higher and lower functioning, it has remained part of neurobiological discourse—at least until recently—to divide the brain in terms of interpreting versus responding to external events. The hippocampal structures, for example, have often been construed as performing interpretative operations and the amygdala bulbs as more reactive; and to be sure, the former may be more involved in cognitively identifying specific threats and the like, while the latter may be more central in generating fear reactions. However, the two tightly interinnervate, and the amygdala bulbs, by facilitating environmental perception and appraisal, also facilitate interpretation, which means that they critically moderate selective attention, among other affective-cognitive-perceptual processes (see chapter 4). In fact, evidence suggests that the hippocampus and amygdala perform interpretive tasks together, in collaboration with a number of other structures.[99] The Kaplans also argue something like this, but from a psychological standpoint[100] (see chapters 3 to 5). They present data suggesting that aesthetic responses are affective appraisals of settings. In their argument, aesthetic responses are also cognitive insofar as they involve immediate evaluations of situations; they are perceptual insofar as they are ways of registering environments through the senses; and they are embodied insofar as they entail assessing what it is possible to do in a setting and to the extent that affective appraisals have visceral components. By virtue of all this, aesthetic responses are also embedded, enacted, and extended.

The experimental psychology of the Kaplans is, in part, an extension of Gibson's ecological framework. As such, it is in the lineage of pragmatism, phenomenology, and Gestalt psychology. It is questionable the extent to which the Kaplans recognize their work as a development in this history, though their 1989 book, *The Experience of Nature: A Psychological Perspective*, exhibits sensitivity to James, specifically in the third part of the work, appropriately titled "Towards a Synthesis." Gibson's work is a stronger example of merging scientific psychology and philosophy. James and Merleau-Ponty achieved something similar by the standards of their day, fusing philosophy, psychology, and biology into an integrated account of human experience. We do not advocate that all scholars do this, and recognize the importance of highly focused empirical research carried out with difficulty in many laboratories, which is critical to the kind of work done in this book. We also appreciate the painstaking rigor of philosophical scholarship that may lack the kind of biological and experimental backing we have attempted to render. At the same time, we endorse the ecological position Dewey expressed when he wrote:

> To see the organism in nature, the nervous system in the organism, the brain in the nervous system, the cortex in the brain is the answer to the problems which haunt philosophy. And when thus seen they will be seen to be in, not as marbles are in a box but as events are in history, in a moving, growing never finished process. Until we have a procedure in actual practice which demonstrates this continuity, we shall continue to engage in appealing to some other specific thing, some other broken off affair.[101]

This suggests that if the goal is to render global and ecological accounts of what it is to be a psychic organism, then it is not sufficient to look at neurons or brain structures in isolation; nor solely experimental accounts measuring behavioral responses; and nor still phenomenology bereft of biological and experimental content. Rather, it is critical to piece all of this together. We have offered a step in this direction, but it goes without saying that our contribution is small and the kind of inquiry we advocate is grueling and without end.

APPENDIX 1

SUBCORTICAL STRUCTURES OF THE BRAIN

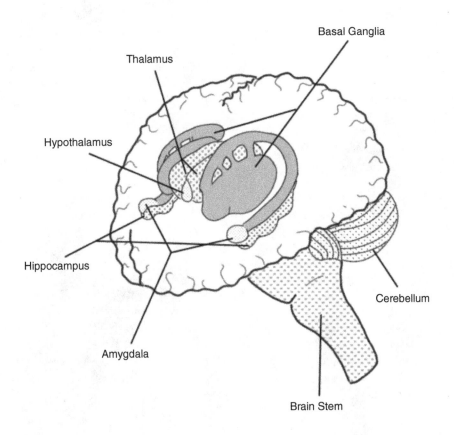

APPENDIX 2

CORTICAL STRUCTURES OF THE BRAIN

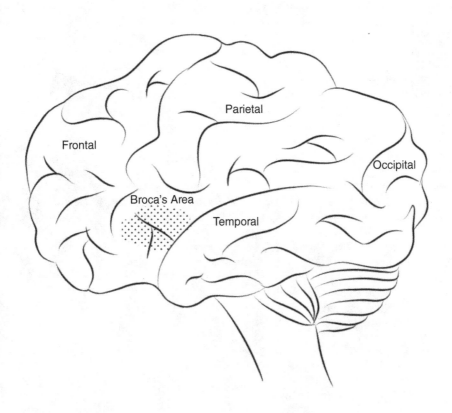

NOTES

INTRODUCTION

1. A. R. Damasio, *Self Comes to Mind: Constructing the Conscious Brain* (New York: Pantheon Books, 2010), 8.
2. See G. Parrott and J. Schulkin, "Neuropsychology and the Cognitive Nature of the Emotions," *Cognition and Emotion* 7 (1993): 43–59; A. R. Damasio, *Descartes' Error: Emotion, Reason, and the Human Brain* (New York: G. P. Putnam, 1994); J. Schulkin, *Roots of Social Sensibility and Neural Function* (Cambridge, MA: MIT Press, 2000); and J. Schulkin, *Bodily Sensibility: Intelligent Action* (New York: Oxford University Press, 2004).
3. For example, T. Rockwell, *Neither Brain nor Ghost: A Nondualist Alternative to the Mind-Brain Identity Theory* (Cambridge, MA: MIT Press, 2005); M. Johnson, *The Meaning of the Body: Aesthetics of Human Understanding* (Chicago: University of Chicago Press, 2007); R. Menary, *Cognitive Integration: Mind and Cognition Unbounded* (New York: Palgrave Macmillan, 2007); R. Menary, "Pragmatism and the Pragmatic Turn in Cognitive Science," in *The Pragmatic Turn: Toward Action-Oriented Views in Cognitive Science*, ed. A. K. Engel, K. J. Friston, and D. Kragic, 215–223 (Cambridge, MA: MIT Press); A. Chemero, *Radical Embodied Cognitive Science* (Cambridge, MA: MIT Press, 2009); S. Gallagher, "Philosophical Antecedents of Situated Cognition," in *The Cambridge Handbook of Situated Cognition*, ed. P. Robbins and M. Aydede, 35–52 (New York: Cambridge University Press, 2009); T. Solymosi and J. Shook, eds., *Neuroscience, Neurophilosophy and Pragmatism: Brains at Work with the World* (New York: Palgrave Macmillan, 2014); Engel, Friston, and Kragic, *The Pragmatic Turn*; M. Crippen, "Dewey, Enactivism and Greek Thought," in *Pragmatism and Embodied Cognitive Science: From Bodily Interaction to Symbolic Articulation*, ed. R. Madzia and M. Jung, 229–246 (Boston: De Gruyter, 2016); M. Crippen, "Embodied Cognition and Perception: Dewey, Science and Skepticism," *Contemporary Pragmatism* 14 (2017): 121–134; and Madzia and Jung, eds., *Pragmatism and Embodied Cognitive Science*.
4. R. G. Collingwood, *An Autobiography* (New York: Oxford University Press, 1939).

5. M. Crippen, "Pragmatic Evolutions of the Kantian *a priori*: From the Mental to the Bodily," in *Pragmatist Kant: Pragmatism, Kant, and Kantianism in the Twenty-First Century*, ed. K. Skowroński and S. Pihlström, 19–40 (Helsinki: Nordic Pragmatism Network, 2019). See also E. Cameron, "From Plato to Socrates: Wittgenstein's Journey on Collingwood's Map," *AE: Canadian Aesthetics Journal* 10 (2004): 1–29, https://www.uqtr.ca/AE/Vol_10/wittgenstein/cameron.htm.

6. See Crippen, "Pragmatic Evolutions."

7. See Crippen, "Dewey, Enactivism and Greek Thought"; and M. Crippen, "Dewey on Arts, Sciences and Greek Philosophy," in *In the Beginning Was the Image: The Omnipresence of Pictures; Time, Truth, Tradition*, ed. A. Benedek and A. Veszelszki, 153–159 (Frankfurt am Main: Peter Lang Press, 2016).

8. J. J. Gibson, *The Ecological Approach to Visual Perception* (Boston: Houghton Mifflin, 1979).

9. M. Merleau-Ponty, *Phenomenology of Perception*, trans. C. Smith (New York: Routledge and Kegan Paul, 1962; first published 1945 in French).

10. M. Rowlands, *The New Science of the Mind: From Extended Mind to Embodied Phenomenology* (Cambridge, MA: MIT Press, 2010), 3.

11. A. Clark, *Supersizing the Mind: Embodiment, Action, and Cognitive Extension* (New York: Oxford University Press, 2008); and A. Clark, *Surfing Uncertainty: Prediction, Action, and the Embodied Mind* (New York: Oxford University Press, 2015).

12. M. Johnson, *The Aesthetics of Meaning and Thought: The Bodily Roots of Philosophy, Science, Morality, and Art* (Chicago: University of Chicago Press, 2018), chap. 1.

13. Rowlands, *New Science*, 3.

14. For example, K. O'Regan and A. Noë, "A Sensorimotor Account of Vision and Visual Consciousness," *Behavior and Brain Sciences* 24 (2001): 939–973; A. Noë, *Action in Perception* (Cambridge, MA: MIT Press, 2004); and A. Noë, *Out of Our Heads: Why You Are Not Your Brain, and Other Lessons from the Biology of Consciousness* (New York: Hill and Wang, 2009).

15. D. Hutto and E. Myin, *Radicalizing Enactivism: Basic Minds Without Content* (Cambridge MA: MIT Press, 2013); D. Hutto and E. Myin, *Evolving Enactivism: Basic Minds Meet Content* (Cambridge MA: MIT Press, 2017); and E. Thompson, *Mind in Life: Biology, Phenomenology, and the Sciences of Mind* (Cambridge, MA: Harvard University Press, 2007).

16. F. Varela, E. Thompson, and E. Rosch, *The Embodied Mind: Cognitive Science and Human Experience* (Cambridge, MA: MIT Press, 1991).

17. We have provided three-dimensional diagrams in the appendixes for brain structures repeatedly discussed in this book. We encourage those with limited familiarity with the brain to consult these as needed.

18. See J. Long, *Darwin's Devices: What Evolving Robots Can Teach Us About the History of Life and the Future of Technology* (New York: Basic Books, 2011), chap. 5.

19. J. Decety et al., "Brain Activity During Observation of Actions: Influence of Action Content and Subject's Strategy," *Brain* 120 (1997): 1763–1777; F. Binkofski et al., "Broca's Region Subserves Imagery of Motion: A Combined Cytoarchitectonic and fMRI Study," *Human Brain Mapping* 11 (2000): 273–285; D. Thoenissen, K. Zilles, and I. Toni, "Differential Involvement of Parietal and Precentral Regions in Movement Preparation and Motor Intention," *Journal of Neuroscience* 22 (2002): 9024–9034; S. Tanaka and T. Inui, "Cortical Involvement for Action Imitation of Hand/Arm Postures Versus Finger Configurations: An fMRI Stury," *NeuroReport* 13 (2002): 1599–1602; F. Hamzei et al., "The Human Action Recognition System and Its Relationship to Broca's Area: An fMRI Study," *NeuroImage* 19 (2003): 637–644; and N. Nishitani et al., "Broca's Region: From Action to Language," *Physiology* 20 (2005): 60–69.

20. B. Maess et al., "Musical Syntax Is Processed in Broca's Area: An MEG study," *Nature Neuroscience* 4 (2001): 540–545; M. Tettamanti and D. Weniger, "Broca's Area: A Supramodal Hierarchical Processor?," *Cortex* 42 (2006): 491–494; and R. Kunert et al., "Music and Language Syntax Interact in Broca's Area: An fMRI Study," *PLoS ONE* 10 (2015): art. e0141069, https://doi.org/10.1371%2Fjournal.pone.0141069.

21. B. Knowlton, J. Mangels, and L. Squire, "A Neostriatal Habit Learning System in Humans," *Science* 273 (1996): 1399–1402; M. T. Ullman et al., "A Neural Dissociation Within Language: Evidence That the Mental Dictionary Is Part of Declarative Memory, and That Grammatical Rules Are Processed by the Procedural System," *Journal of Cognitive Neuroscience* 9 (1997): 266–286; A. J. Calder, A. D. Lawrence, and A. W. Young, "Neuropsychology of Fear and Loathing," *Nature Neuroscience* 2 (2001): 352–363; W. Schultz, "Getting Formal with Dopamine and Reward," *Neuron* 36 (2002): 241–263; J. W. Aldridge, K. C. Berridge, and A. R. Rosen, "Basal Ganglia Neural Mechanisms of Natural Movement Sequences," *Canadian Journal of Physiology and Pharmacology* 82 (2004): 732–739; S. A. Kotz, M. Schwartze, and M. Schmidt-Kassow, "Non-motor Basal Ganglia Functions: A Review and Proposal for a Model of Sensory Predictability in Auditory Language Perception," *Cortex* 45 (2009): 982–990; and S. A. Kotz and M. Schmidt-Kassow, "Basal Ganglia Contribution to Rule Expectancy and Temporal Predictability in Speech," *Cortex* 68 (2015): 48–60.

22. G. Rizzolatti, L. Fogassi, and V. Gallese, "Neuropsychological Mechanisms Underlying the Understanding and Imitation of Action," *Nature Reviews Neuroscience* 2 (2001): 661–670; and M. A. Umiltà et al., "I Know What You Are Doing: A Neurophysiological Study," *Neuron* 31 (2001): 155–165.

23. For example, Parrott and Schulkin, "Neuropsychology"; Damasio, *Descartes' Error*; and A. R. Damasio, *The Feeling of What Happens: Body and Emotion in the Making of Consciousness* (New York: Harcourt Brace, 1999).

24. M. Crippen, "Pragmatism and the Valuative Mind," *Transactions of the Charles S. Peirce Society* 54 (2018): 341–360.

25. H. A. Simon, "Motivational and Emotional Controls of Cognition," *Psychological Review* 74 (1967): 29–39; and G. Loewenstein, "The Psychology of Curiosity: A Review and Reinterpretation," *Psychological Bulletin* 116 (1994): 75–98.

26. H. A. Simon, "Motivational and Emotional Controls"; R. A. Rescorla and A. R. Wagner, "A Theory of Pavlovian Conditioning: Variations in the Effectiveness of Reinforcement and Nonreinforcement," in *Classical Conditioning II: Current Research and Theory*, ed. A. H. Black and W. Prokasy (New York: Appleton-Century-Crofts, 1972); and Loewenstein, "Psychology of Curiosity."

27. For example, R. Kaplan and S. Kaplan, *The Experience of Nature: A Psychological Perspective* (New York: Cambridge University Press, 1989).

28. See J. Schulkin, *Pragmatism and the Search for Coherence in Neuroscience* (New York: Palgrave Macmillan, 2015).

29. See Crippen, "Pragmatism and the Valuative Mind"; and M. Crippen, "Aesthetics and Action: Situations, Emotional Perception and the Kuleshov Effect," *Synthese* [issue unassigned] (2019), https://doi.org/10.1007/s11229-019-02110-2.

30. See Gibson, *Ecological Approach*; E. Reed, *James J. Gibson and the Psychology of Perception* (New Haven, CT: Yale University Press, 1988); H. Heft, *Ecological Psychology in Context: James Gibson, Roger Barker, and the Legacy of William James's Radical Empiricism* (Mahwah, NJ: Lawrence Erlbaum Associates, 2001); and A. Chemero and S. Käufer, "Pragmatism, Phenomenology, and Extended Cognition," in *Pragmatism and Embodied Cognitive Science*, ed. Madzia and Jung, 55–70.

31. See Crippen, "Aesthetics and Action."

32. See C. R. Reid et al., "Slime Mold Uses an Externalized Spatial 'Memory' to Navigate in Complex Environments," *Proceedings of the Natural Academy of Sciences of the United States of America* 109 (2012): 17490–17494.

33. See R. Brooks, *Cambrian Intelligence: The Early History of the New AI* (Cambridge, MA: MIT Press, 1999).

34. Crippen, "Pragmatism and the Valuative Mind"; and Crippen, "Aesthetics and Action."

35. For example, Kaplan and Kaplan, *Experience of Nature.*

36. See J. Dewey, *Experience and Nature* (Chicago: Open Court, 1925), chap. 7.

37. See Gibson, *Ecological Approach*; Kaplan and Kaplan, *Experience of Nature*; H. Dreyfus, *Being-in-the-World: A Commentary on Heidegger's* Being and Time, *Division I* (Cambridge, MA: MIT Press, 2003); G. Böhme, "Atmosphere as the Fundamental Concept of a New Aesthetic," *Thesis Eleven* 36 (1993): 113–126; C. Nussbaum, *The Musical Representation: Meaning, Ontology, and Emotion* (Cambridge, MA: MIT Press, 2007); and J. Slaby, "Emotions and the Extended Mind," in *Collective Emotions: Perspectives from Psychology, Philosophy, and Sociology*, ed. C. von Scheve and M. Salmela (New York: Oxford University Press, 2014).

38. J. P. Webster, C. F. Brunton, and D. W. MacDonald, "Effect of *Toxoplasma gondii* Upon Neophobic Behaviour in Wild Brown Rats, *Rattus norvegicus*," *Parasitology* 108 (1994): 407–411; J. P. Webster, "The Effect of *Toxoplasma gondii* on Animal Behavior: Playing Cat and Mouse," *Schizophrenia Bulletin* 33 (2007): 752–756; J. Flegr, "Effects of *Toxoplasma* on Human Behavior," *Schizophrenia Bulletin* 33 (2007): 757–760; J. Flegr et al., "Increased Incidence of Traffic Accidents in *Toxoplasma*-Infected Military Drivers and Protective Effect RhD Molecule Revealed by a Large-Scale Prospective Cohort Study," *BMC Infectious Diseases* 9 (2009): art. 72, https://doi.org/10.1186/1471-2334-9-72; and M. Clapp et al., "Gut Microbiota's Effect on Mental Health: The Gut-Brain Axis," *Clinics and Practice* 7 (2017): 131–136.

39. See Kaplan and Kaplan, *Experience of Nature*; C. J. Cela-Conde et al., "Activation of the Prefrontal Cortex in the Human Visual Aesthetic Perception," *Proceedings of the National Academy of Sciences of the United States of America* 101 (2004): 6321–6325; H. Kawabata and S. Zeki, "Neural Correlates of Beauty," *Journal of Neurophysiology* 91 (2004): 1699–1705; O. Vartanian and V. Goel, "Neuroanatomical Correlates of Aesthetic Preference for Paintings," *NeuroReport* 15 (2004): 893–897; and Y. Lévêque and D. Schön, "Modulation of the Motor Cortex During Singing-Voice Perception," *Neuropsychologia* 70 (2015): 58–63.

40. For example, Kaplan and Kaplan, *Experience of Nature.*

41. Cela-Conde et al., "Activation of the Prefrontal Cortex"; Kawabata and Zeki, "Neural Correlates of Beauty"; Vartanian and Goel, "Neuroanatomical Correlates"; and Lévêque and Schön, "Modulation of the Motor Cortex."

42. Gibson, *Ecological Approach*; Reed, *James J. Gibson*; Heft, *Ecological Psychology*; and Chemero and Käufer, "Pragmatism, Phenomenology."

43. For example, C. Schiltz et al., "Impaired Face Discrimination in Acquired Prosopagnosia Is Associated with Abnormal Response to Individual Faces in the Right Middle Fusiform Gyrus," *Cerebral Cortex* 16 (2006): 574–586.

44. M. L. Anderson, *After Phrenology: Neural Reuse and the Interactive Brain* (Cambridge, MA: MIT Press, 2014).

45. Maess et al., "Musical Syntax"; Tettamanti and Weniger, "Broca's Area"; and Kunert et al., "Music and Language Syntax."

46. For example, J. G. White et al., "The Structure of the Nervous System of the Nematode *Caenorhabditis elegans*," *Philosophical Transactions of the Royal Society B: Biological Sciences* 314 (1986): 1–340; and B. E. Stein, T. R. Stanford, and B. A. Rowland, "Development

of Multisensory Integration from the Perspective of the Individual Neuron," *Nature Reviews Neuroscience* 15 (2014): 520–535.

47. Crippen, "Dewey, Enactivism and Greek Thought"; M. Crippen, "Body Politics: Revolt and City Celebration," in *Bodies in the Streets: Somaesthetics of City Life*, ed. R. Shusterman, 89–110 (Boston: Brill, 2019).

48. For example, Schulkin, *Bodily Sensibility*; and M. Clapp et al., "Locus Ceruleus Control of Slow-Wave Homeostasis," *Journal of Neuroscience* 25 (2005): 4503–4511.

49. Crippen, "Dewey, Enactivism and Greek Thought." See also W. H. Calvert, "Monarch Butterfly (*Danaus plexippus* L., Nymphalidae) Fall Migration: Flight Behavior and Direction in Relation to Celestial and Physiographic Cues," *Journal of the Lepidopterists' Society* 55 (2001): 162–168.

50. See S. M. Reppert, R. J. Gegear, and C. Merlin, "Navigational Mechanisms of Migrating Monarch Butterflies," *Trends in Neuroscience* 33 (2010): 399–406.

51. Chemero, *Radical Embodied Cognitive Science*; Long, *Darwin's Devices*.

52. See M. Merleau-Ponty, "Film and the New Psychology" [1947], in *Sense and Non-Sense*, ed. H. Dreyfus and P. Dreyfus, 43–59 (Evanston, IL: Northwestern University Press, 1964).

53. For example, W. James, "On Some Omissions of Introspective Psychology" [1884], in *William James: Writings, 1878–1899*, ed. G. E. Myers, 986–1014 (New York: Library of America, 1992); and J. I. Davis et al., "The Effects of Botox Injections on Emotional Experience," *Emotion* 10 (2010): 433–440.

54. See D. Timmann et al., "The Human Cerebellum Contributes to Motor, Emotional and Cognitive Associative Learning: A Review," *Cortex* 46 (2010): 845–857; and O. Baumann et al., "Consensus Paper: The Role of the Cerebellum in Perceptual Processes," *Cerebellum* 14 (2015): 197–220.

55. Noë, *Action in Perception*.

56. V. S. Ramachandran and W. Hirstein, "The Perception of Phantom Limbs: The D. O. Hebb Lecture," *Brain* 121 (1998): 1603–1630; and V. S. Ramachandran and E. Altschuler, "The Use of Visual Feedback, in Particular Mirror Visual Feedback, in Restoring Brain Function," *Brain* 132 (2009): 1693–1710.

57. Merleau-Ponty, *Phenomenology of Perception*.

1. LIFE, EXPERIMENTALISM, AND VALUATION

1. Cf. B. F. Skinner, *Verbal Behavior* (New York: Appleton-Century-Crofts, 1957); and N. A. Chomsky, "A Review of B. F. Skinner's Verbal Behavior," *Language* 35 (1959): 26–58.

2. J. Dewey, *Experience and Nature*, 2nd ed. (1929; repr., New York: Dover, 1958), viii.

3. W. James, *The Principles of Psychology* (New York: Henry Holt, 1890), 1:5.

4. For example, J. Dewey, *Democracy and Education: An Introduction to the Philosophy of Education* (New York: Macmillan, 1916), 361; and G. H. Mead, *Mind, Self, and Society: From the Standpoint of a Social Behaviorist*, ed. C. W. Morris (Chicago: University of Chicago Press, 1934), 112.

5. See J. Schulkin, *Bodily Sensibility: Intelligent Action* (New York: Oxford University Press, 2004); C. D. Green, "Darwinian Theory, Functionalism, and the First American Psychological Revolution," *American Psychologist* 64 (2009): 75–83; R. Frega, "Evolutionary Prolegomena to a Pragmatist Epistemology of Belief," in *Pragmatist Epistemologies*, ed. R. Frega, 127–152 (New York: Lexington Books, 2011); and K. Nyíri, *Meaning and Motoricity: Essays on Image and Time* (New York: Peter Lang: 2014); and M. Crippen, "Embodied Cognition and Perception: Dewey, Science and Skepticism," *Contemporary Pragmatism* 14 (2017): 121–134.

6. H. Calderwood, *The Relations of Mind and Brains* (London: Macmillan, 1879).

7. J. H. Jackson, "Evolution and Dissolution of the Nervous System" [1884], in *Selected Writings of John Hughlings Jackson*, vol. 1, ed. J. Taylor, 45–91 (London: Staples Press, 1958).

8. F. Galton, *Inquiries into Human Faculty and Its Development* (New York: Macmillan, 1883).

9. E. Titchener, *Lectures on the Experimental Psychology of the Thought-Processes* (New York: Macmillan, 1909).

10. M. Washburn, *Movement and Mental Imagery: Outlines of a Motor Theory of the Complexer Mental Processes* (Boston: Houghton Mifflin, 1916).

11. C. Wright, "Evolution of Self-Consciousness," *North America Review* 116 (1873): 245–310.

12. H. Spencer, *The Principles of Psychology* (1855; Westmead, UK: Gregg International, 1970).

13. M. Crippen, "Body Phenomenology, Somaesthetics and Nietzschean Themes in Medieval Art," *Pragmatism Today* 5 (2014): 40–45; and Crippen, "Embodied Cognition."

14. M. Crippen, "Dewey, Enactivism and Greek Thought," in *Pragmatism and Embodied Cognitive Science: From Bodily Interaction to Symbolic Articulation*, ed. R. Madzia and M. Jung, 229–246 (Boston: De Gruyter, 2016); and Crippen, "Embodied Cognition."

15. J. J. Gibson, *The Ecological Approach to Visual Perception* (Boston: Houghton Mifflin, 1979).

16. See E. Reed, *James J. Gibson and the Psychology of Perception* (New Haven, CT: Yale University Press, 1988); H. Heft, *Ecological Psychology in Context: James Gibson, Roger Barker, and the Legacy of William James's Radical Empiricism* (Mahwah, NJ: Lawrence Erlbaum Associates, 2001); A. Chemero and S. Käufer, "Pragmatism, Phenomenology, and Extended Cognition," in *Pragmatism and Embodied Cognitive Science*, ed. Madzia and Jung, 55–70; M. Crippen, "Intuitive Cities: Pre-reflective, Aesthetic and Political Aspects of Urban Design," *Journal of Aesthetics and Phenomenology* 3 (2016): 125–145; M. Crippen, "Pragmatism and the Valuative Mind," *Transactions of the Charles S. Peirce Society* 54 (2018): 341–360; M. Crippen, "Aesthetics and Action: Situations, Emotional Perception and the Kuleshov Effect," *Synthese* [issue unassigned] (2019), https://doi.org/10.1007/s11229-019 -02110-2; and M. Crippen, "Pragmatic Evolutions of the Kantian *a priori*: From the Mental to the Bodily," in *Pragmatist Kant: Pragmatism, Kant, and Kantianism in the Twenty-First Century*, ed. K. Skowroński and S. Pihlström, 19–40 (Helsinki: Nordic Pragmatism Network, 2019).

17. K. O'Regan and A. Noë, "A Sensorimotor Account of Vision and Visual Consciousness," *Behavior and Brain Sciences* 24 (2001): 939–973; A. Noë, *Action in Perception* (Cambridge, MA: MIT Press, 2004); and A. Noë, *Out of Our Heads: Why You Are Not Your Brain, and Other Lessons from the Biology of Consciousness* (New York: Hill and Wang, 2009). See also R. Menary, *Cognitive Integration: Mind and Cognition Unbounded* (New York: Palgrave Macmillan, 2007); Crippen, "Dewey, Enactivism and Greek Thought"; and Crippen, "Embodied Cognition."

18. A. M. Graybiel et al., "The Basal Ganglia and Adaptive Motor Control," *Science* 265 (1994): 1826–1831; B. Knowlton, J. Mangels, and L. Squire, "A Neostriatal Habit Learning System in Humans," *Science* 273 (1996): 1399–1402; M. T. Ullman et al., "A Neural Dissociation Within Language: Evidence That the Mental Dictionary Is Part of Declarative Memory, and That Grammatical Rules Are Processed by the Procedural System," *Journal of Cognitive Neuroscience* 9 (1997): 266–286; M. C. Corballis, *From Hand to Mouth* (Princeton, NJ: Princeton University Press, 2002), chap. 7; and S. A. Kotz, M. Schwartze, and M. Schmidt-Kassow, "Non-motor Basal Ganglia Functions: A Review and Proposal for a Model of Sensory Predictability in Auditory Language Perception," *Cortex* 45 (2009): 982–990.

19. For example, R. Kaplan and S. Kaplan, *The Experience of Nature: A Psychological Perspective* (New York: Cambridge University Press, 1989); A. R. Damasio, *Descartes' Error:*

Emotion, Reason, and the Human Brain (New York: G. P. Putnam, 1994); J. Schulkin, *Roots of Social Sensibility and Neural Function* (Cambridge, MA: MIT Press, 2000); and Schulkin, *Bodily Sensibility*.

20. For an overview, see P. P. Wiener, *Evolution and the Founders of Pragmatism* (Cambridge, MA: Harvard University Press, 1949).

21. Mead, *Mind, Self, and Society*, 337–378. See also E. Thompson, A. Palacios, and F. J. Varela, "Ways of Coloring: Comparative Color Vision as a Case Study for Cognitive Science," *Behavioral and Brain Sciences* 15 (1992): 1–26; and R. Madzia, "Constructive Realism: In Defense of the Objective Reality of Perspectives," *Human Affairs* 23 (2013): 645–657.

22. See R. Richards, *Darwin and the Emergence of Evolutionary Theories of Mind and Behavior* (Chicago: University of Chicago Press, 1987), chap. 9; Green, "Darwinian Theory"; M. Crippen, "William James on Belief: Turning Darwinism Against Empiricistic Skepticism," *Transactions of the Charles S. Peirce Society* 46 (2010): 477–502; M. Crippen, "William James and His Darwinian Defense of Freewill," in *150 Years of Evolution: Darwin's Impact on Contemporary Thought and Culture*, ed. M. Wheeler, 68–89 (San Diego, CA: San Diego State University Press, 2011); and L. McGranahan, *Darwinism and Pragmatism: William James on Evolution and Self-Transformation* (New York: Routledge, 2017).

23. In this book, we follow Stephen Jay Gould's use of the term "neo-Lamarckism," which he employs to connote a reworking of Jean-Baptiste Lamarck's ideas. Neo-Lamarckism overemphasizes Lamarck's principle of inheritance of acquired traits, originally posited as a secondary mechanism. According to Gould, it also abandons Lamarck's "cardinal idea that evolution is an active, creative response by organisms to their felt needs," describing it instead as an outcome of "direct impositions by impressing environments on the passive organism." S. J. Gould, "Shades of Lamarck," in *The Panda's Thumb*, ed. S. J. Gould (New York: Norton, 1979), 77. We would add that Lamarck is important in the history of science and that Darwin himself left room for Lamarckian mechanisms—something that gets lost in many introductions to evolution that pit Darwinian theory against a caricature of Lamarck.

24. See W. James, *Principles of Psychology*, 2:275; W. James, "Philosophical Conceptions and Practical Results" [1898], in *William James: Writings, 1878–1899*, ed. G. E. Myers (New York: Library of America, 1992), 1096.

25. W. James, "Great Men and Their Environment" [1880], in *William James: Writings 1878–1899*, ed. G. E. Myers (New York: Library of America, 1992), 622.

26. W. James, "Brute and Human Intellect" [1878], in *William James: Writings 1878–1899*, ed. G. E. Myers, 910–949 (New York: Library of America, 1992); and James, *Principles of Psychology*, vol. 2, chap. 28.

27. James, *Principles of Psychology*, 2:618.

28. See James, "Brute and Human Intellect," 929; and James, *Principles of Psychology*, 1:402–403.

29. M. Crippen, "Pictures, Experiential Learning and Phenomenology," in *Beyond Words: Pictures, Parables, Paradoxes*, ed. A. Benedek and K. Nyíri, 83–90 (Frankfurt am Main: Peter Lang, 2015); and Crippen, "Pragmatic Evolutions."

30. See James, "Brute and Human Intellect," 929; and James, *Principles of Psychology*, 1:402–403.

31. James, "Brute and Human Intellect," 929; James, "Great Men," 634; James, *Principles of Psychology*, 1:403. Cf. Chomsky, "B. F. Skinner's Verbal Behavior"; and A. N. Chomsky, *Language and Mind* (New York: Cambridge University Press, 2006).

32. James, *Principles of Psychology*, 2:636.

33. James, *Principles of Psychology*, 2:638.

34. James, "Brute and Human Intellect," 929; and James, *Principles of Psychology*, 1:402–403.

35. D. E. Berlyne, *Conflict, Arousal, and Curiosity* (New York: McGraw-Hill, 1960); and T. Wu et al., "The Capacity of Cognitive Control Estimated from a Perceptual Decision Making Task," *Scientific Reports* 6 (2016): art. 34025, https://doi.org/10.1038/srep34025.

36. James, *Principles of Psychology*, 1:139. Italics in the original.

37. James, *Principles of Psychology*, 1:288.

38. James, *Principles of Psychology*, 1:403. Italics in the original. See also W. James, "Are We Automata?," *Mind* 4 (1879): 1–22.

39. W. James, "Remarks on Spencer's Definition of Mind as Correspondence" [1878], in *William James: Writings, 1878–1899*, ed. G. E. Myers (New York: Library of America, 1992), 908.

40. James, "Spencer's Definition of Mind," 908.

41. James, "Great Men," 622.

42. James, "Great Men," 641.

43. James, *Principles of Psychology*, 2:636.

44. James, *Principles of Psychology*, 2:636.

45. James, "Are We Automata?," 12; James, *Principles of Psychology*, 1:287, 2:chap. 28.

46. S. J. Gould, "Shades of Lamarck," in *The Panda's Thumb*, ed. S. J. Gould, 76–84 (New York: W. W. Norton , 1979).

47. See Crippen, "William James and His Darwinian Defense."

48. See James, *Principles of Psychology*, 1:403. See also James, "Brute and Human Intellect," 930.

49. Crippen, "William James and His Darwinian Defense."

50. Crippen, "Pragmatic Evolutions."

51. James, *Principles of Psychology*, 2:618.

52. James, "Spencer's Definition of Mind," 897n.

53. W. James to Henry Holt, letter, November 22, 1878, quoted in R. B. Perry, *The Thought and Character of William James, as Revealed in Unpublished Correspondence and Notes, Together with His Published Writings*, vol. 1 (Boston: Little, Brown, 1935), 35.

54. Crippen, "Pragmatism and the Valuative Mind."

55. M. Crippen, "Art and Pragmatism: James and Dewey on the Reconstructive Presuppositions of Experience" (PhD diss., York University, 2010).

56. R. G. Collingwood, *An Essay on Metaphysics* (Oxford: Oxford University Press, 1940), 4.

57. See W. James, *The Letters of William James*, ed. H. James, vol. 1 (Boston: Atlantic Monthly Press, 1920), chap. 3; and Perry, *Thought and Character of William James*, chap. 12.

58. C. S. Peirce, "How to Make Our Ideas Clear," *Popular Science Monthly* 12 (1878): 286–302.

59. Peirce, "How to Make Our Ideas Clear," 293.

60. Peirce, "How to Make Our Ideas Clear," 293.

61. W. James, *Pragmatism: A New Name for Some Old Ways of Thinking* [1907], in *William James: Writings, 1902–1910*, ed. B. Kuklick (New York: Library of America, 1987), 507.

62. James, *Pragmatism*, 507–508.

63. James, *Pragmatism*, 508.

64. James, *Pragmatism*, 507.

65. See Crippen, "Pragmatic Evolutions."

66. W. James, "The Psychology of Belief" [1889], in *William James: Writings, 1878–1899*, ed. G. E. Myers (New York: Library of America, 1992), 1021; and James, *Principles of Psychology*, 1:301.

67. W. James, "Rationality, Activity and Faith," *Princeton Review* 2 (1882): 65.

68. M. Merleau-Ponty, *Phenomenology of Perception*, trans. C. Smith (New York: Routledge and Kegan Paul, 1962; first published 1945 in French), 159.

69. James, "Rationality, Activity and Faith," 66.

70. James, "Rationality, Activity and Faith," 70.

71. W. James, "The Will to Believe" [1896], in *William James: Writings, 1878–1899*, ed. G. E. Myers (New York: Library of America, 1992), 458.
72. James, "Psychology of Belief," 1021; and James, *Principles of Psychology*, 2:283.
73. James, *Pragmatism*, 574.
74. James, *Principles of Psychology*, 2:321.
75. Crippen, "William James on Belief."
76. James, *Pragmatism*, 579.
77. Crippen, "Pragmatic Evolutions."
78. J. Dewey, *Art as Experience* (New York: Minton, Balch, 1934), 53.
79. See G. H. Mead, *The Philosophy of the Act*, ed. C. W. Morris (Chicago: University of Chicago Press, 1938), chap. 1; and Merleau-Ponty, *Phenomenology of Perception*, 367–368.
80. Crippen, "Body Phenomenology," 40–45.
81. For example, J. Dewey, "Experience and Objective Idealism," *Philosophical Review* 15 (1906): 469–475; J. Dewey, *Reconstruction in Philosophy* (New York: Henry Holt, 1920), 81–91; J. Dewey, "Syllabus: Types of Philosophical Thought" [1922–1923], in *The Middle Works, 1899–1922*, vol. 13, ed. J. A. Boydston (Carbondale, IL: Southern Illinois University Press, 1983), 351. See also Crippen, "Dewey, Enactivism and Greek Thought"; and Crippen, "Pragmatic Evolutions."
82. J. Dewey, *Human Nature and Conduct: An Introduction to Social Psychology* (New York: Henry Holt, 1922).
83. Dewey, *Human Nature and Conduct*, 30–31.
84. Dewey, *Human Nature and Conduct*, 31.
85. Dewey, *Human Nature and Conduct*, 31.
86. Dewey, *Human Nature and Conduct*, 32.
87. Dewey, *Experience and Nature* (Chicago: Open Court, 1925), 277.
88. Dewey, *Human Nature and Conduct*, 52.
89. Dewey, *Experience and Nature*, 1925 ed., 277. See also J. Dewey, *The Quest for Certainty* (New York: Minton, Balch, 1929), 208–222.
90. Crippen, "Pragmatic Evolutions."
91. Dewey, *Reconstruction in Philosophy*, 91.
92. E. Thompson, *Mind in Life: Biology, Phenomenology, and the Sciences of Mind* (Cambridge, MA: Harvard University Press, 2007), 74–75; and Noë, *Out of Our Heads*, 43.
93. Crippen, "Dewey, Enactivism and Greek Thought."
94. Dewey, *Experience and Nature*, 1925 ed., 354.
95. Aristotle, *Posterior Analytics*, trans. G. R. G. Mure, in *The Basic Works of Aristotle*, ed. R. McKeon (New York: Random House, 1941), 99b35–100a8.
96. Dewey, "Syllabus"; and J. Dewey, "Unfinished Introduction" [1951], in *The Later Works, 1925–1953*, vol. 1, ed. J. A. Boydston, 361–364 (Cardondale, IL: Southern Illinois University Press, 1981).
97. J. Lennon, "The Notion of Experience," *Thomist: A Speculative Quarterly Review* 23 (1960): 316.
98. Dewey, *Experience and Nature*, 1925 ed., 354.
99. Dewey, *Reconstruction in Philosophy*, 79.
100. J. Dewey, *Philosophy and Education in Their Historic Relation*, ed. J. J. Chambliss (Boulder, CO: Westview, 1910), 47–48.
101. Dewey, *Experience and Nature*, 1925 ed., 354.
102. Dewey, *Experience and Nature*, 1925 ed., 354.
103. O'Regan and Noë, "Sensorimotor Account of Vision."
104. See Aristotle, *Metaphysics*, trans. W. D. Ross, in *The Basic Works of Aristotle*, ed. R. McKeon (New York: Random House, 1941), 981a5–11.
105. Crippen, "Dewey, Enactivism and Greek Thought."

106. Plato, *Laws*, trans. A. E. Taylor, in *Plato: The Collected Dialogues*, ed. E. Hamilton and H. Cairns (Princeton, NJ: Princeton University Press, 1963), 720b.

107. Plato, *Gorgias*, trans. W. D. Woodhead, in *Plato: The Collected Dialogues*, ed. E. Hamilton and H. Cairns (Princeton, NJ: Priceton University Press, 1963), 463b.

108. Plato, *Gorgias*, 465a.

109. Dewey, *Democracy and Education*, 263.

110. Dewey, *Democracy and Education*, 263.

111. Dewey, *Democracy and Education*, 263.

112. Dewey, *Democracy and Education*, 263.

113. Dewey, *Reconstruction in Philosophy*, 92.

114. Dewey, *Democracy and Education*, 311.

115. Dewey, *Reconstruction in Philosophy*, 92.

116. Dewey, *Human Nature and Conduct*, 72.

117. See Dewey, *Experience and Nature*, 1925 ed., 358–361.

118. J. Dewey, "The Practical Character of Reality" [1908], in *Philosophy and Civilization* (New York: Minton, Balch, 1931), 48.

119. Dewey, "Practical Character of Reality," 47.

120. Dewey, "Practical Character of Reality," 48.

121. Crippen, "Art and Pragmatism."

122. Dewey, *Quest for Certainty*, 204.

123. Dewey, *Quest for Certainty*, 204–205.

124. Dewey, *Quest for Certainty*, 204.

125. Dewey, *Quest for Certainty*, 87.

126. Dewey, *Quest for Certainty*, 84.

127. Crippen, "Art and Pragmatism."

128. Dewey, *Reconstruction in Philosophy*, 114–115.

129. Dewey, *Reconstruction in Philosophy*, 114–115.

130. Dewey, *Reconstruction in Philosophy*, 114–115.

131. Dewey, *Reconstruction in Philosophy*, 114–115.

132. Dewey, *Reconstruction in Philosophy*, 114–115.

133. C. I. Lewis, *Mind and the World-Order* (New York: Charles Scribner's Sons, 1929), 286.

134. Mead, *Philosophy of the Act*, 22.

135. F. Nietzsche, *The Will to Power*, ed. W. Kaufmann, trans. W. Kaufmann and R. J. Hollingdale (New York: Vintage Books, 1968), §557.

136. Crippen, "William James on Belief"; and Crippen, "Aesthetics and Action."

137. Dewey, *Quest for Certainty*, 128–129.

138. Crippen, "William James on Belief"; and Crippen, "Aesthetics and Action."

139. Mead, *Philosophy of the Act*, 21.

140. Dewey, *Experience and Nature*, 1925 ed., 259. See also M. Mackenzie, "Dewey, Enactivism, and the Qualitative Dimension," *Humana Mente: Journal of Philosophical Studies* 31 (2016): 21–36.

141. Dewey, "Practical Character of Reality," 45.

142. Dewey, "Practical Character of Reality," 45.

143. Dewey, "Practical Character of Reality," 45.

144. Dewey, "Practical Character of Reality," 45. See also G. H. Mead, *The Philosophy of the Present*, with introductory remarks by J. Dewey and A. Murphy (LaSalle, IL: Open Court, 1932), 4; and S. Hurley, *Consciousness in Action* (Cambridge, MA: Harvard University Press, 1998), 333.

145. See Crippen, "Embodied Cognition"; and P. McReynolds, "Autopoiesis and Transaction," *Transactions of the Charles S. Peirce Society* 53 (2017): 312–334.

146. Mead, *Mind, Self, and Society*, 245.

147. J. von Uexküll, *Umwelt und Innenwelt der Tiere* (Berlin: Julius Springer, 1909); and J. von Uexküll, "A Stroll Through the Worlds of Animals and Men: A Picture Book of Invisible Worlds" [1934], in *Instinctive Behavior: The Development of a Modern Concept*, ed. and trans. C. H. Schiller, 5–80 (New York: International Universities Press, 1957).

148. James, *Pragmatism*.

149. See C. S. Peirce, "Fixation of Belief," *Popular Science Monthly* 12 (1877): 6.

150. James, "Will to Believe," 464.

151. Crippen, "William James on Belief."

152. Crippen, "William James on Belief."

153. James, *Pragmatism*, 580.

154. James, *Pragmatism*, 580–581.

155. W. James, "On Some Omissions of Introspective Psychology" [1884], in *William James: Writings 1878–1899*, ed. G. E. Myers, 986–1014 (New York: Library of America, 1992); and W. James, "The Stream of Thought," chap. 9 in *Principles of Psychology*, vol. 1.

156. This paraphrases Mark Johnson's account of James. See M. Johnson, *The Meaning of the Body: Aesthetics of Human Understanding* (Chicago: University of Chicago Press, 2007), chap. 5; and M. Johnson, "Keeping the Pragmatism in Neuropragmatism," in *Neuroscience, Neurophilosophy and Pragmatism: Brains at Work with the World*, ed. T. Solymosi and J. Shook, 37–56 (New York: Palgrave Macmillan, 2014).

157. Johnson, *Meaning of the Body*, 96. See also Johnson, "Keeping the Pragmatism," 45–47.

158. For example, James, "Brute and Human Intellect"; James, "Spencer's Definition of Mind"; James, "Are We Automata?"; W. James, "The Sentiment of Rationality" [1879], in *William James: Writings, 1878–1899*, ed. G. E. Myers, 950–985 (New York: Library of America, 1992); and James, "Great Men."

159. For example, Parrott and Schulkin, "Neuropsychology"; Damasio, *Descartes' Error*; and L. Pessoa, *The Cognitive-Emotional Brain: From Interactions to Integration* (Cambridge, MA: MIT Press, 2013).

160. James, "Sentiment of Rationality," 952.

161. Crippen, "William James on Belief."

162. Crippen, "Pragmatism and the Valuative Mind."

163. James, "Sentiment of Rationality," 952.

164. James, *Principles of Psychology*, 2:335–336.

165. James, "Sentiment of Rationality," 952.

166. James, *Pragmatism*, 606.

167. W. James, *Some Problems of Philosophy* [1911], in *William James: Writings, 1902–1910*, ed. B. Kuklick (New York: Library of America, 1987), 1013.

168. James, *Some Problems of Philosophy*, 1020.

169. See Crippen, "Art and Pragmatism."

170. James, *Some Problems of Philosophy*, 1015.

171. James, *Some Problems of Philosophy*, 1015.

172. James, *Some Problems of Philosophy*, 1015–1016.

173. See Crippen, "Art and Pragmatism."

174. James, "Sentiment of Rationality."

175. James, "Sentiment of Rationality," 950.

176. See James, "Sentiment of Rationality," 954–956.

177. See L. Festinger, *A Theory of Cognitive Dissonance* (Evanston, IL: Row, Peterson, 1957); Berlyne, *Conflict, Arousal, and Curiosity*; D. E. Berlyne, "Novelty, Complexity, and Hedonic Values," *Perception and Psychophysics* 8 (1970): 279–285; I. McGregor, I. R. Newby-Clark, and M. P. Zanna, "'Remembering' Dissonance: Simultaneous Accessibility of Inconsistent Cognitive Elements Moderates Epistemic Discomfort," in *Cognitive Dissonance: Progress on a Pivotal Theory in Social Psychology*, ed. E. Harmon-Jones and J. Mills,

325–353 (Washington, DC: American Psychological Association, 1999); E. Harmon-Jones, "Cognitive Dissonance and Experienced Negative Affect: Evidence That Dissonance Increases Experienced Negative Affect Even in the Absence of Aversive Consequences," *Personality and Social Psychology Bulletin* 26 (2000): 1490–1501; and I. McGregor et al., "Compensatory Conviction in the Face of Personal Uncertainty: Going to Extremes and Being Oneself," *Journal of Personality and Social Psychology* 80 (2001): 472–488.

178. Dewey, *Experience and Nature*, 1925 ed., 378.

179. Dewey, *Experience and Nature*, 1925 ed., 379.

180. Dewey, *Experience and Nature*, 1925 ed., 378.

181. Dewey, *Experience and Nature*, 1925 ed., 378.

182. For example, G. Böhme, "Atmosphere as the Fundamental Concept of a New Aesthetic," *Thesis Eleven* 36 (1993): 113–126; M. Ratcliffe, *Feelings of Being: Phenomenology, Psychiatry and the Sense of Reality* (New York: Oxford University Press, 2008); J. Krueger, "Extended Cognition and the Space of Social Interaction," *Consciousness and Cognition* 20 (2011): 643–655; and J. Slaby, "Emotions and the Extended Mind," in *Collective Emotions: Perspectives from Psychology, Philosophy, and Sociology*, ed. C. von Scheve and M. Salmela (New York: Oxford University Press, 2014).

183. Dewey, *Art as Experience*.

184. M. Merleau-Ponty, "Film and the New Psychology" [1947], in *Sense and Non-Sense*, trans. H. Dreyfus and P. Dreyfus, 43–59 (Evanston, IL: Northwestern University Press, 1964).

185. Merleau-Ponty, "Film," 52.

186. Merleau-Ponty, "Film," 52.

187. Dewey, *Art as Experience*, 42.

188. Dewey, *Art as Experience*, 42. See also Dewey, *Human Nature and Conduct*, 140.

189. R. C. Solomon, *True to Our Feelings: What Our Emotions Are Really Telling Us* (New York: Oxford University Press, 2007).

190. H. Dreyfus, *Being-in-the-World: A Commentary on Heidegger's Being and Time, Division I* (Cambridge, MA: MIT Press, 2003), 171.

191. K. Koffka, *Principles of Gestalt Psychology* (New York: Harcourt, Brace, 1935), 326. See also Crippen, "Aesthetics and Action."

192. Dewey, *Art as Experience*, 30.

193. Dewey, *Art as Experience*, 67.

194. Dewey, *Art as Experience*, 16.

195. M. Bhalla and D. R. Proffitt, "Visual-Motor Recalibration in Geographical Slant Perception," *Journal of Experimental Psychology: Human Perception and Performance* 25 (1999): 1076–1096; and D. R. Proffitt, "Embodied Perception and the Economy of Action," *Perspectives on Psychological Science* 1 (2006): 110–122.

196. Dewey, *Art as Experience*, 67.

197. Dewey, *Art as Experience*, 67.

198. Dewey, *Human Nature and Conduct*, 62.

199. Dewey, *Human Nature and Conduct*, 16.

200. Dewey, *Human Nature and Conduct*, 62.

201. Dewey, *Art as Experience*, 177.

202. Dewey, *Art as Experience*, 250.

203. Dewey, *Art as Experience*, 251.

204. Crippen, "Aesthetics and Action."

205. J. Dewey, "The Need for a Recovery of Philosophy," in *Creative Intelligence: Essays in the Pragmatic Attitude*, by J. Dewey et al. (New York: Henry Holt, 1917), 11.

206. I. Kant, *Critique of Pure Reason* [synthesis of 1781 and 1787 eds.], trans. P. Guyer and A. W. Wood (New York: Cambridge University Press, 1998), Bxvi.

207. Kant, *Critique of Pure Reason*, Bxviii.

208. Kant, *Critique of Pure Reason*, Bxvi.
209. See Kant, *Critique of Pure Reason*, Bxvi–Bxix.
210. Kant, *Critique of Pure Reason*, Bxii–Bxiv.
211. Kant, *Critique of Pure Reason*, B161–166.

2. PRAGMATISM AND EMBODIED COGNITIVE SCIENCE

1. See V. Kestenbaum, *The Phenomenological Sense of John Dewey: Habit and Meaning* (Atlantic Highlands, NJ: Humanities Press, 1977).
2. F. Varela, E. Thompson, and E. Rosch, *The Embodied Mind: Cognitive Science and Human Experience* (Cambridge, MA: MIT Press, 1991).
3. For example, T. Rockwell, *Neither Brain nor Ghost: A Nondualist Alternative to the Mind-Brain Identity Theory* (Cambridge, MA: MIT Press, 2005); R. Menary, *Cognitive Integration: Mind and Cognition Unbounded* (New York: Palgrave Macmillan, 2007); R. Menary, "Pragmatism and the Pragmatic Turn in Cognitive Science," in *The Pragmatic Turn: Toward Action-Oriented Views in Cognitive Science*, ed. A. K. Engel, K. J. Friston, and D. Kragic, 215–223 (Cambridge, MA: MIT Press, 2015); P. Steiner, "Sciences cognitives, tournant pragmatique et horizons pragmatists," *Tracés: Revue de Sciences humaines* 15 (2008): 85–108; P. Steiner, "Philosophie, technologie et cognition: État des lieux et perspectives," *Intellectica* 53 (2010): 7–40; A. Chemero, *Radical Embodied Cognitive Science* (Cambridge, MA: MIT Press, 2009); S. Gallagher, "Philosophical Antecedents of Situated Cognition," in *The Cambridge Handbook of Situated Cognition*, ed. P. Robbins and M. Aydede, 35–52 (New York: Cambridge University Press, 2009); S. Gallagher, *Enactivist Interventions: Rethinking the Mind* (New York: Oxford University Press, 2017); J. Fingerhut, "Sensorimotor Signature, Skill, and Synaesthesia: Two Challenges to Enactive Theories of Perception," in *Habitus in Habitat III: Synaesthesia and Kinaesthetics*, ed. J. Fingerhut, S. Flach, and J. Söffner, 101–120 (New York: Peter Lang, 2011); R. Madzia, "Constructive Realism: In Defense of the Objective Reality of Perspectives," *Human Affairs* 23 (2013): 645–657; M. Crippen, "Dewey, Enactivism and Greek Thought," in *Pragmatism and Embodied Cognitive Science: From Bodily Interaction to Symbolic Articulation*, ed. R. Madzia and M. Jung, 229–246 (Boston: De Gruyter, 2016); and M. Crippen, "Embodied Cognition and Perception: Dewey, Science and Skepticism," *Contemporary Pragmatism* 14 (2017): 121–134.
4. See J. Long, *Darwin's Devices: What Evolving Robots Can Teach Us About the History of Life and the Future of Technology* (New York: Basic Books, 2011), chap. 5.
5. Crippen, "Dewey, Enactivism and Greek Thought." See also W. H. Calvert, "Monarch Butterfly (*Danaus plexippus* L., Nymphalidae) Fall Migration: Flight Behavior and Direction in Relation to Celestial and Physiographic Cues," *Journal of the Lepidopterists' Society* 55 (2001): 162–168.
6. See S. M. Reppert, R. J. Gegear, and C. Merlin, "Navigational Mechanisms of Migrating Monarch Butterflies," *Trends in Neuroscience* 33 (2010): 399–406.
7. R. Pfeifer and J. Bongard, *How the Body Shapes the Way We Think: A New View of Intelligence* (Cambridge, MA: MIT Press, 2007), 81.
8. A. Noë, *Action in Perception* (Cambridge, MA: MIT Press, 2004); A. Noë, *Out of Our Heads: Why You Are Not Your Brain, and Other Lessons from the Biology of Consciousness* (New York: Hill and Wang, 2009); and L. Barrett, *Beyond the Brain: How Body and Environment Shape Animal and Human Minds* (Princeton, NJ: Princeton University Press, 2011).
9. See A. Dietrich, "Neurocognitive Mechanism Underlying the Experience of Flow," *Consciousness and Cognition* 13 (2004): 746–761.
10. Dietrich, "Neurocognitive Mechanism"; S. Brown and L. M. Parsons, "The Neuroscience of Dance," *Scientific American* 299 (2008): 78–83; and A. M. Graybiel, "Habits, Rituals, and the Evaluative Brain," *Annual Review of Neuroscience* 31 (2008): 359–387.

11. Varela, Thompson, and Rosch, *Embodied Mind*, 9.
12. Noë, *Out of Our Heads*, 5.
13. See Noë, *Action in Perception*, 2.
14. A. Clark, *Supersizing the Mind: Embodiment, Action, and Cognitive Extension* (New York: Oxford University Press, 2008); A. Clark, *Surfing Uncertainty: Prediction, Action, and the Embodied Mind* (New York: Oxford University Press, 2015); Menary, *Cognitive Integration*; E. Thompson, *Mind in Life: Biology, Phenomenology, and the Sciences of Mind* (Cambridge, MA: Harvard University Press, 2007); Varela, Thompson, and Rosch, *Embodied Mind*.
15. J. Daugman, "Brain Metaphor and Brain Theory," in *Philosophy and the Neurosciences: A Reader*, ed. W. Bechtel et al. (Oxford: Blackwell, 2001), 24. See also J. Schulkin, "Foraging for Coherence in Neuroscience: A Pragmatist Orientation," *Contemporary Pragmatism* 13 (2016): 8.
16. D. Hoffman, "The Interface Theory of Perception: Natural Selection Drives True Perception to Swift Extinction," in *Object Categorization: Computer and Human Vision Perspectives*, ed. S. Dickinson et al., 148–165 (New York: Cambridge University Press, 2009); D. Hoffman, "The Construction of Visual Reality," in *Hallucination: Research and Practice*, ed. J. D. Blom and I. Sommer, 7–15 (New York: Springer, 2011); and D. Hoffman and C. Prakash, "Objects of Consciousness," *Frontiers in Psychology* 5 (2014): art. 577, https://doi .org/10.3389/fpsyg.2014.00577.
17. Hoffman, "Construction of Visual Reality," 12.
18. Hoffman, "Construction of Visual Reality," 11.
19. Hoffman, "Construction of Visual Reality," 12.
20. R. Dawkins, *The God Delusion* (Boston: Houghton Mifflin, 2006), 88.
21. Dawkins, *God Delusion*, 88–89.
22. A. R. Damasio and G. B. Carvalho, "The Nature of Feelings: Evolutionary and Neurobiological Origins," *Nature Reviews Neuroscience* 14 (2013): 143–152; and A. R. Damasio, *The Strange Order of Things: Life, Feeling, and the Making of Cultures* (New York: Pantheon Books, 2018), chap. 4.
23. Dawkins, *God Delusion*, 361.
24. Dawkins, *God Delusion*, 361.
25. Dawkins, *God Delusion*, 367.
26. Dawkins, *God Delusion*, 361–374.
27. Dawkins, *God Delusion*, 361–362.
28. Dawkins, *God Delusion*, 361.
29. H. Dreyfus, "Existential Phenomenology and the Brave New World of *The Matrix*," *Harvard Review of Philosophy* 11 (2003): 18–31.
30. J. Dewey, *Experience and Nature* (Chicago: Open Court, 1925), 172. See also R. Rorty, *Philosophy and the Mirror of Nature* (Princeton, NJ: Princeton University Press, 1979).
31. Dewey, *Experience and Nature*, 1925 ed., 230.
32. Dewey, *Experience and Nature*, 1925 ed., 230.
33. K. O'Regan and A. Noë, "A Sensorimotor Account of Vision and Visual Consciousness," *Behavior and Brain Sciences* 24 (2001): 943.
34. O'Regan and Noë, "Sensorimotor Account of Vision," 943, 946.
35. D. R. Finkelstein, "Physical Process and Physical Law," in *Physics and Whitehead: Quantum, Process, and Experience*, ed. T. E. Eastman and H. Keeton (Albany, NY: State University of New York Press, 2004), 182.
36. Finkelstein, "Physical Process," 182. Emphasis added. See also M. Crippen, "The Totalitarianism of Therapeutic Philosophy: Reading Wittgenstein Through Critical Theory," *Essays in Philosophy* 8 (2007): 29–55; and M. Crippen, "William James on Belief: Turning

Darwinism Against Empiricistic Skepticism," *Transactions of the Charles S. Peirce Society* 46 (2010): 477–502.

37. Finkelstein, "Physical Process."

38. Dewey, *Experience and Nature*, 1925 ed., 381.

39. J. Dewey, *Essays in Experimental Logic* (Chicago: University of Chicago Press, 1916).

40. Dewey, *Essays in Experimental Logic*, 14.

41. O'Regan and Noë, "Sensorimotor Account of Vision"; and Noë, *Action in Perception*.

42. Noë, *Action in Perception*, 2.

43. Noë, *Action in Perception*, 194.

44. Clark, *Supersizing the Mind*; Clark, *Surfing Uncertainty*; and Thompson, *Mind in Life*.

45. J. Dewey, *Democracy and Education: An Introduction to the Philosophy of Education* (New York: Macmillan, 1916), 361.

46. T. Solymosi, "A Reconstruction of Freedom in the Age of Neuroscience: A View from Neuropragmatism," *Contemporary Pragmatism* 8 (2011): 153–171; and T. Solymosi and J. Shook, eds., *Neuroscience, Neurophilosophy and Pragmatism: Brains at Work with the World* (New York: Palgrave Macmillan, 2014).

47. M. Merleau-Ponty, *Phenomenology of Perception*, trans. C. Smith (New York: Routledge and Kegan Paul, 1962; first published 1945 in French), 272.

48. Dewey, *Democracy and Education*, 361.

49. Dewey, *Democracy and Education*, 361–362.

50. O'Regan and Noë, "Sensorimotor Account of Vision," 947.

51. K. Takakusaki, "Functional Neuroanatomy for Posture and Gait Control," *Journal of Movement Disorders* 10 (2017): 1–17.

52. Schulkin, "Foraging for Coherence."

53. E. Titchener, *Lectures on the Experimental Psychology of the Thought-Processes* (New York: Macmillan, 1909), 176.

54. M. Washburn, *Movement and Mental Imagery: Outlines of a Motor Theory of the Complexer Mental Processes* (Boston: Houghton, 1916), xiii.

55. F. Galton, *Inquiries into Human Faculty and Its Development* (New York: Macmillan, 1883), 61. Kristof Nyíri quoted the foregoing passages from Galton's, Titchener's, and Washburn's books in his plenary talk, "Towards a Theory of Common-Sense Realism," at the 6th Budapest Visual Learning Conference (*Visual Learning: Trust, Time, Tradition*, Budapest University of Technology and Economics, November 13, 2015).

56. J. H. Jackson, "Evolution and Dissolution of the Nervous System" [1884], in *Selected Writings of John Hughlings Jackson*, vol. 1, ed. J. Taylor (London: Staples Press, 1958), 41.

57. A. D. Friederici, "Towards a Neural Basis of Auditory Sentence Processing," *Trends in Cognitive Sciences* 6 (2002): 78–84; M. T. Ullman, "Is Broca's Area Part of a Basal Ganglia Thalamocortical Circuit?," *Cortex* 42 (2006): 480–485; and S. Koelsch et al., "Investigating Emotion with Music: An fMRI Study," *Human Brain Mapping* 27 (2006): 239–250.

58. B. Knowlton, J. Mangels, and L. Squire, "A Neostriatal Habit Learning System in Humans," *Science* 273 (1996): 1399–1402; A. J. Calder, A. D. Lawrence, and A. W. Young, "Neuropsychology of Fear and Loathing," *Nature Neuroscience* 2 (2001): 352–363; W. Schultz, "Getting Formal with Dopamine and Reward," *Neuron* 36 (2002): 241–263; J. W. Aldridge, K. C. Berridge, and A. R. Rosen, "Basal Ganglia Neural Mechanisms of Natural Movement Sequences," *Canadian Journal of Physiology and Pharmacology* 82 (2004): 732–739; Ullman, "Broca's Area"; S. A. Kotz, M. Schwartze, and M. Schmidt-Kassow, "Non-motor Basal Ganglia Functions: A Review and Proposal for a Model of Sensory Predictability in Auditory Language Perception," *Cortex* 45 (2009): 982–990; and S. A. Kotz and M. Schmidt-Kassow, "Basal Ganglia Contribution to Rule Expectancy and Temporal Predictability in Speech," *Cortex* 68 (2015): 48–60.

59. See J. Schulkin, *Sport: A Biological, Philosophical, and Cultural Perspective* (New York: Columbia University Press, 2016), 23–24.

60. Varela, Thompson, and Rosch, *Embodied Mind*.

61. R. Held and A. Hein, "Movement-Produced Stimulation in the Development of Visually Guided Behavior," *Journal of Comparative and Physiological Psychology* 56 (1963): 872–876.

62. Varela, Thompson, and Rosch, *Embodied Mind*, 175.

63. See J. J. Prinz, *The Conscious Brain: How Attention Engenders Experience* (New York: Oxford Universty Press, 2012), 179.

64. For example, Varela, Thompson, and Rosch, *Embodied Mind*; Noë, *Action in Perception*; Thompson, *Mind in Life*; and E. Myin and J. Degenaar, "Enactive Vision," in *The Routledge Handbook of Embodied Cognition*, ed. L. Shapiro, 90–98 (New York: Routledge, 2014).

65. P. Bach-y-Rita, "Tactile Vision Substitution: Past and Future," *International Journal of Neuroscience* 19 (1983): 29–36; P. Bach-y-Rita, "The Relationship Between Motor Processes and Cognition in Tactile Vision Substitution," in *Cognition and Motor Processes*, ed. A. F. Sanders and W. Prinz, 150–159 (Berlin: Springer, 1984); and P. Bach-y-Rita and S. Kercel, "Sensory Substitution and Augmentation: Incorporating Humans-in-the-Loop," *Intellectica* 2 (2002): 287–297.

66. Noë, *Action in Perception*, 112.

67. M. Ptito et al., "Cross-Modal Plasticity Revealed by Electrotactile Stimulation of the Tongue in the Congenitally Blind," *Brain* 128 (2005): 606–614; M. Ptito and R. Kupers, "Cross-Modal Plasticity in Early Blindness," *Journal of Integrative Neuroscience* 4 (2005): 479–488; R. Kupers and M. Ptito, "Insights from Darkness: What the Study of Blindness Has Taught Us About Brain Structure and Function," *Progress in Brain Research* 192 (2011): 17–31; and M. C. Murphy et al., "Top-Down Influence on the Visual Cortex of the Blind During Sensory Substitution," *NeuroImage* 125 (2016): 932–940.

68. A. G. De Volder et al., "Changes in Occipital Cortex Activity in Early Blind Humans Using a Sensory Substitution Device," *Brain Research* 826 (1999): 128–134.

69. Varela, Thompson, and Rosch, *Embodied Mind*.

70. Varela, Thompson, and Rosch, *Embodied Mind*.

71. For example, L. S. Petro, A. T. Paton, and L. Muckli, "Contextual Modulation of Primary Visual Cortex by Auditory Signals," *Philosophical Transactions of the Royal Society B: Biological Sciences* 372 (2017): art. 20160104, https://doi.org/10.1098/rstb.2016.0104.

72. A. Amedi et al., "Convergence of Visual and Tactile Shape Processing in the Human Lateral Occipital Complex," *Cerebral Cortex* 12 (2002): 1202–1212.

73. See P. Arno et al., "Occipital Activation by Pattern Recognition in the Early Blind Using Auditory Substitution for Vision," *NeuroImage* 13 (2001): 632–645; H. Burton, "Visual Cortex Activity in Early and Late Blind People," *Journal of Neuroscience* 23 (2003): 4005–4011; and P. Voss et al., "Early- and Late-Onset Blind Individuals Show Supra-Normal Auditory Abilities in Far-Space," *Current Biology* 14 (2004): 1734–1738.

74. For example, Bach-y-Rita, "Tactile Vision Substitution"; P. Bach-y-Rita, "Brain Plasticity as a Basis of Sensory Substitution," *Journal of Neurologic Rehabilitation* 2 (1987): 67–71; and Bach-y-Rita and Kercel, "Sensory Substitution and Augmentation."

75. C. Büchel et al., "Different Activation Patterns in the Visual Cortex of Late and Congenitally Blind Subjects," *Brain* 121 (1998): 409–419; and N. Sadato et al., "Neural Networks for Braille Reading by the Blind," *Brain* 121 (1998): 1213–1229.

76. L. B. Merabet et al., "Rapid and Reversible Recruitment of Early Visual Cortex for Touch," *PLoS ONE* 3 (2008): art. e3046, https://doi.org/10.1371%2Fjournal.pone.0003046. See also O. Kauffmann, "Brain Plasticity and Phenomenal Consciousness," *Journal of Consciousness Studies* 18 (2011): 46–70.

77. Merabet et al., "Rapid and Reversible Recruitment."

78. J. Dewey, *Reconstruction in Philosophy* (New York: Henry Holt, 1920), 86–87. See also J. Dewey, *Art as Experience* (New York: Minton, Balch, 1934), 43–45.

79. J. Dewey, "The Reflex Arc Concept in Psychology," *Psychological Review* 3 (1896): 359.

80. Dewey, "Reflex Arc Concept," 358–359.

81. H. S. Langfeld, "Fifty Volumes of the *Psychological Review*," *Psychological Review* 50 (1943): 141–155.

82. Merleau-Ponty, *Phenomenology of Perception*; M. Merleau-Ponty, "Film and the New Psychology" [1947], in *Sense and Non-Sense*, trans. H. Dreyfus and P. Dreyfus, 43–59 (Evanston, IL: Northwestern University Press, 1964).

83. Merleau-Ponty, "Film."

84. Merleau-Ponty, *Phenomenology of Perception*, 164–166.

85. Merleau-Ponty, *Phenomenology of Perception*, 367–368.

86. Myin and Degenaar, "Enactive Vision," 91.

87. Merleau-Ponty, *Phenomenology of Perception*; and D. Hutto and E. Myin, *Radicalizing Enactivism: Basic Minds Without Content* (Cambridge MA: MIT Press, 2013).

88. Noë, *Action in Perception*, 73.

89. Mead, *Philosophy of the Act*, 4.

90. Mead, *Philosophy of the Act*, 3.

91. Crippen, "Pragmatic Evolutions."

92. Dewey, *Art as Experience*, 246.

93. Dewey, *Reconstruction in Philosophy*, 86.

94. Noë, *Action in Perception*, 11.

95. Dewey, *Experience and Nature*, 1925 ed., 259. Merleau-Ponty, *Phenomenology of Perception*; and O'Regan and Noë, "Sensorimotor Account of Vision," 960.

96. Dewey, *Art as Experience*, 177.

97. Crippen, "Art and Pragmatism."

98. Mead, *Philosophy of the Act*, 3–4.

99. R. M. Pritchard, "Stabilized Images on the Retina," *Scientific American* 204 (1961): 72–78.

100. B. Laeng and U. Sulutvedt, "The Eye Pupil Adjusts to Imaginary Light," *Psychological Science* 1 (2014): 188–197, cited in Clark, *Surfing Uncertainty*, 98.

101. See Dewey, *Reconstruction in Philosophy*, 113.

102. S. Gallagher, *Enactivist Interventions*, 38; and F. E. Zajac, "Muscle Coordination of Movement: A Perspective," *Journal of Biomechanics* 26 (1993): 109–124.

103. Dewey, *Art as Experience*, 122.

104. Jackson, "Evolution and Dissolution."

105. Dewey, *Art as Experience*, 100.

106. O'Regan and Noë, "Sensorimotor Account of Vision," 947.

107. Noë, *Action in Perception*, 2.

108. Myin and Degenaar, "Enactive Vision," 91.

109. Noë, *Action in Perception*, 73.

110. Dewey, *Art as Experience*, 49–50.

111. Dewey, *Art as Experience*, 175; O'Regan and Noë, "Sensorimotor Account of Vision"; and Noë, *Action in Perception*, 73. See Crippen, "Dewey, Enactivism and Greek Thought."

112. Noë, *Action in Perception*, 90.

113. O'Regan and Noë, "Sensorimotor Account of Vision." Also see, Myin and Degenaar, "Enactive Vision."

114. O'Regan and Noë, "Sensorimotor Account of Vision," 953.

115. J. Dewey, *Human Nature and Conduct: An Introduction to Social Psychology* (New York: Henry Holt, 1922), 31.

116. Dewey, *Art as Experience*, 104.

117. Merleau-Ponty, *Phenomenology of Perception*, 381, 175.
118. See R. Lickliter, "The Integrated Development of Sensory Organization," *Clinics in Perinatology* 38 (2011): 591–603.
119. Dewey, *Art as Experience*, 98.
120. Dewey, *Art as Experience*, 100–101.
121. For example, Bhalla and D. R. Proffitt, "Visual-Motor Recalibration in Geographical Slant Perception," *Journal of Experimental Psychology: Human Perception and Performance* 25 (1999): 1076–1096; and D. R. Proffitt, "Embodied Perception and the Economy of Action," *Perspectives on Psychological Science* 1 (2006): 110–122.
122. Dewey, *Art as Experience*, 255.
123. Dewey, *Art as Experience*, 256. Emphasis added.
124. J. J. Gibson, *The Ecological Approach to Visual Perception* (Boston: Houghton Mifflin, 1979), 210–211.
125. Gibson, *Ecological Approach*, 210.
126. Noë, *Action in Perception*, 111. Emphasis added.
127. Noë, *Action in Perception*, 109.
128. Noë, *Action in Perception*, 90.
129. Dewey, *Art as Experience*.
130. Dewey, *Art as Experience*, 175.
131. S. Doucet et al., "The 'Smellscape' of Mother's Breast: Effects of Odor Masking and Selective Unmasking on Neonatal Arousal, Oral, and Visual Responses," *Developmental Psychobiology* 49 (2007): 2470–2473.
132. Varela, Thompson, and Rosch, *Embodied Mind*, 163; O'Regan and Noë, "Sensorimotor Account of Vision," 959; and S. Gallagher, *Enactivist Interventions*, 179.
133. See Dewey, *Art as Experience*, 22–53; and Merleau-Ponty, *Phenomenology of Perception*. See also M. Crippen, "Body Phenomenology, Somaesthetics and Nietzschean Themes in Medieval Art," *Pragmatism Today* 5 (2014): 40–41.
134. M. Zampini and C. Spence, "The Role of Auditory Cues in Modulating the Perceived Crispiness and Staleness of Potato Chips," *Journal of Sensory Studies* 19 (2004): 347–363.
135. J. Dewey, *Affective Thought* [1926], in *Philosophy and Civilization* (New York: Minton, Balch, 1931), 122.
136. Merleau-Ponty, *Phenomenology of Perception*, 371–372.
137. See Crippen, "Body Phenomenology."
138. For example, R. M. Pangborn, H. W. Berg, and B. Hansen, "The Influence of Color on Discrimination of Sweetness in Dry Table-Wine," *American Journal of Psychology* 76 (1963): 492–495; G. Morrot, F. Brochet, and D. Dubourdieu, "The Color of Odors," *Brain and Language* 79 (2001): 309–320; and W. V. Parr, K. G. White, and D. Heatherbell, "The Nose Knows: Influence of Colour on Perception of Wine Aroma," *Journal of Wine Research* 14 (2003): 79–101.
139. Merleau-Ponty, *Phenomenology of Perception*, 371.
140. See A. Maravita and A. Iriki, "Tools for the Body (Schema)," *Trends in Cognitive Sciences* 8 (2004): 79–86. See also Kauffmann, "Brain Plasticity."
141. L. Yu, B. E. Stein, and B. A. Rowland, "Adult Plasticity in Multisensory Neurons: Short-Term Experience-Dependent Changes in the Superior Colliculus," *Journal of Neuroscience* 29 (2009): 15910–15922; and B. E. Stein, T. R. Stanford, and B. A. Rowland, "Development of Multisensory Integration from the Perspective of the Individual Neuron," *Nature Reviews Neuroscience* 15 (2014): 520–535.
142. See Dewey, *Art as Experience*, 126; Merleau-Ponty, *Phenomenology of Perception*, 371–371; and Crippen, "Art and Pragmatism."
143. Merleau-Ponty, *Phenomenology of Perception*, 266.
144. See Merleau-Ponty, *Phenomenology of Perception*, 272.

145. For example, M. H. Giard and F. Peronnet, "Auditory-Visual Integration During Multi-modal Object Recognition in Humans: A Behavioral and Electrophysiological Study," *Journal of Cognitive Neuroscience* 11 (1999): 473–490; E. Macaluso, C. D. Frith, and J. Driver, "Modulation of Human Visual Cortex by Crossmodal Spatial Attention," *Science* 289 (2000): 1206–1208; K. O. Bushara, J. Grafman, and M. Hallett, "Neural Correlates of Auditory-Visual Stimulus Onset Asnchrony Detection," *Journal of Neuroscience* 21 (2001): 300–304; and J. Driver and T. Noesselt, "Multisensory Interplay Reveals Crossmodal Influences on 'Sensory-Specific' Brain Regions, Neural Responses, and Judgments," *Neuron* 57 (2008): 11–23.

146. K. J. Wheaton et al., "Viewing the Motion of Human Body Parts Activates Different Regions of Premotor, Temporal, and Parietal Cortex," *NeuroImage* 22 (2004): 277–288.

147. G. A. Calvert et al., "Activation of Auditory Cortex During Silent Lipreading," *Science* 276 (1997): 593–596.

148. See Merleau-Ponty, *Phenomenology of Perception*, 272.

149. For example, H. Calderwood, *The Relations of Mind and Brain* (London: Macmillan, 1879); and Jackson, "Evolution and Dissolution."

150. B. E. Stein and T. R. Stanford, "Multisensory Integration: Current Issues from the Perspective of the Individual Neuron," *Nature Reviews Neuroscience* 15 (2014): 520–535; and Stein, Stanford, and Rowland, "Development of Multisensory Integration."

151. Stein, Stanford, and Rowland, "Development of Multisensory Integration."

152. Yu, Stein, and Rowland, "Adult Plasticity."

153. Dewey, *Art as Experience*, 175.

154. W. Whyte, *The Social Life of Small Urban Spaces* (Washington, DC: Conservation Foundation, 1980), 48.

155. For example, D. N. Lee and J. R. Lishman, "Visual Proprioceptive Control of Stance," *Journal of Human Movement Studies* 1 (1975): 87–95. See also Merleau-Ponty, "Film," 51–52.

156. Merleau-Ponty, *Phenomenology of Perception*, 266–267.

157. Dewey, *Art as Experience*, 123.

158. Dewey, *Art as Experience*, 124.

159. Aristotle, *On Dreams*, trans. J. I. Beare, in *The Basic Works of Aristotle*, ed. R. McKeon (New York: Random House, 1941), 460b20.

160. Merleau-Ponty, *Phenomenology of Perception*, 205. Emphasis added.

161. Crippen, "Embodied Cognition."

162. H. McGurk and J. MacDonald, "Hearing Lips and Seeing Voices," *Nature* 264 (1976): 746–748.

163. D. Senkowski et al., "Look Who's Talking: The Deployment of Visuo-Spatial Attention During Multisensory Speech Processing Under Noisy Environmental Conditions," *NeuroImage* 43 (2008): 379–387; and D. Senkowski et al., "Crossmodal Binding Through Neural Coherence: Implications for Multisensory Processing," *Trends in Neurosciences* 31 (2008): 401–409.

164. See J. Keil et al., "On the Variability of the McGurk Effect: Audiovisual Integration Depends on Prestimulus Brain States," *Cerebral Cortex* 22 (2011): 221–231; and A. R. Nath and M. S. Beauchamp, "A Neural Basis for Interindividual Differences in the McGurk Effect, a Multisensory Speech Illusion," *NeuroImage* 59 (2012): 781–787.

165. Crippen, "William James on Belief"; and Crippen, "Aesthetics and Action."

166. Crippen, "Embodied Cognition."

167. Dawkins, *God Delusion*, 89, 371, 361.

168. Hoffman, "Interface Theory of Perception"; Hoffman, "Construction of Visual Reality"; and Hoffman and Prakash, "Objects of Consciousness."

169. Dawkins, *God Delusion*, 372, 373.

170. Dawkins, *God Delusion*, 373.

171. Dawkins, *God Delusion*, 372.
172. Dawkins, *God Delusion*, 372.
173. Dewey, *Experience and Nature*, 1925 ed., 259.
174. Crippen, "William James on Belief"; Crippen, "Embodied Cognition"; and Crippen, "Aesthetics and Action."
175. Dawkins, *God Delusion*, 368.
176. Crippen, "Art and Pragmatism"; and Crippen, "Embodied Cognition."
177. Crippen, "Art and Pragmatism"; and Crippen, "Embodied Cognition."
178. Crippen, "Art and Pragmatism"; and Crippen, "Embodied Cognition."
179. G. Lakoff and M. Johnson, *Metaphors We Live By* (Chicago: University of Chicago Press, 1980); G. Lakoff and M. Johnson, *Philosophy in the Flesh* (New York: Basic Books, 1999); M. Johnson, *The Body in the Mind: The Bodily Basis of Meaning, Imagination, and Reason* (Chicago: University of Chicago Press, 1987); and M. Johnson, *Embodied Mind, Meaning, and Reason: How Our Bodies Give Rise to Understanding* (Chicago: University of Chicago Press, 2017).
180. Merleau-Ponty, *Phenomenology of Perception*.
181. Dawkins, *God Delusion*, 368–372.
182. Dawkins, *God Delusion*, 371.
183. Dewey, *Reconstruction in Philosophy*.
184. Dewey, *Reconstruction in Philosophy*, 154.
185. See Crippen, "Embodied Cognition"; and H. Jackman, "William James's Naturalistic Account of Concepts and His 'Rejection of Logic,'" in *Philosophy of Mind in the Nineteenth Century*, ed. S. Lapointe, 133–146 (New York: Routledge, 2018).
186. Dawkins, *God Delusion*, 361.
187. Dewey, *Reconstruction in Philosophy*, 154.
188. Dewey, *Reconstruction in Philosophy*, 154.
189. See Hoffman and Prakash, "Objects of Consciousness," 1–22.
190. Hoffman, "Interface Theory of Perception," 153.
191. Crippen, "Embodied Cognition"; Crippen, "Aesthetics and Action"; and Crippen, "Pragmatic Evolutions."
192. D. R. Proffitt and S. A. Linkenauger, "Perception Viewed as a Phenotypic Expression," in *Action Science: Foundations of an Emerging Discipline*, ed. W. Prinz, M. Beisert, and A. Herwig, 171–197 (Cambridge, MA: MIT Press, 2013).
193. J. K. Witt, D. R. Proffitt, and W. Epstein, "Tool Use Affects Perceived Distance, but Only When You Intend to Use It," *Journal of Experimental Psychology: Human Perception and Performance* 31 (2005): 80–88; and J. K. Witt and D. R. Proffitt, "Action-Specific Influences on Distance Perception: A Role for Motor Simulation," *Journal of Experimental Psychology: Human Perception and Performance* 34 (2008): 1479–1492.
194. D. A. Lessard, S. A. Linkenauger, and D. R. Proffitt, "Look Before You Leap: Jumping Ability Affects Distance Perception," *Perception* 38 (2009): 1863–1866.
195. J. K. Stefanucci and M. N. Geuss, "Duck! Scaling the Height of a Horizontal Barrier to Body Height," *Attention, Perception, and Psychophysics* 72 (2010): 1338–1349.
196. J. K. Stefanucci and M. N. Guess, "Big People, Little World: The Body Influences Size Perception," *Perception* 38 (2009): 1782–1795.
197. D. R. Proffitt et al., "Perceiving Geographical Slant," *Psychonomic Bulletin and Review* 16 (1995): 970–972; Bhalla and Proffitt, "Visual-Motor Recalibration"; S. Schnall, J. Zadra, and D. R. Proffitt, "Direct Evidence for the Economy of Actions: Glucose and the Perception of Geographical Slant," *Perception* 39 (2010): 464–482; and J. Zadra et al., "Direct Physiological Evidence for the Economy of Action: Bioenergetics and the Perception of Spatial Layout," *Journal of Vision* 10 (2010): 54.
198. Hoffman, "Interface Theory of Perception," 151.

199. P. Lewicki, T. Hill, and M. Czyzewska, "Nonconscious Acquisition of Information," *American Psychologist* 47 (1992): 796.
200. R. G. Brown and C. D. Marsden, "How Common Is Dementia in Parkinson's Disease?," *Lancet* 324 (1984): 1262–1265.
201. A. Tessitore et al., "Dopamine Modulates the Response of the Human Amygdala: A Study in Parkinson's Disease," *Journal of Neuroscience* 22 (2002): 9099–9103.
202. J. Schulkin, "Cognitive Functions, Bodily Sensibility and the Brain," *Phenomenology and the Cognitive Sciences* 5 (2006): 341–349; Schulkin, *Sport*, chap. 2; and Crippen, "Dewey, Enactivism and Greek Thought."
203. M. Csikszentmihalyi, *Flow and the Foundations of Positive Psychology: The Collected Works of Mihaly Csikszentmihalyi* (New York: Springer, 2014).
204. Dietrich, "Neurocognitive Mechanism."
205. Dietrich, "Neurocognitive Mechanism."
206. Clark, *Supersizing the Mind*, 14.
207. Dewey, *Reconstruction in Philosophy*, 90–91.
208. Dewey, *Reconstruction in Philosophy*, 91.
209. Dewey, *Reconstruction in Philosophy*, 91.
210. Noë, *Out of Our Heads*, 14. See also Thompson, *Mind in Life*, 74–75.
211. Dewey, *Experience and Nature*, 1925 ed., chap. 7.
212. Long, *Darwin's Devices*, 95.
213. Long, *Darwin's Devices*, 95–96.
214. Noë, *Out of Our Heads*, 8.
215. Long, *Darwin's Devices*, 97.
216. Thompson, *Mind in Life*, chap. 8.
217. Thompson, *Mind in Life*, chap. 13.
218. See T. Fuchs, *Ecology of the Brain: The Phenomenology and Biology of the Embodied Mind* (New York: Oxford University Press, 2018).
219. T. Fong, I. Nourbakhsh, and K. Dautenhahn, "A Survey of Socially Interactive Robots," *Robotics and Autonomous Systems* 42 (2003): 143–166.
220. Clark, *Supersizing the Mind*.
221. Chemero, *Radical Embodied Cognitive Science*, 27.
222. Long, *Darwin's Devices*, 104.
223. Knowlton, Mangels, and Squire, "Neostriatal Habit Learning System."
224. M. T. Ullman et al., "A Neural Dissociation Within Language: Evidence That the Mental Dictionary Is Part of Declarative Memory, and That Grammatical Rules Are Processed by the Procedural System," *Journal of Cognitive Neuroscience* 9 (1997): 266–286; and Kotz and Schmidt-Kassow, "Basal Ganglia Contribution," 48–60.
225. Kotz and Schmidt-Kassow, "Basal Ganglia Contribution," 48–60.
226. A. M. Graybiel et al., "The Basal Ganglia and Adaptive Motor Control," *Science* 265 (1994): 1826–1831.
227. O. Hauk, I. Johnsrude, and F. Pulvermüller, "Somatotopic Representation of Action Words in Human Motor and Premotor Cortex," *Neuron* 41 (2004): 301–307.
228. M. Merleau-Ponty, *Consciousness and the Acquisition of Language*, trans. H. J. Silverman (Evanston, IL: Northwestern University Press, 1991; first published 1964 in French).
229. For example, S. Goldin-Meadow, "The Role of Gesture in Communication and Thinking," *Trends in Cognitive Sciences* 3 (1999): 419–429; and M. W. Alibali, R. Boncoddo, and A. B. Hostetter, "Gesture in Reasoning: An Embodied Perspective," in *The Routledge Handbook of Embodied Cognition*, ed. L. Shapiro, 150–159 (New York: Routledge, 2014).
230. M. J. Farah, "The Neurobiological Basis of Visual Imagery: A Componential Analysis," *Cognition* 18 (1984): 245–272; S. M. Kosslyn, *Image and Mind* (Cambridge, MA: Harvard University Press, 1986); S. M. Kosslyn, *Image and Brain: The Resolution of the Imagery*

Debate (Cambridge, MA: MIT Press, 1994), chap. 9; and J. Decety, "Do Imagined and Executed Actions Share the Same Neural Substrate?," *Cognitive Brain Research* 3 (1996): 87–93.

231. For example, V. Gallese and A. Goldman, "Mirror Neurons and the Simulation Theory of Mind-Reading," *Trends in Cognitive Science* 2 (1998): 493–501; and G. Rizzolatti, L. Fogassi, and V. Gallese, "Neuropsychological Mechanisms Underlying the Understanding and Imitation of Action," *Nature Reviews Neuroscience* 2 (2001): 661–670.

232. M. Iacoboni et al., "Cortical Mechanisms of Human Imitation," *Science* 286 (1999): 2526–2528.

233. M. Jeannerod, *The Neural and Behavioural Organization of Goal-Directed Movements* (New York: Oxford University Press, 1988); M. Jeannerod, "The Representing Brain: Neural Correlates of Motor Intention and Imagery," *Behavioral and Brain Sciences* 17 (1994): 187–201; M. Jeannerod, *The Cognitive Neuroscience of Action* (Oxford: Blackwell, 1997); and M. Jeannerod, "To Act or Not to Act: Perspectives on the Representation of Action," *Quarterly Journal of Experimental Psychology* 52 (1999): 1–29.

234. Kosslyn, *Image and Mind*.

235. G. Leisman, A. Moustafa, and T. Shafir, "Thinking, Walking, Talking: Integratory Motor and Cognitive Brain Function," *Frontiers in Public Health* 4 (2016): art. 94, https://doi.org/10.3389/fpubh.2016.00094.

236. Dewey, *Human Nature and Conduct*, 22. See also M. Costantini and M. Stapleton, "How the Body Narrows the Interaction with the Environment," in *Foundations of Embodied Cognition: Perceptual and Emotional Embodiment*, ed. Y. Coello and M. H. Fischer, 181–197 (New York: Routledge, 2016).

237. L. Wittgenstein, *Philosophical Investigations* [written 1929–1949], trans. G. E. M. Anscombe (Oxford: Basil Blackwell, 1953), §43.

238. Wittgenstein, *Philosophical Investigations*, §150, §183.

239. Dewey, *Experience and Nature*, 1925 ed., 186.

240. A. Clark, *Being There: Putting Brain, Body, and World Together Again* (Cambridge, MA: MIT Press, 1997).

3. SOCIAL COHESION, EXPERIENCE, AND AESTHETICS

1. Aristotle, *Politics*, trans. B. Jowett, in *The Basic Works of Aristotle*, ed. R. McKeon, 1113–1316 (New York: Random House, 1941), 1253a.

2. For example, C. S. Peirce, "Fixation of Belief," *Popular Science Monthly* 12 (1877): 1–15; J. Dewey, *Reconstruction in Philosophy* (New York: Henry Holt, 1920); J. Dewey, *Human Nature and Conduct: An Introduction to Social Psychology* (New York: Henry Holt, 1922); J. Dewey, *Experience and Nature* (Chicago: Open Court, 1925); and G. H. Mead, *Mind, Self, and Society: From the Standpoint of a Social Behaviorist*, ed. C. W. Morris (Chicago: University of Chicago Press, 1934).

3. J. Dewey, "Syllabus: Types of Philosophical Thought" [1922–1923], in *The Middle Works, 1899–1922*, vol. 13, ed. J. A. Boydston (Carbondale, IL: Southern Illinois University Press, 1983), 351; and J. Dewey, "Unfinished Introduction" [1951], in *The Later Works, 1925–1953*, vol. 1, ed. J. A. Boydston (Cardondale, IL: Southern Illinois University Press, 1981), 363.

4. See M. Crippen, "Group Cognition in Pragmatism, Developmental Psychology and Aesthetics," *Pragmatism Today* 8 (2017): 185–197.

5. For example, A. R. Damasio, *Descartes' Error: Emotion, Reason, and the Human Brain* (New York: G. P. Putnam, 1994); P. Brown and C. D. Maarsden, "What Do the Basal Ganglia Do?," *Lancet* 351 (1998): 1801–1804; S. Baron-Cohen et al., "Social Intelligence in the Normal and Autistic Brain: An fMRI Study," *European Journal of Neuroscience* 11 (1999):

1891–1898; and T. Canli et al., "Amygdala Response to Happy Faces as a Function of Extraversion," *Science* 296 (2002): 2191.

6. For example, A. M. Graybiel et al., "The Basal Ganglia and Adaptive Motor Control," *Science* 265 (1994): 1826–1831; B. Knowlton, J. Mangels, and L. Squire, "A Neostriatal Habit Learning System in Humans," *Science* 273 (1996): 1399–1402; M. T. Ullman et al., "A Neural Dissociation Within Language: Evidence That the Mental Dictionary Is Part of Declarative Memory, and That Grammatical Rules Are Processed by the Procedural System," *Journal of Cognitive Neuroscience* 9 (1997): 266–286; and S. A. Kotz, M. Schwartze, and M. Schmidt-Kassow, "Non-motor Basal Ganglia Functions: A Review and Proposal for a Model of Sensory Predictability in Auditory Language Perception," *Cortex* 45 (2009): 982–990.

7. K. C. Berridge, "Measuring Hedonic Impact in Animals and Infants: Microstructure of Affective Taste Reactivity Patterns," *Neuroscience and Biobehavioral Reviews* 24 (2000): 173–198; A. J. Calder, A. D. Lawrence, and A. W. Young, "Neuropsychology of Fear and Loathing," *Nature Neuroscience* 2 (2001): 352–363; and W. Schultz, "Getting Formal with Dopamine and Reward," Neuron 36 (2002): 241–263.

8. J. Aggleton, *The Amygdala* (New York: Oxford University Press, 2000), chap. 1.

9. L. Brothers, "Neurophysiology of the Perception of Intentions in Primates," in *The Cognitive Neurosciences*, ed. S. Gazzaniga, 1107–1115 (Cambridge, MA: MIT Press, 1994); D. I. Perret and N. J. Emery, "Understanding the Intentions of Others from Visual Signals: Neurophysiological Evidence," *Cahiers de psychologie cognitive* 13 (1994): 683–694; and E. T. Rolls, *The Brain and Emotion* (New York: Oxford University Press, 1999).

10. See K. Nakayama, "Introduction: Vision Going Social," in *The Science of Social Vision*, ed. R. N. Adams et al., 3–17 (New York: Oxford University Press, 2010).

11. J. Dewey, *Art as Experience* (New York: Minton, Balch, 1934).

12. W. James, "Are We Automata?," *Mind* 4 (1879): 1–22.

13. F. Nietzsche, *The Will to Power*, ed. W. Kaufmann, trans. W. Kaufman and R. J. Hollingdale (New York: Vintage Books, 1968); M. Heidegger, *Being and Time*, trans. J. Macquarrie and E. Robinson (New York: Harper and Row, 1962; first published 1927 in German). See M. Crippen, "Art and Pragmatism: James and Dewey on the Reconstructive Presuppositions of Experience" (PhD diss., York University, 2010); and M. Crippen and M. Dixon, "Echoes of Past and Present," in *Tom Petty and Philosophy: We Need to Know*, ed. R. Auxier and M. Volpert, 16–25 (Chicago: Open Court, 2019).

14. A. Chemero, "Sensorimotor Empathy," *Journal of Consciousness Studies* 23 (2016): 138–152.

15. M. Merleau-Ponty, *Phenomenology of Perception*, trans. C. Smith (New York: Routledge and Kegan Paul, 1962; first published 1945 in French), 143.

16. See C. W. Reynolds, "Flocks, Herds, and Schools: A Distributed Behavioral Model," *Computer Graphics* 21 (1987): 25–34. See also M. J. Richardson and A. Chemero, "Complex Dynamical Systems and Embodiment," in *The Routledge Handbook of Embodied Cognition*, ed. L. Shapiro (New York: Routledge, 2014), 39–50.

17. S. J. Portugal et al., "Upwash Exploitation and Downwash Avoidance by Flap Phasing in Ibis Formation Flight," *Nature* 505 (2014): 399–402.

18. M. Crippen, "Group Cognition."

19. S. Sonea and M. Panisset, *A New Bacteriology* (Boston: Jones and Bartlett, 1983), 8, quoted in E. Thompson, *Mind in Life: Biology, Phenomenology, and the Sciences of Mind* (Cambridge, MA: Harvard University Press, 2007), 119.

20. See M. Crippen, "Body Politics: Revolt and City Celebration," in *Bodies in the Streets: Somaesthetics of City Life*, ed. R. Shusterman, 89–110 (Boston: Brill, 2019); Crippen "Group Cognition"; and M. Crippen, "The Soma in City Life: Cultural, Political and Bodily Aesthetics of Mandalay's Water Festival," *Pragmatism Today* 9 (2018): 29–40.

21. For example, R. Zazzo, "Le problème de l'imitation chez le nouveau-né," *Enfance* 10 (1957): 135–142; A. Kendon, "Movement Coordination in Social Interactions: Some Examples Described," *Acta Psychologica* 32 (1970): 101–125; A. N. Meltzoff and M. K. Moore, "Imitation of Facial and Manual Gestures by Human Neonates," *Science* 198 (1977): 75–78; A. N. Meltzoff and M. K. Moore, "Imitation in Newborn Infants: Exploring the Range of Gestures Imitated and the Underlying Mechanisms," *Developmental Psychology* 25 (1989): 954–962; A. N. Meltzoff and M. K. Moore, "Imitation, Memory, and the Representation of Persons," *Infant Behavior and Development* 17 (1994): 83–99; A. N. Meltzoff and M. K. Moore, "Explaining Facial Imitation: A Theoretical Model," *Early Development and Parenting* 6 (1997): 179–192; O. Maratos, "Trends in Development of Imitation in Early Infancy," in *Regressions in Mental Development: Basis Phenomena and Theories*, ed. T. G. Bever, 81–101 (Hillsdale, NJ: Erlbaum, 1982); U. Dimberg, "Facial Relations to Facial Expressions," *Psychophysiology* 19 (1982): 643–647; J. N. Cappella, "The Biological Origins of Automated Patterns of Human Interaction," *Communication Theory* 1 (1991): 4–35; J. L. Chartrand and J. A. Bargh, "The Chameleon Effect: The Perception-Behavior Link and Social Interaction," *Journal of Personality and Social Psychology* 76 (1999): 893–910; G. Kugiumutzakis, "Genesis and Development of Early Infant Mimesis to Facial and Vocal Models," in *Imitation in Infancy*, ed. J. Nadel and G. Butterworth, 127–185 (New York: Cambridge University Press, 1999); K. G. Niederhoffer and J. W. Pennebaker, "Linguistic Style Matching in Social Interaction," *Journal of Language and Social Psychology* 10 (2002): 59–65; N. A. Harrison et al., "Pupillary Contagion: Central Mechanisms Engaged in Sadness Processing," *Social Cognitive and Affective Neuroscience* 1 (2006): 5–17; and P. R. Cannon, A. E. Hayes, and S. P. Tipper, "An Electromyographic Investigation of the Impact of Task Relevance on Facial Mimicry," *Cognition and Emotion* 23 (2009): 918–929.

22. E. J. Charney, "Postural Configuration in Psychotherapy," *Psychosomatic Medicine* 28 (1966): 305–315.

23. P. M. Niedenthal et al., "When Did Her Smile Drop? Facial Mimicry and the Influences of Emotional State on the Detection of Change in Emotional Expression," *Cognition and Emotion* 15 (2001): 853–864; M. Stel and A. van Knippenberg, "The Role of Facial Mimicry in the Recognition of Affect," *Psychological Science* 19 (2008): 984–985; and D. T. Neal and T. Chartrand, "Embodied Emotion Perception: Amplifying and Dampening Facial Feedback Modulates Emotion Perception Accuracy," *Social Psychology and Personality Science* 2 (2011): 673–678.

24. See Chemero, "Sensorimotor Empathy."

25. J. Steinbeck, *The Grapes of Wrath* (New York: Viking, 1939), 264–265.

26. Steinbeck, *Grapes of Wrath*, 266.

27. Steinbeck, *Grapes of Wrath*, 266–267.

28. For example, J. Dewey, *Philosophy and Education in Their Historic Relation*, ed. J. J. Chambliss from 1910–1911 lectures transcribed by E. R. Clapp (Boulder, CO: Westview, 1910); Dewey, *Reconstruction in Philosophy*; and Dewey, *Experience and Nature*, 1925 ed.

29. R. Descartes to Mersenne, April 1, 1640, in *The Philosophical Writings of Descartes*, trans. J. Cottingham et al., vol. 3, *Correspondence*, 145–146 (New York: Cambridge University Press, 1991).

30. T. H. Huxley, "On the Hypothesis That Animals Are Automata, and Its History," *Nature* 10 (1874): 362–366.

31. Dewey, "Unfinished Introduction," 361.

32. B. Smuts, "Encounters with Animal Minds," *Journal of Consciousness Studies* 8 (2001): 293–309.

33. Smuts, "Encounters with Animal Minds," 395.

34. Dewey, *Human Nature and Conduct*, 104.

35. Dewey, *Human Nature and Conduct*, 62–63.
36. Dewey, *Human Nature and Conduct*, 62.
37. Dewey, *Reconstruction in Philosophy*, 92.
38. Zazzo, "Problème de l'imitation"; Maratos, "Trends in Development"; Meltzoff and Moore, "Facial and Manual Gestures"; Meltzoff and Moore, "Imitation in Newborn Infants"; Meltzoff and Moore, "Imitation, Memory"; Meltzoff and Moore, "Explaining Facial Imitation"; and Kugiumutzakis, "Genesis and Development."
39. L. W. Swanson, "The Cerebral Hemisphere Regulation of Motivated Behavior," *Brain Research* 836 (2000): 113–164; I. Kokal et al., "Synchronized Drumming Enhances Activity in the Caudate and Facilitates Prosocial Commitment—if the Rhythm Comes Easily," *PLoS ONE* 6 (2011): art. e27272, https://doi.org/10.1371/journal.pone.0027272; and J. Schulkin, *Pragmatism and the Search for Coherence in Neuroscience* (New York: Palgrave Macmillan, 2015), chap. 6.
40. G. Rizzolatti, L. Fogassi, and V. Gallese, "Neuropsychological Mechanisms Underlying the Understanding and Imitation of Action," *Nature Reviews Neuroscience* 2 (2001): 661–670.
41. M. A. Umiltà et al., "I Know What You Are Doing: A Neurophysiological Study," *Neuron* 31 (2001): 155–165.
42. E. Kohler et al., "Hearing Sounds, Understanding Actions: Action Representation in Mioor Neurons," *Science* 297 (2002): 846–484.
43. P. F. Ferrari, "Mirror Neurons Responding to the Observation of Ingestive and Communicative Mouth Actions in the Monkey Ventral Premotor Cortex," *European Journal of Neuroscience* 17 (2003): 1703–1714. For an overview, see L. Craighero, "The Role of the Motor System in Cognitive Functions," in *The Routledge Handbook of Embodied Cognition*, ed. L. Shapiro, 51–58 (New York: Routledge, 2014).
44. T. Singer et al., "Empathy for Pain Involves the Affective but Not Sensory Components of Pain," Science 303 (2004): 1157–1162. See also N. I. Eisenberg, "Identifying the Neural Correlates Underlying Social Pain: Implications for Developmental Processes," *Human Development* 49 (2006): 273–293.
45. For example, M. Tomasello, A. C. Kruger, and H. H. Ratner, "Cultural Learning," *Behavioral and Brain Sciences* 16 (1993): 495–511; M. Tomasello, *The Cultural Origins of Human Cognition* (Cambridge, MA: Harvard University Press, 1999); M. Tomasello, *Origins of Human Communication* (Cambridge, MA: MIT Press, 2008); C. Trevarthen, "The Intersubjective Psychobiology of Human Meaning: Learning of Culture Depends on Interest for Co-operative Practical Work—and Affection for the Joyful Art of Good Company," *Psychoanalytic Dialogues* 19 (2009): 507–518; C. Trevarthen, "Embodied Human Intersubjectivity: Imaginative Agency, to Share Meaning," *Journal of Cognitive Semiotics* 4 (2009): 6–56; C. Trevarthen, "What Is It Like to Be a Person Who Knows Nothing? Defining the Active Intersubjective Mind of a Newborn Human Being," *Infant and Child Development* 20 (2011): 119–135; C. Trevarthen, "In Praise of a Doctor Who Welcomes the Newborn Infant Person," *Journal of Child and Adolescent Psychiatric Nursing* 26 (2013): 204–213; C. Trevarthen, "Awareness of Infants: What Do They, and We, Seek?," *Psychoanalytic Inquiry* 35 (2015): 395–416; and C. Trevarthen, "Infant Semiosis: The Psychobiology of Action and Shared Experience from Birth," *Cognitive Development* 36 (2015): 130–141.
46. Trevarthen, "What Is It Like"; Trevarthen, "Awareness of Infants."
47. E. Nagy and P. Molnár, "Homo Imitans or Homo Provocans? The Phenomenon of Neonatal Initiation," *Infant Behavior and Development* 27 (2004): 57–63. See also E. Nagy, "Innate Intersubjectivity: Newborns' Sensitivity to Communication Disturbance," *Developmental Psychology* 44 (2008): 1779–1784; C. Trevarthen et al., "Collaborative Regulations of Vitality in Early Childhood: Stress in Intimate Relationships and Postnatal

Psychopathology," in *Developmental Psychopathology*, ed. D. Cichetti and D. J. Cohen, 65–126 (New York: Wiley, 2015).

48. T. B. Brazelton et al., "Early Mother-Infant Reciprocity," in *Parent-Infant Interaction*, ed. M. Hofer, 137–154 (New York: Elsevier, 1975); C. Trevarthen, "Early Attempts at Speech," in *Child Alive: New Insights into the Development of Young Children*, ed. R. Lewin, 62–80 (London: Temple Smith, 1975); C. Trevarthen, "Descriptive Analyses of Infant Communication Behavior," in *Studies in Mother-Infant Interaction: The Loch Lomond Symposium*, ed. H. R. Schaffer, 227–270 (New York: Academic Press, 977); C. Trevarthen, "Communication and Cooperation in Early Infancy. A Description of Primary Intersubjectivity," in *Before Speech: The Beginning of Human Communication*, ed. M. Bullowa, 321–347 (New York: Cambridge University Press, 1979); H. Papoušek and M. Papoušek, "Mothering and Cognitive Head Start: Psychobiological Considerations," in *Studies in Mother-Infant Interaction*, ed. Schaffer, 63–85; H. Papoušek and M. Papoušek, "Intuitive Parenting: A Dialectic Counterpart to the Infant's Integrative Competence," in *Handbook of Infant Development*, ed. J. D. Osofsky, 669–720 (New York: Wiley, 1987); and E. Z. Tronick et al., "The Infant's Response to Entrapment Between Contradictory Messages in Face-to-Face Interaction," *Journal of the American Academy of Child Psychiatry* 17 (1978): 1–13.

49. Dewey, *Reconstruction in Philosophy*, 84–85.

50. Dewey, *Reconstruction in Philosophy*, 85.

51. A. N. Whitehead, *Science and the Modern World* (1926; New York: Cambridge University Press, 1953), 140.

52. Trevarthen, "What Is It Like."

53. Trevarthen, "What Is It Like."

54. Trevarthen, "What Is It Like," 121.

55. Trevarthen, "Intersubjective Psychobiology," 507.

56. Trevarthen, "Intersubjective Psychobiology," 507.

57. Trevarthen, "Embodied Human Intersubjectivity," 25–26.

58. Trevrathen, "Embodied Human Intersubjectivity," 33.

59. Dewey, *Art as Experience*, chap. 3.

60. For example, Dewey, *Experience and Nature*, 1925 ed., 358; J. Dewey, *Affective Thought* [1926], in *Philosophy and Civilization* (New York: Minton, Balch, 1931), 121; and Dewey, *Art as Experience*, 139.

61. For example, D. E. Berlyne, "Novelty, Complexity, and Hedonic Values," *Perception and Psychophysics* 8 (1970): 279–285; D. E. Berlyne, "Ends and Means of Experimental Aesthetics," *Canadian Journal of Psychology* 26 (1972): 303–325; R. Kaplan and S. Kaplan, *The Experience of Nature: A Psychological Perspective* (New York: Cambridge University Press, 1989); and C. Kidd, S. T. Piantadosi, and R. N. Aslin, "The Goldilocks Effect: Human Infants Allocate Attention to Visual Sequences That Are Neither Too Simple nor Too Complex," *PLoS ONE* 7 (2012): art. e36399, https://doi.org/10.1371%2Fjournal.pone.0036399.

62. See C. E. Shannon, "The Mathematical Theory of Communication," in *The Mathematical Theory of Communication*, ed. C. E. Shannon and W. Weaver, 25–115 (Chicago: University of Illinois Press, 1949); and W. Weaver, "Recent Contributions to the Mathematical Theory of Communication," in *Mathematical Theory of Communication*, ed. Shannon and Weaver, 1–28.

63. See R. S. Ulrich, "Aesthetic and Affective Response to Natural Environment," in *Behavior and the Natural Environment*, ed. I. Altman and J. F. Wohlwill (New York: Plenum, 1983); S. Kaplan, "Aesthetics, Affect, and Cognition: Environmental Preference from an Evolutionary Perspective," in *Environment and Behavior* 1 (1987): 4–32; and S. Kaplan, "Environmental Preference in a Knowledge-Seeking, Knowledge-Using Organism," in *The Adapted Mind*, ed. J. Barkow, L. Cosmides, and J. Tooby, 581–598 (New York: Oxford University Press, 1992).

64. For example, A. M. Graybiel and S. T. Grafton, "The Striatum: Where Skills and Habits Meet," *Cold Spring Harbor Perspectives in Biology* 7 (2015): 1–13.
65. I. Fried et al., "Increased Dopamine Release in the Human Amygdala During Performance of Cognitive Tasks," *Nature Neuroscience* 4 (2001): 201–206; and S. Koelsch, T. Fritz, and G. Schlaug, "Amygdala Activity Can Be Modulated by Unexpected Chord Functions During Music Listening," *NeuroReport* 19 (2008): 1815–1819.
66. Dewey, *Experience and Nature*, 1925 ed.; and Dewey, *Art as Experience*.
67. Dewey, *Art as Experience*, 35.
68. Dewey, *Art as Experience*, 35.
69. Dewey, *Art as Experience*, 35.
70. Dewey, *Art as Experience*, 174.
71. Dewey, *Affective Thought*, 121.
72. Dewey, *Art as Experience*, 175. See also chap. 2.
73. Crippen, "Art and Pragmatism."
74. P. Sterling and J. Eyer, "Allostasis: A New Paradigm to Explain Arousal Pathology," in *Handbook of Life Stress, Cognition, and Health*, eds. S. Fisher and J. Reason (New York: John Wiley and Sons, 1988), 629–649.
75. For example, P. Lee, M. M. Swarbrick, and K. K. Ho, "Brown Adipose Tissue in Adult Humans: A Metabolic Renaissance," *Endocrine Reviews* 34 (2013): 413–438; and A. A. van der Lans et al., "Cold Acclimation Recruits Human Brown Fat and Increases Nonshivering Thermogenesis," *Journal of Clinical Investigation* 123 (2013): 3395–3340.
76. T. A. Bedrosian and R. J. Nelson, "Timing of Light Exposure Affects Mood and Brain Circuits," *Translational Psychiatry* 7 (2017): art. e1017, https://doi.org/10.1038%2Ftp.2016.262.
77. See W. Deng, J. B. Aimone, and F. H. Gage, "New Neurons and New Memories: How Does Adult Hippocampal Neurogenesis Affect Learning and Memory?," *Nature Reviews Neuroscience* 11 (2010): 339–350.
78. A. Vyas et al., "Chronic Stress Induces Contrasting Patterns of Dendritic Remolding in Hippocampal and Amygdaloid Neurons," *Journal of Neuroscience* 22 (2002): 6810–6818; and A. Vyas, A. Pillai, and S. Chattarji, "Recovery After Chronic Stress Fails to Reverse Amygdaloid Neuronal Hypertrophy and Enhanced Anxiety-like Behavior," *Neuroscience* 128 (2004): 667–673.
79. A. D. Tramontin and E. A. Brenowitz, "Seasonal Plasticity in the Adult Brain," *Trends in Neuroscience* 23 (2000): 251–258.
80. P. Sterling, "Principles of Allostasis: Optimal Design, Predictive Regulation, Pathophysiology, and Rational Therapeutics," in *Allostasis, Homeostasis, and the Costs of Physiological Adaptation*, ed. J. Schulkin (New York: Cambridge University Press, 2004), 18.
81. M. L. Power, "Viability as Opposed to Stability: An Evolutionary Perspective on Physiological Regulation," in *Allostasis, Homeostasis*, ed. Schulkin, 343–364.
82. M. Crippen, "Dewey on Arts, Sciences and Greek Philosophy," in *In the Beginning Was the Image: The Omnipresence of Pictures; Time, Truth, Tradition*, ed. A. Benedek and A. Veszelszki, 153–159 (Frankfurt am Main: Peter Lang Press, 2016); and E. Chudoba, "What Controls and What Is Controlled? Deweyan Aesthetic Experience and Shusterman's Somatic Experience," *Contemporary Pragmatism* 14 (2017): 112–134.
83. See L. Mealey and P. Theis, "The Relationship Between Mood and Preferences Among Natural Landscapes: An Evolutionary Perspective," *Ethology and Sociobiology* 16 (1995): 247–256.
84. See C. J. Cela-Conde et al., "Activation of the Prefrontal Cortex in the Human Visual Aesthetic Perception," *Proceedings of the National Academy of Sciences of the United States of America* 101 (2004): 6321–6325; H. Kawabata and S. Zeki, "Neural Correlates of Beauty," *Journal of Neurophysiology* 91 (2004): 1699–1705; O. Vartanian and V. Goel, "Neuroanatomical Correlates of Aesthetic Preference for Paintings," *NeuroReport* 15 (2004): 893–897;

and Y. Lévêque and D. Schön, "Modulation of the Motor Cortex During Singing-Voice Perception," *Neuropsychologia* 70 (2015): 58–63.

85. A. E. Stahl and L. Feigenson, "Observing the Unexpected Enhances Infants' Learning and Exploration," *Science* 348 (2015): 91–94. See also L. Schulz, "Infants Explore the Unexpected," *Science* 348 (2015): 42–43.

86. Tomasello, *Cultural Origins*.

87. Trevarthen, "In Praise of a Doctor"; and Trevarthen, "Awareness of Infants."

88. L. W. Sander, *Living Systems, Evolving Consciousness, and the Emerging Person: A Selection of Papers from the Life Work of Louis Sander*, ed. G. Amadei and I. Bianchi (New York: Analytic Press, 2008).

89. For example, E. Z. Tronick, "Emotions and Emotional Communication in Infants," *American Psychologist* 44 (1989): 112–126; E. Z. Tronick, "Why Is Connection with Others So Critical? The Formation of Dyadic States of Consciousness: Coherence-Governed Selection and the Co-creation of Meaning Out of Messy Meaning Making," in *Emotional Development: Recent Research Advances*, ed. J. Nadel and D. Muir, 293–315 (New York: Oxford University Press, 2005); Tronick et al., "Infant's Response to Entrapment"; E. Z. Tronick et al., "Dyadically Expanded States of Consciousness and the Process of Therapeutic Change," *Infant Mental Health Journal* 19 (1999): 290–299; M. K. Weinberg and E. Z. Tronick, "Beyond the Face: An Empirical Study of Infant Affective Configurations of Facial, Vocal, Gestural, and Regulatory Behaviors," *Child Development* 65 (1994): 1495–1507.

90. Tronick et al., "Dyadically Expanded States," 296.

91. J. A. Mennella and P. L. Garcia, "Children's Hedonic Response to the Smell of Alcohol: Effects of Parental Drinking Habits," *Alcohol: Clinical and Experimental Research* 24 (2000): 1167–1171; C. A. Forestell and J. A. Mennella, "Children's Hedonic Judgments of Cigarette Smoke Odor: Effect of Parental Smoking and Maternal Mood," *Psychology of Addiction and Behavior* 19 (2005): 423–432; and J. A. Mennella and C. A. Forestell, "Children's Hedonic Responses to the Odors of Alcoholic Beverages: A Window to Emotions," *Alcohol* 24 (2008): 249–260. See also B. Schaal and K. Durand, "The Role of Olfaction in Human Multisensory Development," in *Multisensory Development*, ed. A. J. Bremner, D. J. Lewkowicz, and C. Spence, 29–62 (New York: Oxford University Press, 2012).

92. J. T. Delafield-Butt and C. Trevarthen, "The Ontogenesis of Narrative: From Moving to Meaning," *Frontiers in Psychology* 6 (2015): art. 1157, https://doi.org/10.3389/fpsyg.2015.01157.

93. Trevarthen, "What Is It Like," 129.

94. Trevarthen, "Infant Semiosis."

95. Trevarthen, "Intersubjective Psychobiology," 512.

96. For example, S. Malloch, "Mother and Infants and Communicative Musicality," in "Rhythm, Musical Narrative, and the Origins of Human Communication," ed. C. Trevarthen, supplement, *Musicae Scientiae* 3, no. S1 (1999–2000): 29–57; and S. Malloch and C. Trevarthen, "The Neuroscience of Emotion in Music," in *Communicative Musicality: Exploring the Basis of Human Companionship*, ed. S. Malloch and C. Trevarthen, 105–146 (New York: Oxford University Press, 2009).

97. For example, A. D. Patel, "Syntactic Processing in Language and Music: Different Cognitive Operations, Similar Neural Resources," *Music Perception* 16 (1998): 27–42; A. D. Patel, "Language, Music, Syntax and the Brain," *Nature Neuroscience* 6 (2003): 674–681; A. D. Patel, "The Relationship of Music to the Melody of Speech and to Syntactic Processing Disorders in Aphasia," *Annals of the New York Academy of Sciences* 1060 (2005): 59–70; A. D. Patel, "Talk of the Tone," *Nature* 453 (2008): 726–727; A. D. Patel et al., "Processing Syntactic Relations in Language and Music: An Event-Related Potential Study," *Journal of Cognitive Neuroscience* 10 (1998): 717–733; A. D. Patel et al., "Musical

Syntactic Processing in Agrammatic Broca's Aphasia," *Aphasiology* 22 (2008): 776–789; B. Maess et al., "Musical Syntax Is Processed in Broca's Area: An MEG study," *Nature Neuroscience* 4 (2001): 540–545; S. Koelsch et al., "Bach Speaks: A Cortical 'Language-Network' Serves the Processing of Music," *NeuroImage* 17 (2002): 956–966; S. Koelsch, "Significance of Broca's Area and Ventral Premotor Cortex for Music-Syntactic Processing," *Cortex* 42 (2006): 518–520; and S. Koelsch and N. Steinbeis, "Shared Neural Resources Between Music and Language Indicate Semantic Processing of Musical Tension-Resolution Patterns," *Cerebral Cortex* 18 (2008): 1169–1178.

98. See N. Cook, *Music: A Very Short Introduction* (Oxford: Oxford University Press, 1998); and Patel, "Language, Music, Syntax."

99. Koelsch, "Significance of Broca's Area"; and J. A. Grahn and J. B. Rowe, "Feeling the Beat: Premotor and Striatal Interactions in Musicians and Nonmusicians During Beat Perception," *Journal of Neuroscience* 29 (2009): 7540–7548.

100. T. C. Zhao and P. K. Kuhl, "Musical Intervention Enhances Infants' Neural Processing of Temporal Structure in Music and Speech," *Proceedings of the National Academy of Sciences of the United States of America* 113 (2016): 5212–5217.

101. See U. Hasson et al., "Brain-to-Brain Coupling: A Mechanism for Creating and Sharing a Social World," *Trends in Cognitive Sciences* 16 (2012): 114–121.

102. C. Darwin, *The Descent of Man, and Selection in Relation to Sex*, 2 vols. (London: J. Murray, 1871).

103. M. L. Power and J. Schulkin, *The Evolution of the Human Placenta* (Baltimore, MD: Johns Hopkins University Press, 2012), chap. 7.

104. C. F. Zink et al., "Vasopressin Modulates Social Recognition-Related Activity in the Left Temporoparietal Junction in Humans," *Translational Psychiatry* 1 (2011): art. e3, https://doi.org/10.1038/tp.2011.2.

105. C. F. Zink and A. Meyer-Lindenberg, "Human Neuroimaging of Oxytocin and Vasopressin in Social Cognition," *Hormones and Behavior* 61 (2012): 400–409; and S. B. Algoe, L. E. Kurtz, and K. Grewen, "Oxytocin and Social Bonds: The Role of Oxytocin in Perceptions of Romantic Partners' Bonding Behavior," *Psychological Science* 28 (2017): 1763–1772.

106. E. Savaskan et al., "Post-Learning Intranasal Oxytocin Modulates Human Memory for Facial Identity," *Psychoneuroendocrinology* 33 (2008): 368–374.

107. P. J. Zak, R. Kurzban, and W. T. Matzner, "Oxytocin Is Associated with Human Trustworthiness," *Hormones and Behavior* 48 (2005): 522–527; and M. Kosfeld et al., "Oxytocin Increases Trust in Humans," *Science* 435 (2005): 673–676.

108. J. R. Keeler, "The Neurochemistry and Social Flow of Singing: Bonding and Oxytocin," *Frontiers in Human Neuroscience* 9 (2015): art. 518, https://doi.org/10.3389/fnhum.2015 .00518; and Y. Ooishi et al., "Increase in Salivary Oxytocin and Decrease in Salivary Cortisol After Listening to Relaxing Slow-Tempo and Exciting Fast-Tempo Music," *PLoS ONE* 12 (2017): art. e0189075, https://doi.org/10.1371%2Fjournal.pone.0189075.

109. Keeler, "Neurochemistry and Social Flow."

110. I. Gordon et al., "Oxytocin and the Development of Parenting in Humans," *Biological Psychiatry* 68 (2010): 377–382.

111. Dewey, *Art as Experience*, 246.

112. Dewey, *Art as Experience*, 44.

113. Dewey, *Experience and Nature*, 1925 ed., 67.

114. J. Dewey, *Experience and Nature*, 2nd ed. (1929; repr., New York: Dover, 1958), viii–ix. See also Crippen, "Art and Pragmatism"; M. Crippen, "Dewey, Enactivism and Greek Thought," in *Pragmatism and Embodied Cognitive Science: From Bodily Interaction to Symbolic Articulation*, ed. R. Madzia and M. Jung, 229–246 (Boston: De Gruyter, 2016); S. Vera and D. Schoeller, "Cognition as a Transformative Process: Re-affirming a

Classical Pragmatist Understanding," *European Journal of Pragmatism and American Philosophy* 10 (2018): 1–21, https://doi.org/10.4000/ejpap.1211.

115. J. Dewey, "The Reflex Arc Concept in Psychology," *Psychological Review* 3 (1896): 66–70. See also Dewey, *Experience and Nature*, 1925 ed., 379; and Dewey, *Art as Experience*, 37–38.

116. Dewey, *Reconstruction in Philosophy*, 2–3.

117. See Dewey, *Art as Experience*, 171.

118. L. de Bruin and S. de Haan, "Enactivism and Social Cognition: In Search of the Whole Story," *Journal of Cognitive Semiotics* 4 (2009): 225–250.

119. De Bruin and De Haan, "Enactivism and Social Cognition," 236–237.

120. Dewey, "Reflex Arc Concept"; and Dewey, *Reconstruction in Philosophy*, 86–87.

121. Meltzoff and Moore, "Explaining Facial Imitation."

122. Dewey, *Reconstruction in Philosophy*, 86.

123. Dewey, *Reconstruction in Philosophy*, 87.

124. For example, G. Lakoff and M. Johnson, *Metaphors We Live By* (Chicago: University of Chicago Press, 1980); G. Lakoff and M. Johnson, *Philosophy in the Flesh* (New York: Basic Books, 1999); M. Johnson, *The Body in the Mind: The Bodily Basis of Meaning, Imagination, and Reason* (Chicago: University of Chicago Press, 1987); and M. Johnson, *Embodied Mind, Meaning, and Reason: How Our Bodies Give Rise to Understanding* (Chicago: University of Chicago Press, 2017).

125. See De Bruin and De Haan, "Enactivism and Social Cognition," 238.

126. See Thompson, *Mind in Life*; and G. Colombetti, *The Feeling Body: Affective Science Meets the Enactive Mind* (Cambridge, MA: MIT Press, 2014).

127. Dewey, *Art as Experience*.

128. A. Noë, *Out of Our Heads: Why You Are Not Your Brain, and Other Lessons from the Biology of Consciousness* (New York: Hill and Wang, 2009), 31.

129. Trevarthen, "What Is It Like," 129.

130. M. Merleau-Ponty, *Consciousness and the Acquisition of Language*, trans. H. J. Silverman (Evanston, IL: Northwestern University Press, 1991; first published 1964 in French), 34.

131. Trevarthen, "What Is It Like," 129.

132. See E. Reed, *James J. Gibson and the Psychology of Perception* (New Haven, CT: Yale University Press, 1988); H. Heft, *Ecological Psychology in Context: James Gibson, Roger Barker, and the Legacy of William James's Radical Empiricism* (Mahwah, NJ: Lawrence Erlbaum Associates, 2001); and A. Chemero and S. Käufer, "Pragmatism, Phenomenology, and Extended Cognition," in *Pragmatism and Embodied Cognitive Science*, ed. Madzia and Jung, 55–70.

133. M. Crippen, "Pragmatism and the Valuative Mind," *Transactions of the Charles S. Peirce Society* 54 (2018): 341–360; and M. Crippen, "Aesthetics and Action: Situations, Emotional Perception and the Kuleshov Effect," *Synthese* [issue unassigned] (2019), https://doi.org /10.1007/s11229-019-02110-2.

134. For example, A. Still and J. Good, "The Ontology of Mutualism," *Ecological Psychology* 10 (1998): 39–63; F. Clément and L. Kaufmann, "How Culture Comes to Mind: From Social Affordances to Cultural Analogies," *Intellectica* 46 (2007): 221–250; J. Krueger, "Extended Cognition and the Space of Social Interaction," *Consciousness and Cognition* 20 (2011): 643–655; and M. Crippen, "Contours of Cairo Revolt: Semiology and Political Affordances in Street Discourses," *Topoi* [issue unassigned] (2019), https://doi.org/10.1007/s11245-019 -09650-9.

135. Kaplan and Kaplan, *Experience of Nature*; S. Kaplan, R. Kaplan, and J. S. Wendt, "Rated Preference and Complexity for Natural and Urban Visual Material," *Perception and Psychophysics* 12 (1972): 354–356; S. Kaplan, "Perception and Landscape: Conceptions and Misconceptions" [1979], in *Environmental Aesthetics*, ed. J. Nasar, 45–55 (New York:

Cambridge University Press, 1988); S. Kaplan, "Where Cognition and Affect Meet: A Theoretical Analysis of Preference" [1982], in *Environmental Aesthetics*, ed. Nasar, 56–63; S. Kaplan, "Aesthetics, Affect, and Cognition"; and S. Kaplan, "Environmental Preference."

136. M. Crippen, "Intuitive Cities: Pre-reflective, Aesthetic and Political Aspects of Urban Design," *Journal of Aesthetics and Phenomenology* 3 (2016): 125–145.

137. Kaplan and Kaplan, *Experience of Nature*, 58.

138. For example, Kaplan and Kaplan, *Experience of Nature*.

139. Shannon, "Mathematical Theory of Communication"; and Weaver, "Recent Contributions."

140. For example, Berlyne, "Novelty, Complexity, and Hedonic Values"; and Berlyne, "Ends and Means."

141. See S. Kaplan, "Aesthetics, Affect, and Cognition"; S. Kaplan, "Environmental Preferences"; and Kaplan and Kaplan, *Experience of Nature*.

142. Kaplan and Kaplan, *Experience of Nature*, 37.

143. See S. Kaplan, "Aesthetics, Affect, and Cognition"; and Kaplan and Kaplan, *Experience of Nature*, 32. See also I. Xenakis and A. Arnellos, "The Relation Between Interaction Aesthetics and Affordances," *Design Studies* 34 (2013): 57–73.

144. Krueger, "Extended Cognition."

145. See Krueger, "Extended Cognition," 650.

146. Still and Good, "Ontology of Mutualism," 56.

147. Krueger, "Extended Cognition."

148. Dewey's aesthetics stress the near-total engagement and the dramatic, culminating qualities that pull experience into something that stands out as a whole. This can happen with both the ugly and the beautiful. Dewey illustrated the point: "There is that storm one went through in crossing the Atlantic—the storm that seemed in its fury, as it was experienced, to sum up in itself all that a storm can be, complete in itself, standing out" and "marked out from what went before and what came after." *Art as Experience*, 36. For Dewey, "an experience has a unity that gives it its name, that meal, that storm, that rupture of friendship." *Art as Experience*, 37.

149. Krueger, "Extended Cognition," 652.

150. See Crippen, "Pragmatism and the Valuative Mind"; and M. Crippen, "Pragmatic Evolutions of the Kantian *a priori*: From the Mental to the Bodily," in *Pragmatist Kant: Pragmatism, Kant, and Kantianism in the Twenty-First Century*, ed. K. Skowroński and S. Pihlström, 19–40 (Helsinki: Nordic Pragmatism Network, 2019).

151. Crippen, "Art and Pragmatism"; Crippen, "Dewey, Enactivism and Greek Thought"; and M. Crippen, "Embodied Cognition and Perception: Dewey, Science and Skepticism," *Contemporary Pragmatism* 14 (2017): 121–134.

152. De Bruin and De Haan, "Enactivism and Social Cognition," 231.

4. PRAGMATISM AND AFFECTIVE COGNITION

1. D. Hume, *A Treatise of Human Nature*, ed. D. Norton and M. Norton (1740; New York: Oxford University Press, 2000).

2. F. Nietzsche, *Twilight of the Idols* [1888], trans. W. Kaufman, in *The Portable Nietzsche*, ed. W. Kaufman, 463–563 (New York: Penguin, 1954).

3. See Plato, *Republic*, trans. P. Shorey, in *Plato: The Collected Dialogues*, ed. E. Hamilton and H. Cairns (Princeton, NJ: Princeton University Press, 1963), bks. 2–4.

4. For example, W. James, "The Sentiment of Rationality" [1879], in *William James: Writings, 1878–1899*, ed. G. E. Myers, 950–985 (New York: Library of America, 1992).

5. For example, J. Dewey, *Reconstruction in Philosophy* (New York: Henry Holt, 1920); and J. Dewey, *Art as Experience* (New York: Minton, Balch, 1934).

6. C. S. Peirce, "Fixation of Belief," *Popular Science Monthly* 12 (1877): 1–15.

7. G. H. Mead, "The Relations of Psychology and Philology," *Psychological Bulletin* 1 (1904): 375–391; and G. H. Mead, "Social Psychology as Counterpart to Physiological Psychology," *Psychological Bulletin* 6 (1909): 401–408.

8. For example, G. Parrott and J. Schulkin, "Neuropsychology and the Cognitive Nature of the Emotions," *Cognition and Emotion* 7 (1993): 43–59; A. R. Damasio, *Descartes' Error: Emotion, Reason, and the Human Brain* (New York: G. P. Putnam, 1994); A. R. Damasio, *The Feeling of What Happens: Body and Emotion in the Making of Consciousness* (New York: Harcourt Brace, 1999); and A. R. Damasio, *Self Comes to Mind: Constructing the Conscious Brain* (New York: Pantheon Books, 2010).

9. M. Crippen, "Pragmatism and the Valuative Mind," *Transactions of the Charles S. Peirce Society* 54 (2018): 341–360.

10. Crippen, "Pragmatism and the Valuative Mind."

11. G. Gigerenzer, *Adaptive Thinking: Rationality in the Real World* (New York: Oxford University Press, 2000); and G. Gigerenzer, *Gut Feelings: The Intelligence of the Unconscious* (New York: Viking, 2007).

12. For example, K. J. Friston, "The Free-Energy Principle: A Rough Guide to the Brain?," *Trends in Cognitive Sciences* 13 (2009): 293–301; A. Clark, "Whatever Next? Predictive Brains, Situated Agents, and the Future of Cognitive Science," *Behavioral and Brain Sciences* 36 (2013): 181–204; A. Clark, *Surfing Uncertainty: Prediction, Action, and the Embodied Mind* (New York: Oxford University Press, 2015); and A. Linson et al., "The Active Inference Approach to Ecological Perception: General Information Dynamics for Natural and Artificial Embodied Cognition," *Frontiers in Robotics and AI* 5 (2018): art. 21, https://doi.org/10.3389/frobt.2018.00021.

13. For example, H. A. Simon, *The Sciences of the Artificial* (Cambridge, MA: MIT Press, 1969); and R. A. Rescorla and A. R. Wagner, "A Theory of Pavlovian Conditioning: Variations in the Effectiveness of Reinforcement and Nonreinforcement," in *Classical Conditioning II: Current Research and Theory*, ed. A. H. Black and W. Prokasy (New York: Appleton-Century-Crofts, 1972); and G. Loewenstein, "The Psychology of Curiosity: A Review and Reinterpretation," *Psychological Bulletin* 116 (1994): 75–98.

14. See J. Schulkin, *Pragmatism and the Search for Coherence in Neuroscience* (New York: Palgrave Macmillan, 2015).

15. W. James, *The Principles of Psychology* (New York: Henry Holt, 1890), 2:449–450.

16. James, *Principles of Psychology*, 2:440. See also W. James, "What Is an Emotion?," *Mind* 9 (1884): 190.

17. James, "What Is an Emotion?"

18. J. J. Prinz, *Gut Reactions: A Perceptual Theory of Emotion* (New York: Oxford University Press, 2004); J. J. Prinz, "Is Emotion a Form of Perception?," *Canadian Journal of Philosophy* 36 (2006): 137–160; Damasio, *Descartes' Error*; Damasio, *Feeling of What Happens*; and Damasio, *Self Comes to Mind*.

19. R. McDermott, "The Feeling of Emotionality: The Meaning of Neuroscientific Advances for Political Science," *Perspectives on Politics* 2 (2004): 691.

20. James, "What Is an Emotion?"

21. Crippen, "Pragmatism and the Valuative Mind." See also J. Schulkin, *Bodily Sensibility: Intelligent Action* (New York: Oxford University Press, 2004), chap. 1.

22. K. Koffka, *Principles of Gestalt Psychology* (New York: Harcourt, Brace, 1935), 326.

23. See D. K. Deady et al., "Examining the Effect of Spinal Cord Injury on Emotional Awareness, Expressivity and Memory for Emotional Material," *Psychology, Health and Medicine* 15 (2010): 406–419.

24. M. Crippen, "Digital Fabrication and Its Meaning for Film," in *Conceiving Virtuality: From Art to Technology*, ed. J. Braga (New York: Springer, 2019), 120–122; Crippen,

"Pragmatism and the Valuative Mind," 343–344; M. Merleau-Ponty, preface to *Phenomenology of Perception*, trans. C. Smith (New York: Routledge and Kegan Paul, 1962; first published 1945 in French).

25. James, "What Is an Emotion?," 189.

26. James, "What Is an Emotion?," 200.

27. See R. de Sousa, *The Rationality of Emotion* (Cambridge, MA: MIT Press, 1987).

28. See M. Gallagher and P. C. Holland, "The Amygdala Complex: Multiple Roles in Associative Learning and Emotion," *Proceedings of the National Academy of Sciences of the United States of America* 91 (1994): 11771–11776; J. B. Rosen and J. Schulkin, "From Normal Fear to Pathological Anxiety," *Psychological Review* 105 (1998): 325–350; J. Schulkin, B. L. Thompson, and J. B. Rosen, "Demythologizing the Emotions: Adaptation, Cognition, and Visceral Representations of Emotion in the Nervous System," *Brain and Cognition* 52 (2003): 15–23; and L. Pessoa, *The Cognitive-Emotional Brain: From Interactions to Integration* (Cambridge, MA: MIT Press, 2013), chap. 2.

29. Damasio, *Descartes' Error*, 70–72.

30. Damasio, *Feeling of What Happens*.

31. See A. Venkatraman, B. L. Edlow, and M. H. Immordino-Yang, "The Brainstem in Emotion: A Review," *Frontiers in Neuroanatomy* 11 (2017): art. 15, https://doi.org/10.3389/fnana.2017.00015.

32. Damasio, *Feeling of What Happens*, 273. See also L. Kühle, "William James and the Embodied Mind," *Contemporary Pragmatism* 14 (2017): 51–75.

33. For example, Damasio, *Descartes' Error*; Damasio, *Feeling of What Happens*; and R. Reisenzein, "What Is an Emotion in the Belief-Desire Theory of Emotion?," in *The Goals of Cognition: Essays in Honour of Cristiano Castelfranchi*, ed. F. Paglieri et al., 193–223 (London: College Publications, 2012).

34. For example, Dewey, *Experience and Nature* (Chicago: Open Court, 1925); Dewey, *Art as Experience*; Schulkin, Thompson, and Rosen, "Demythologizing the Emotions"; Pessoa, *Cognitive-Emotional Brain*; and M. Miceli and C. Castelfranchi, *Expectancy and Emotion* (New York: Oxford University Press, 2015).

35. See L. W. Swanson and G. D. Petrovich, "What Is the Amygdala?," *Trends in Neural Science* 21 (1998): 323–331; L. W. Swanson, "The Cerebral Hemisphere Regulation of Motivated Behavior," *Brain Research* 836 (2000): 113–164; and L. W. Swanson, "What Is the Brain?," *Trends in Neural Science* 23 (2000): 519–527.

36. See J. F. Thayer and R. D. Lane, "A Model of Neurovisceral Integration in Emotion Regulation and Dysregulation," *Journal of Affective Disorders* 61 (2000): 201–216; and Schulkin, *Bodily Sensibility*, chap. 2.

37. Damasio, *Feeling of What Happens*, 273–274.

38. Damasio, *Feeling of What Happens*, 274.

39. L. Gregory et al., "Cognitive Modulation of the Cerebral Processing of Human Oesophageal Sensation Using Functional Magnetic Resonance Imaging," *Gut* 52 (2003): 1671–1677; and E. Matthias et al., "On the Relationship Between Interoceptive Awareness and the Attentional Processing of Visual Stimuli," *International Journal of Psychophysiology* 72 (2009): 154–159.

40. Gigerenzer, *Gut Feelings*.

41. See M. Crippen, "William James on Belief: Turning Darwinism Against Empiricistic Skepticism," *Transactions of the Charles S. Peirce Society* 46 (2010): 477–502; and M. Crippen, "William James and His Darwinian Defense of Freewill," in *150 Years of Evolution: Darwin's Impact on Contemporary Thought and Culture*, ed. M. Wheeler, 68–89 (San Diego, CA: San Diego State University Press, 2011).

42. [W. James], review of *Lectures on the Elements of Comparative Anatomy*, by T. H. Huxley, *North American Review* 100 (1865): 290–298.

43. [W. James], review of "The Origin of Human Races and the Antiquity of Man Deduced from the Theory of 'Natural Selection,'" by A. R. Wallace, *North American Review* 101 (1865): 261–263.

44. A. R. Wallace, "The Origin of Human Races and the Antiquity of Man Deduced from the Theory of 'Natural Selection,'" *Journal of the Anthropological Society of London* 2 (1864): clviii–clxxxvii.

45. Wallace, "Origin of Human Races," clxix.

46. W. James, "Remarks on Spencer's Definition of Mind as Correspondence" [1878], in *William James: Writings, 1878–1899*, ed. G. E. Myers (New York: Library of America, 1992), 899.

47. James, "Spencer's Definition of Mind," 899.

48. See R. B. Perry, *The Thought and Character of William James, as Revealed in Unpublished Correspondence and Notes, Together with His Published Writings*, vol. 1 (Boston: Little, Brown, 1935), 323.

49. W. James, "Great Men and Their Environment" [1880], in *William James: Writings 1878–1899*, ed. G. E. Myers (New York: Library of America, 1992), 641.

50. See James, *Principles of Psychology*, 2:636.

51. W. James, "Brute and Human Intellect" [1878], in *William James: Writings 1878–1899*, ed. G. E. Myers, 910–949 (New York: Library of America, 1992); James, "Spencer's Definition of Mind"; W. James, "Are We Automata?," *Mind* 4 (1879): 1–22; James, "Sentiment of Rationality"; and James, *Principles of Psychology*, vol. 1, chaps. 5 and 11.

52. James, "Great Men," 620.

53. James, "Spencer's Definition of Mind," 921–922.

54. James, "Spencer's Definition of Mind," 929; James, *Principles of Psychology*, 1:402–403.

55. H. A. Simon, "Motivational and Emotional Controls of Cognition," *Psychological Review* 74 (1967): 29–39.

56. D. Moffat, "On the Positive Value of Affect" (paper presented at the AISB 2001 Symposium on Emotion, Cognition, and Affective Computing; York, UK, March 23–24, 2001).

57. W. James, "On Some Omissions of Introspective Psychology" [1884], in *William James: Writings 1878–1899*, ed. G. E. Myers, 986–1014 (New York: Library of America, 1992); and James, *Principles of Psychology*, vol. 1, chap. 9.

58. M. Johnson, *The Meaning of the Body: Aesthetics of Human Understanding* (Chicago: University of Chicago Press, 2007); and M. Johnson, "Keeping the Pragmatism in Neuropragmatism," in *Neuroscience, Neurophilosophy and Pragmatism: Brains at Work with the World*, ed. T. Solymosi and J. Shook, 37–56 (New York: Palgrave Macmillan, 2014).

59. J. L. Austin, "A Plea for Excuses," *Proceedings of the Aristotelian Society* 57 (1956–1957): 1–30.

60. Gigerenzer, *Gut Feelings*, 97–98.

61. Mead, "Relations of Psychology"; and Mead, "Social Psychology."

62. C. S. Peirce, "How to Make Our Ideas Clear," *Popular Science Monthly* 12 (1878): 286–302.

63. James, "Sentiment of Rationality."

64. Gigerenzer, *Gut Feelings*.

65. A. R. Luria, *The Mind of a Mnemonist: A Little Book About a Vast Memory*, trans. L. Solotaroff (New York: Basic Books, 1968).

66. James, *Principles of Psychology*, 1:680, quoted in Gigerenzer, *Gut Feelings*, 23.

67. James, *Principles of Psychology*, 1:680.

68. James, *Principles of Psychology*, vol. 1, chap. 16; Luria, *Mind of a Mnemonist*; and Gigerenzer, *Gut Feelings*.

69. See Pessoa, *Cognitive-Emotional Brain*, chap. 3.

70. S. B. Hamann et al., "Amygdala Activity Related to Enhanced Memory for Pleasant and Aversive Stimuli," *Nature Neuroscience* 2 (1999): 289–294. See also M. G. Packard, L. Cahill,

and J. L. McGaugh, "Amygdala Modulation of Hippocampal-Dependent and Caudate Nucleus-Dependent Memory Processes," *Proceedings of the National Academy of Sciences of the United States of America* 91 (1994): 8477–8481; and E. A. Phelps, "Human Emotion and Memory: Interactions of the Amygdala and Hippocampal Complex," *Current Opinion in Neurobiology* 14 (2004): 198–202.

71. Damasio, *Descartes' Error*, 197.
72. Crippen, "Pragmatism and the Valuative Mind."
73. Damasio, *Descartes' Error*.
74. See Damasio, *Descartes' Error*, 73.
75. See Crippen, "Pragmatism and the Valuative Mind."
76. Damasio, *Descartes' Error*, 51.
77. Damasio, *Descartes' Error*, 70.
78. For example, Parrott and Schulkin, "Neuropsychology"; Schulkin, Thompson, and Rosen, "Demythologizing the Emotions"; and Pessoa, *Cognitive-Emotional Brain*.
79. Pessoa, *Cognitive-Emotional Brain*. See also J. Aggleton, *The Amygdala* (New York: Oxford University Press, 2000), chap. 1.
80. Pessoa, *Cognitive-Emotional Brain*, 38.
81. Pessoa, *Cognitive-Emotional Brain*, 38.
82. See R. Kaplan and S. Kaplan, *The Experience of Nature: A Psychological Perspective* (New York: Cambridge University Press, 1989).
83. Dewey, *Reconstruction in Philosophy*, 143.
84. W. Weaver, "Recent Contributions to the Mathematical Theory of Communication," in *The Mathematical Theory of Communication*, ed. C. E. Shannon and W. Weaver (Chicago: University of Illinois Press, 1949), 11.
85. Dewey, *Art as Experience*, 15.
86. Miceli and Castelfranchi, *Expectancy and Emotion*, 47–58.
87. Kaplan and Kaplan, *Experience of Nature*, chap. 2.
88. See C. E. Shannon, "The Mathematical Theory of Communication," in *Mathematical Theory of Communication*, ed. Shannon and Weaver, 25–115; Weaver, "Recent Contributions"; and Kaplan and Kaplan, *Experience of Nature*.
89. Aristotle, *Rhetorica*, trans. W. R. Roberts, in *The Basic Works of Aristotle*, ed. R. McKeon (New York: Random House, 1941), 2.5.1382a19–27.
90. James, *Principles of Psychology*, 1:8.
91. D. C. Dennett, *Consciousness Explained* (New York: Little, Brown, 1991), 177.
92. H. Helmholtz, *Handbook of Physiological Optics* (1867; repr., New York: Dover, 1963); Rescorla and Wagner, "Theory of Pavlovian Conditioning"; Clark, *Surfing Uncertainty*; and Friston, "Free-Energy Principle."
93. A. Bechara et al., "Deciding Advantageously Before Knowing the Advantageous Strategy," *Science* 275 (1997): 1293–1295.
94. Bechara et al., "Deciding Advantageously," 1294.
95. Loewenstein, "Psychology of Curiosity."
96. See also De Sousa, *Rationality of Emotion*.
97. See Schulkin, *Pragmatism*, 22.
98. Loewenstein, "Psychology of Curiosity."
99. Dewey, *Reconstruction in Philosophy*, 142.
100. Dewey, *Reconstruction in Philosophy*, 142.
101. J. Dewey, *Affective Thought* [1926], in *Philosophy and Civilization* (New York: Minton, Balch, 1931), 121.
102. Rescorla and Wagner, "Theory of Pavlovian Conditioning."
103. Peirce, "Fixation of Belief"; and Peirce, "How to Make Our Ideas Clear."
104. Peirce, "Fixation of Belief," 6.

105. See R. Menary, *Cognitive Integration: Mind and Cognition Unbounded* (New York: Palgrave Macmillan, 2007).
106. Peirce, "How to Make Our Ideas Clear," 154.
107. Peirce, "How to Make Our Ideas Clear," 150.
108. Peirce, "How to Make Our Ideas Clear," 154.
109. Peirce, "How to Make Our Ideas Clear," 150.
110. J. Dewey, "The Theory of Emotions II: The Significance of Emotions," *Psychological Review* 2 (1895): 13–32.
111. Dewey, "Theory of Emotions II," 13.
112. Dewey, "Theory of Emotions II," 15.
113. Mead, "Relations of Psychology"; and Mead, "Social Psychology."
114. L. G. Ward and R. Throop, "The Dewey-Mead Analysis of Emotions," *Social Science Journal* 26 (1989): 465–479.
115. See R. C. Solomon, "Emotions and Choice," *Review of Metaphysics* 27 (1973): 20–41; and R. C. Solomon, *True to Our Feelings: What Our Emotions Are Really Telling Us* (New York: Oxford University Press, 2007).
116. F. Strack, L. L. Martin, and S. Stepper, "Inhibiting and Facilitating Conditions of the Human Smile: A Nonobtrusive Test of the Facial Feedback Hypothesis," *Journal of Personality and Social Psychology* 54 (1988): 768–777; J. I. Davis, A. Senghas, and K. N. Ochsner, "How Does Facial Feedback Modulate Emotional Experience?," *Journal of Research in Personality* 43 (2009): 822–829; and J. I. Davis et al., "The Effects of Botox Injections on Emotional Experience," *Emotion* 10 (2010): 433–440.
117. See E. J. Wagenmakers et al., "Registered Replication Report," *Perspectives on Psychological Science* 11 (2016): 917–928.
118. See W. James, "Rationality, Activity and Faith," *Princeton Review* 2 (1882): 74–75 [most of this article is incorporated into James, "The Sentiment of Rationality"]; W. James, "The Will to Believe" [1896], in *William James: Writings, 1878–1899*, ed. G. E. Myers, 457–479 (New York: Library of America, 1992); and W. James, *A Pluralistic Universe* [1909], in *William James: Writings, 1902–1910*, ed. B. Kuklick (New York: Library of America, 1987), 773–780.
119. James, "Rationality, Activity and Faith"; James, "Will to Believe"; James, *Pluralistic Universe*. See also Crippen, "William James on Belief"; and M. Crippen, "Art and Pragmatism: James and Dewey on the Reconstructive Presuppositions of Experience" (PhD diss., York University, 2010).
120. See S. Hawking, ed., *On the Shoulders of Giants: The Great Works of Physics and Astronomy* (Philadelphia: Running Press, 2004), chap. 3; and D. Love, "Who Was Johannes Kepler?," *Astronomy and Geophysics* 50 (2009): 6.15–6.17.
121. Crippen, "Art and Pragmatism."
122. See Dewey, *Reconstruction in Philosophy*; and J. Dewey, *The Quest for Certainty* (New York: Minton, Balch, 1929).
123. See Aristotle, *De Anima*, trans. J. A. Smith, in *Basic Works of Aristotle*, ed. McKeon, 533–603, bk. 2.
124. G. Lakoff and M. Johnson, *Philosophy in the Flesh* (New York: Basic Books, 1999); and M. Johnson, *Embodied Mind, Meaning, and Reason: How Our Bodies Give Rise to Understanding* (Chicago: University of Chicago Press, 2017).
125. Crippen, "Art and Pragmatism"; and M. Crippen, "Embodied Cognition and Perception: Dewey, Science and Skepticism," *Contemporary Pragmatism* 14 (2017): 121–134.
126. See Lakoff and Johnson, *Philosophy in the Flesh*; Loewenstein, "Psychology of Curiosity"; and Johnson, *Embodied Mind*.
127. See Miceli and Castelfranchi, *Expectancy and Emotion*, 85–88.
128. R. Pelagio-Flores et al., "Serotonin, a Tryptophan-Derived Signal Conserved in Plants and Animals, Regulates Root System Architecture Probably Acting as a Natural Auxin

Inhibitor in *Arabidopsis thaliana*," *Plant and Cell Physiology* 52 (2011): 504. See also R. Madzia, "Root-Brains: The Frontiers of Cognition in the Light of John Dewey's Philosophy of Nature," *Contemporary Pragmatism* 14 (2017): 93–111.

129. Pelagio-Flores et al., "Serotonin."

130. C. T. Robbins et al., "Role of Tannins in Defending Plants Against Ruminants: Reduction in Dry Matter Digestion?," *Ecology* 68 (1987): 1606–1615; P. Yam, "Acacia Trees Kill Antelope in the Transvaal," *Scientific American* 263 (1990): 28; M. Heil, "Indirect Defence via Tritrophic Interactions," *New Phytologist* 178 (2008): 41–61; C. J. Frost et al., "Plant Defense Priming Against Herbivores: Getting Ready for a Different Battle," *Plant Physiology* 146 (2008): 818–824; G. I. Arimura, K. Matsui, and J. Takabayashi, "Chemical and Molecular Ecology of Herbivore-Induced Plant Volatiles: Proximate Factors and Their Ultimate Functions," *Plant and Cell Physiology* 50 (2009): 911–923; C. R. Rodriguez-Saona, L. E. Rodriguez-Saona, and C. J. Frost, "Herbivore-Induced Volatiles in the Perennial Shrub, *Vaccinium corymbosum*, and Their Role in Inter-branch Signaling," *Journal of Chemical Ecology* 35 (2009): 163–175; A. R. War et al., "Mechanisms of Plant Defense Against Insect Herbivores," *Plant Signaling and Behavior* 7 (2012): 1306–1320; and P. Wohlleben, *The Hidden Life of Trees—What They Feel, How They Communicate: Discoveries from a Secret World*, trans. J. Billinghurst (Vancouver, BC: Greystone Books, 2016).

131. P. D. Coley, "Costs and Benefits of Defense by Tannins in a Neotropical Tree," *Oecologia* 70 (1986): 238–241.

132. Yam, "Acacia Trees"; and Wohlleben, *Hidden Life of Trees*.

133. H. Spencer, *The Principles of Psychology* (1855; Westmead, UK: Gregg International, 1970), §127–173.

134. See M. Ben-Attia et al., "Blooming Rhythms of Cactus *Cereus peruvianus* with Nocturnal Peak at Full Moon During Seasons of Prolonged Daytime Photoperiod," *Chronobiology International* 33 (2016): 419–430.

135. See K. L. Gamble et al., "Circadian Clock Control of Endocrine Factors," *Nature Reviews Endocrinology* 10 (2014): 466–475.

136. M. C. Olianas et al., "Corticotropin-Releasing Hormone Stimulates Adenylyl Cyclase Activity in the Retinas of Different Animal Species," *Regulatory Peptides* 47 (1993): 127–132; and M. C. Olianas and P. Onali, "G Protein-Coupled Corticotropin-Releasing Hormone Receptors in Rat Retina," *Regulatory Peptides* 56 (1995): 61–70.

137. See C. P. Richter, *Biological Clocks in Medicine and Psychiatry* (Springfield, IL: Charles C. Thomas, 1965), chaps. 2–4; Olianas et al., "Corticotropin-Releasing Hormone"; Olianas and Onali, "G Protein-Coupled"; A. S. Fisk et al., "Light and Cognition: Roles for Circadian Rhythms, Sleep, and Arousal," *Frontiers in Neurology* 9 (2018): art. 56, https://doi.org/10.3389/fneur.2018.00056.

138. D. W. Pfaff, I. Phillips, and R. T. Rubin, *Principles of Hormone/Behavior Relations* (Boston: Academic Press, 2004), 57.

139. S. Peciña, J. Schulkin, and K. C. Berridge, "Nucleus Accumbens Corticotropin-Releasing Factor Increases Cue-Triggered Motivation for Sucrose Reward: Paradoxical Positive Incentive Effects in Stress?," *BMC Biology* 4 (2006): art, 8. https://doi.org/10.1186/1741-7007-4-8.

140. See Fisk et al., "Light and Cognition."

141. D. W. King et al., "Confirmatory Factor Analysis of the Clinician-Administered PTSD Scale: Evidence for the Dimensionality of Posttraumatic Stress Disorder," *Psychological Assessment* 10 (1998): 90–96; S. B. Norman, M. B. Stein, and J. R. Davidson, "Profiling Posttraumatic Functional Impairment," *Journal of Nervous and Mental Disease* 195 (2007): 48–53; T. Jovanovic and S. D. Norrholm, "Neural Mechanisms of Impaired Fear Inhibition in Posttraumatic Stress Disorder," *Frontiers in Behavioral Neuroscience* 5 (2011): art. 44, https://doi.org/10.3389/fnbeh.2011.00044; and B. E. Depue et al., "Reduced Amygdala

Volume Is Associated with Deficits in Inhibitory Control: A Voxel- and Surface-Based Morphometric Analysis of Comorbid PTSD/Mild TBI," *BioMed Research International* (2014): art. 691505, https://doi.org/10.1155/2014/691505.

142. G. B. Raglan, L. A. Schmidt, and J. Schulkin, "The Role of Glucocorticoids and Corticotropin-Releasing Hormone Regulation on Anxiety Symptoms and Response to Treatment," *Endocrine Connections* 6 (2017): R1–R7.

143. J. Heller, *Catch-22* (New York: Simon and Schuster, 1961), 165.

144. N. E. Rosenthal et al., "Seasonal Affective Disorder: A Description of the Syndrome and Preliminary Findings with Light Therapy," *Archives of General Psychiatry* 41 (1984): 72–80; and A. Magnusson and T. Partonen, "The Diagnosis, Symptomatology, and Epidemiology of Seasonal Affective Disorder," *CNS Spectrums* 10 (2005): 625–634.

145. See L. Mealey and P. Theis, "The Relationship Between Mood and Preferences Among Natural Landscapes: An Evolutionary Perspective," *Ethology and Sociobiology* 16 (1995): 247–256.

146. C. R. Riener et al., "An Effect of Mood on the Perception of Geographical Slant," *Cognition and Emotion* 25 (2011): 174–182.

147. C. R. Gallistel, *The Organization of Learning* (Cambridge, MA: MIT Press, 1992).

148. Crippen, "Art and Pragmatism"; and Crippen, "Embodied Cognition and Perception."

149. Gigerenzer, *Adaptive Thinking*; and Gigerenzer, *Gut Feelings*.

150. See Schulkin, Thompson, and Rosen, "Demythologizing the Emotions"; and Crippen, "Pragmatism and the Valuative Mind."

151. Nietzsche, *Twilight of the Idols*, 478.

152. Gigerenzer, *Adaptive Thinking*; and Gigerenzer, *Gut Feelings*.

153. See M. Crippen, "Asleep at the Press: Thoreau, Egyptian Revolt and Nuances of Democracy," *Arab Media and Society*, no. 20 (2015), https://www.arabmediasociety.com/wp-content/uploads/2017/12/20150302085925_Crippen_AsleepAtThePress.pdf; and M. Crippen, "Egypt and the Middle East: Democracy, Anti-Democracy and Pragmatic Faith," *Saint Louis University Public Law Review* 35 (2016): 281–302.

154. T. Solymosi, "Dewey on the Brain: Dopamine, Digital Devices, and Democracy," *Contemporary Pragmatism* 14 (2017): 5–34.

155. Crippen, "Pragmatism and the Valuative Mind"; M. Crippen, "Pragmatic Evolutions of the Kantian *a priori*: From the Mental to the Bodily," in *Pragmatist Kant: Pragmatism, Kant, and Kantianism in the Twenty-First Century*, ed. K. Skowroński and S. Pihlström, 19–40 (Helsinki: Nordic Pragmatism Network, 2019); H. Jackman, "William James's Naturalistic Account of Concepts and His 'Rejection of Logic,'" in *Philosophy of Mind in the Nineteenth Century*, ed. S. Lapointe, 133–146 (New York: Routledge, 2018).

156. Crippen, "Pragmatism and the Valuative Mind."

157. See Schulkin, *Pragmatism*, chap. 1.

158. See James, *Pragmatism*, chaps. 5–6.

159. Crippen, "William James on Belief"; and Crippen, "Art and Pragmatism."

160. James, *Pragmatism*, chap. 6.

161. H. Putnam, *The Collapse of the Fact/Value Dichotomy and Other Essays* (Cambridge, MA: Harvard University Press, 2002), 30–31. See also J. Dewey, *Theory of Valuation* (Chicago: University of Chicago Press, 1939); R. C. Neville, *The Cosmology of Freedom* (New Haven, CT: Yale University Press, 1974); and D. Weissman, *Truth's Debt to Value* (New Haven, CT: Yale University Press, 1993).

162. James, "Rationality, Activity and Faith"; and James, "Will to Believe."

163. See Crippen, "Pragmatism and the Valuative Mind"; and M. Crippen, "Aesthetics and Action: Situations, Emotional Perception and the Kuleshov Effect," *Synthese* [issue unassigned] (2019), https://doi.org/10.1007/s11229-019-02110-2.

164. W. Craig, "Appetites and Aversions as Constituents of Instinct," *Biological Bulletin* 34 (1918): 91–107; and Dewey, *Experience and Nature*, 1925 ed.

5. PERCEPTION, AFFECT, WORLD

1. See J. J. Gibson, *The Ecological Approach to Visual Perception* (Boston: Houghton Mifflin, 1979), chap. 8; B. H. Hodges and R. M. Baron, "Values as Constraints on Affordances: Perceiving and Acting Properly," *Journal for the Theory of Social Behaviour* 22 (1992): 263–294; M. Crippen, "Intuitive Cities: Pre-reflective, Aesthetic and Political Aspects of Urban Design," *Journal of Aesthetics and Phenomenology* 3 (2016): 125–145; M. Crippen, "Pragmatism and the Valuative Mind," *Transactions of the Charles S. Peirce Society* 54 (2018): 341–360; M. Crippen, "Aesthetics and Action: Situations, Emotional Perception and the Kuleshov Effect," *Synthese* [issue unassigned] (2019), https://doi.org/10.1007/s11229-019 -02110-2; M. Crippen, "Contours of Cairo Revolt: Semiology and Political Affordances in Street Discourses," *Topoi* [issue unassigned] (2019), https://doi.org/10.1007/s11245-019 -09650-9; and J. Schulkin, *Pragmatism and the Search for Coherence in Neuroscience* (New York: Palgrave Macmillan, 2015).

2. See throughout J. Schulkin, *Bodily Sensibility: Intelligent Action* (New York: Oxford University Press, 2004).

3. J. Dewey, *Reconstruction in Philosophy* (New York: Henry Holt, 1920).

4. Gibson, *Ecological Approach*.

5. See Crippen, "Aesthetics and Action."

6. For example, J. Dewey, *Art as Experience* (New York: Minton, Balch, 1934), chap. 1.

7. See M. Crippen, "Embodied Cognition and Perception: Dewey, Science and Skepticism," *Contemporary Pragmatism* 14 (2017): 121–134.

8. J. Dewey, *Experience and Nature* (Chicago: Open Court, 1925), chap. 7.

9. A. R. Luria, *The Mind of a Mnemonist: A Little Book About a Vast Memory*, trans. L. Solotaroff (New York: Basic Books, 1968).

10. W. James, "Remarks on Spencer's Definition of Mind as Correspondence" [1878], in *William James: Writings, 1878–1899*, ed. G. E. Myers (New York: Library of America, 1992).

11. Crippen, "Pragmatism and the Valuative Mind"; and Crippen, "Aesthetics and Action."

12. W. James, *The Principles of Psychology* (New York: Henry Holt, 1890), 2:335–336.

13. Crippen, "Pragmatism and the Valuative Mind."

14. K. Koffka, *Principles of Gestalt Psychology* (New York: Harcourt, Brace, 1935), 345.

15. C. M. Brendl, A. B. Markman, and C. Messner, "The Devaluation Effect: Activating a Need Devalues Unrelated Objects," *Journal of Consumer Research* 29 (2003): 463–473; and M. Veltkamp, H. Aarts, and R. Custers, "Perception in the Service of Goal Pursuit: Motivation to Attain Goals Enhances the Perceived Size of Goal-Instrumental Objects," *Social Cognition* 26 (2008): 720–736.

16. See M. I. Posner and S. E. Peterson, "The Attention System of the Human Brain," *Annual Review of Neuroscience* 13 (1990): 25–42; R. Desimone and J. Duncan, "Neural Mechanisms of Selective Visual Attention," *Annual Review of Neuroscience* 18 (1995): 193–222; S. E. Peterson and M. I. Posner, "The Attention System of the Human Brain: 20 Years After," *Annual Review of Neuroscience* 35 (2012): 73–89; and L. Pessoa, *The Cognitive-Emotional Brain: From Interactions to Integration* (Cambridge, MA: MIT Press, 2013), chap. 6.

17. See Gibson, *Ecological Approach*, chap. 8.

18. For example, H. Werner, "On Physiognomic Modes of Perception and Their Experimental Investigation" [1927], in *Developmental Processes: Heinz Werner's Selected Writings*, ed. S. S. Barten and M. B. Franklin, vol. 1, *General Theory and Perceptual Experience*, 149–152 (New York: International University Press, 1978).

19. Gibson, *Ecological Approach*, 138.
20. Gibson, *Ecological Approach*, 138. Bracketed ellipses added.
21. Gibson, *Ecological Approach*, chap. 8.
22. See Crippen, "Aesthetics and Action."
23. N. Frijda, *The Emotions* (New York: Cambridge University Press, 1986), 188.
24. M. Heidegger, *Being and Time*, trans. J. Macquarrie and E. Robinson (New York: Harper and Row, 1962; first published 1927 in German); M. Merleau-Ponty, "Film and the New Psychology" [1947], in *Sense and Non-Sense*, trans. H. Dreyfus and P. Dreyfus, 43–59 (Evanston, IL: Northwestern University Press, 1964); J.-P. Sartre, *The Emotions: Outline of a Theory*, trans. B. Frechtman (New York: Carol Publishing Group, 1993; first published 1948 in French).
25. Koffka, *Principles of Gestalt Psychology*, 326; and G. Böhme, "Atmosphere as the Fundamental Concept of a New Aesthetic," *Thesis Eleven* 36 (1993): 113–126.
26. Dewey, *Art as Experience*, 67.
27. Merleau-Ponty, "Film."
28. Dewey, *Art as Experience*.
29. Dewey, *Art as Experience*, 42.
30. C. S. Peirce, "Some Consequences of Four Incapacities," *Journal of Speculative Philosophy* 2 (1868): 150.
31. H. Dreyfus, *Being-in-the-World: A Commentary on Heidegger's* Being and Time, *Division I* (Cambridge, MA: MIT Press, 2003), 171.
32. Crippen, "Aesthetics and Action."
33. Frjida, *Emotions*.
34. H. Schmitz, R. O. Müllan, and J. Slaby, "Emotions Outside the Box—the New Phenomenology of Feeling and Corporeality," *Phenomenology and the Cognitive Sciences* 10 (2011): 241–259.
35. Schmitz, Müllan, and Slaby, "Emotions Outside the Box," 246.
36. J. Slaby, "Emotions and the Extended Mind," in *Collective Emotions: Perspectives from Psychology, Philosophy, and Sociology*, ed. C. von Scheve and M. Salmela (New York: Oxford University Press, 2014), 36.
37. Slaby, "Emotions and the Extended Mind," 36.
38. See Böhme, "Atmosphere."
39. J. J. Prinz, *Gut Reactions: A Perceptual Theory of Emotion* (New York: Oxford University Press, 2004), 235.
40. Crippen, "Aesthetics and Action."
41. Dewey, *Art as Experience*.
42. Dewey, *Art as Experience*, 177.
43. Gibson, *Ecological Approach*, 129.
44. Dewey, *Art as Experience*, 250–251.
45. See Crippen, "Aesthetics and Action."
46. C. R. Riener et al., "An Effect of Mood on the Perception of Geographical Slant," *Cognition and Emotion* 25 (2011): 174–182.
47. See N. Frijda, "The Laws of Emotion," *American Psychologist* 43 (1988): 349–358; N. Frijda, "Emotion Experiences and Its Varieties," *Emotion Review* 1 (2009): 264–271; and N. Frijda, "Impulsive Action and Motivation," *Biological Psychology* 84 (2010): 570–579.
48. R. Brooks, *Cambrian Intelligence: The Early History of the New AI* (Cambridge, MA: MIT Press, 1999), viii.
49. Brooks, *Cambrian Intelligence*, chaps. 7–8.
50. Brooks, *Cambrian Intelligence*, 115.
51. Brooks, *Cambrian Intelligence*, 115.

52. J. Long, *Darwin's Devices: What Evolving Robots Can Teach Us About the History of Life and the Future of Technology* (New York: Basic Books, 2011).
53. Merleau-Ponty, "Film."
54. Prinz, *Gut Reactions*.
55. Prinz, *Gut Reactions*, 61.
56. C. Nussbaum, *The Musical Representation: Meaning, Ontology, and Emotion* (Cambridge, MA: MIT Press, 2007), 193.
57. See Crippen, "Aesthetics and Action."
58. Merleau-Ponty, "Film."
59. Frijda, *Emotions*, 188.
60. Dewey, *Art as Experience*, 177.
61. Dewey, *Experience and Nature*, 1925 ed., 259.
62. Gibson, *Ecological Approach*, 129.
63. See G. Parrott and J. Schulkin, "Neuropsychology and the Cognitive Nature of the Emotions," *Cognition and Emotion* 7 (1993): 43–59.
64. J. E. LeDoux, "Cognitive-Emotional Interactions in the Brain," *Cognition and Emotion* 3 (1989): 267–289.
65. See J. E. LeDoux, *Anxious: Using the Brain to Understand and Treat Fear and Anxiety* (New York: Viking, 2015).
66. Gibson, *Ecological Approach*, 142.
67. See Crippen, "Aesthetics and Action."
68. Dewey, *Art as Experience*, 16.
69. Dewey, *Art as Experience*, 209.
70. Dewey, *Experience and Nature*, 1925 ed., 42.
71. Dewey, *Experience and Nature*, 1925 ed., 42.
72. Gibson, *Ecological Approach*.
73. Dewey, *Experience and Nature*, 1925 ed., 42.
74. Dewey, *Art as Experience*, 16.
75. See Crippen, "Pragmatism and the Valuative Mind."
76. Gibson, *Ecological Approach*, 140.
77. See A. G. Taylor et al., "Top-Down and Bottom-Up Mechanisms in Mind-Body Medicine: Development of an Integrative Framework for Psychophysiological Research," *Explore* 6 (2010): 29–41; H. D. Critchley and N. Harrison, "Visceral Influences on Brain and Behavior," *Neuron* 77 (2013): 624–638; M. Mather and J. F. Thayer, "How Heart Rate Variability Affects Emotion Regulation Brain Networks," *Current Opinion in Behavioral Sciences* 19 (2018): 98–104; and D. Azzallini, I. Rebollo, and C. Tallon-Baudry, "Visceral Signals Shape Brain Dynamics and Cognition," *Trends in Cognitive Sciences* 23 (2019): 488–509.
78. See L. Mealey and P. Theis, "The Relationship Between Mood and Preferences Among Natural Landscapes: An Evolutionary Perspective," *Ethology and Sociobiology* 16 (1995): 247–256; and D. R. Proffitt et al., "Perceiving Geographical Slant," *Psychonomic Bulletin and Review* 16 (1995): 970–972.
79. For example, C. R. Reid et al., "Slime Mold Uses an Externalized Spatial 'Memory' to Navigate in Complex Environments," *Proceedings of the Natural Academy of Sciences of the United States of America* 109 (2012): 17490–17494.
80. See Reid et al., "Slime Mold."
81. T. Latty and M. Beekman, "Food Quality Affects Search Strategy in the Acellular Slime Mould, *Physarum polycephalum*," *Behavioral Ecology* 20 (2009): 1160–1167.
82. Reid et al., "Slime Mold," 17490. See also T. Ueda, T. Hirose, and Y. Kobatake, "Membrane Biophysics of Chemoreception and Taxis in the Plasmodium of *Physarum polycephalum*," *Biophysical Chemistry* 11 (1980): 461–473; and Latty and Beekman, "Food Quality."

83. Reid et al., "Slime Mold," 17490. See also Ueda, Hirose, and Kobatake, "Membrane Biophysics."

84. E. Guttes and S. Guttes, "Mitotic Synchrony in the Plasmodia of *Physarum polycephalum* and Mitotic Synchronization by Coalescence of Microplasmodia," *Methods in Cell Biology* 1 (1964): 43–54.

85. T. Nakagaki, H. Yamada, and Á. Tóth, "Maze-Solving by an Amoeboid Organism," *Nature* 407 (2000): 470.

86. For example, M. J. Carlile, "Nutrition and Chemotaxis in the Myxomycete *Physarum polycephalum*: The Effect of Carbohydrates on the Plasmodium," *Journal of General Microbiology* 63 (1970): 221–226; I. Chet, A. Naveh, and Y. Henis, "Chemotaxis of *Physarum polycephalum* Towards Carbohydrates, Amino Acids and Nucleotides," *Journal of General Microbiology* 102 (1977): 145–148; R. L. Kincaid and T. E. Mansour, "Chemotaxis Toward Carbohydrates and Amino Acids in *Physarum polycephalum*," *Experimental Cell Research* 116 (1978): 377–385; and A. Dussutour et al., "Amoeboid Organism Solves Complex Nutritional Challenges," *Proceedings of the National Academy of Sciences of the United States of America* 107 (2010): 4607–4611.

87. Nakagaki, Yamada, and Tóth, "Maze-Solving"; T. Nakagaki, H. Yamada, and Á. Tóth, "Path Finding by Tube Morphogenesis in an Amoeboid Organism," *Biophysical Chemistry* 92 (2001): 47–52; T. Nakagaki et al., "Obtaining Multiple Separate Food Sources: Behavioural Intelligence in the *Physarum* Plasmodium," *Proceedings of the Royal Society B: Biological Sciences* 27 (2004): 2305–2310; T. Nakagaki et al., "Minimum-Risk Path Finding by an Adaptive Amoebal Network," *Physical Review Letters* 99 (2007): art. 068104, https://doi.org/10.1103/PhysRevLett.99.068104; and T. Nakagaki, H. Yamada, and M. Hara, "Smart Network Solutions in an Amoeboid Organism," *Biophysical Chemistry* 107 (2004): 1–5.

88. Reid et al., "Slime Mold"; C. R. Reid et al., "Amoeboid Organism Uses Extracellular Secretions to Make Smart Foraging Decisions," *Behavioral Ecology* 24 (2013): 812–818.

89. Crippen, "Aesthetics and Action."

90. N. R. Franks and C. R. Fletcher, "Spatial Patterns in Army Ant Foraging and Migration: *Eciton burchelli* on Barro Colorado Island, Panama," *Behavioral Ecology and Sociobiology* 12 (1983): 261–270; T. Nakashima, M. Teshiba, and Y. Hirose, "Flexible Use of Patch Marks in an Insect Predator: Effect of Sex, Hunger State, and Patch Quality," *Ecological Entomology* 27 (2002): 581–587; and S. K. Willson et al., "Spatial Movement Optimization in Amazonian *Eciton burchellii* Army Ants," *Insectes Sociaux: International Journal for the Study of Social Arthropods* 58 (2011): 325–334.

91. C. Bernstein and G. Driessen, "Patch-Marking and Optimal Search Patterns in the Parasitoid *Venturia canescens*," *Journal of Animal Ecology* 65 (1996): 211–219; and T. S. Hoffmeister and B. D. Roitberg, "To Mark the Host or the Patch: Decisions of a Parasitoid Searching for Concealed Host Larvae," *Evolutionary Ecology* 11 (1997): 145–168.

92. J. D. Henry, "The Use of Urine Marking in Scavenging Behavior of Red Fox (*Vulpes vulpes*)," *Behaviour* 61 (1977): 82–106; F. H. Harrington, "Urine-Marking and Caching Behavior in the Wolf," *Behaviour* 76 (1981): 280–288; L. Devenport, J. Devenport, and C. Kokesh, "The Role of Urine Marking in the Foraging Behaviour of Least Chipmunks," *Animal Behaviour* 57 (1999): 557–563. See also Reid et al., "Slime Mold"; and Reid et al., "Amoeboid Organism."

93. A. Clark and D. J. Chalmers, "The Extended Mind," *Analysis* 58 (1998): 7–19. See Crippen, "Aesthetics and Action."

94. For example, R. Kaplan and S. Kaplan, *The Experience of Nature: A Psychological Perspective* (New York: Cambridge University Press, 1989).

95. R. S. Ulrich, "Aesthetic and Affective Response to Natural Environment," in *Behavior and the Natural Environment*, ed. I. Altman and J. F. Wohlwill (New York: Plenum, 1983).

96. M. Johnson, *Embodied Mind, Meaning, and Reason: How Our Bodies Give Rise to Understanding* (Chicago: University of Chicago Press, 2017), 162.

97. R. Gibbs, "Embodied Experience and Linguistic Meaning," *Brain and Language* 84 (2003): 9.

98. Critchley and Harrison, "Visceral Influences."

99. Critchley and Harrison, "Visceral Influences," 625.

100. Critchley and Harrison, "Visceral Influences," 625. See also W. W. Blessing, *The Lower Brainstem and Bodily Homeostasis* (New York: Oxford University Press, 1997); L. E. Goehler et al., "Vagal Immune-to-Brain Communication: A Visceral Chemosensory Pathway," *Autonomic Neuroscience* 85 (2000): 49–59; and C. B. Saper, "The Central Autonomic Nervous System: Conscious Visceral Perception and Autonomic Pattern Generation," *Annual Review of Neuroscience* 25 (2002): 433–469.

101. R. Norgren, "Gustatory System," in *The Rat Nervous System*, ed. G. Paxinos, 751–771 (San Diego, CA: Academic Press, 1995); and T. L. Powley et al., "Gastrointestinal Projection Maps of the Vagus Nerve Are Specified Permanently in the Perinatal Period," *Developmental Brain Research* 129 (2001): 570–572.

102. See Schulkin, *Bodily Sensibility*.

103. J. T. Fitzsimmons, "Angiotensin, Thirst, and Sodium Appetite," *Physiological Review* 78 (1999): 583–686; D. A. Denton, M. J. McKinley, and R. S. Wessinger, "Hypothalamic Integration of Body Fluid Regulation," *Proceedings of the National Academy of Sciences of the United States of America* 93 (1996): 7397–7404; D. A. Denton et al., "Neuroimaging of Genesis and Satiation of Thirst and an Interoceptor-Driven Theory of Origins of Primary Consciousness," *Proceedings of the National Academy of Sciences of the United States of America* 96 (1999): 5304–5309. See also Schulkin, *Bodily Sensibility*.

104. For example, J. Rzoska, "Bait Shyness, a Study in Rat Behaviour," *British Journal of Animal Behaviour* 1 (1953): 128–135; J. Garcia, W. G. Hankins, and K. W. Rusiniak, "Behavioral Regulation of the Milieu Interne in Man and Rat," *Science* 185 (1974): 824–831; and P. Rozin, "The Selection of Foods by Rats, Humans, and Other Animals," in *Advances in the Study of Behavior*, vol. 6, ed. J.S. Rosenblatt et al. (New York: Academic Press, 1976).

105. Norgren, "Gustatory System"; L. W. Swanson and G. D. Petrovich, "What Is the Amygdala?," *Trends in Neural Science* 21 (1998): 323–331; and G. D. Petrovich, N. W. Canteras, and L. W. Swanson, "Combinatorial Amygdalar Inputs to Hippocampal Domains and the Hypothalamic Behavior Systems," *Brain Research Reviews* 38 (2001): 247–289.

106. For example, F. Bermúdez-Rattoni et al., "Flavor-Illness Aversions: The Role of the Amygdala in the Acquisition of Taste-Potentiated Odor Aversions," *Physiology and Behavior* 38 (1986): 503–508; A. C. Spector, R. Norgren, and H. J. Grill, "Parabrachial Gustatory Lesions Impair Taste Aversion Learning in Rats," *Behavioral Neuroscience* 106 (1992): 147–161; A. C. Spector, "Gustatory Function in Parabrachial Nuclei: Implications from Lesion Studies," *Revolutions in Neuroscience* 6 (1995): 143–175; and K. Touzani and A. Sclafani, "Critical Role of Amygdala in Flavor but Not Taste Preference Learning in Rats," *European Journal of Neuroscience* 22 (2005): 1767–1774.

107. D. H. Zald and J. V. Pardo, "Emotion, Olfaction, and the Human Amygdala: Amygdala Activation During Aversive Olfactory Stimulation," *Proceedings of the National Academy of Sciences of the United States of America* 94 (1997): 4119–4124.

108. See J. M. Fuster, *The Prefrontal Cortex* (New York: Elsevier, 2008).

109. Swanson and Petrovich, "What Is the Amygdala?"; L. W. Swanson, "The Cerebral Hemisphere Regulation of Motivated Behavior," *Brain Research* 836 (2000): 113–164; L. W. Swanson, *Brain Architecture* (New York: Oxford University Press, 2003).

110. J. Herbert, "Peptides in the Limbic System: Neurochemical Codes for Co-ordinated Adaptive Responses to Behavioral and Physiological Demand," *Progress in Neurobiology* 41 (1993): 723–791.

111. See R. Leproult et al., "Transition from Dim to Bright Light in the Morning Induces an Immediate Elevation of Cortisol Levels," *Journal of Clinical Endocrinology and Metabolism* 86 (2001): 151–157; L. L. Wellman, L. Yang, and L. D. Sanford, "Effects of Corticotropin Releasing Factor (CRF) on Sleep and Temperature Following Predictable Controllable and Uncontrollable Stress in Mice," *Frontiers in Neuroscience* 9 (2015): art. 258, https://doi.org/10.3389/fnins.2015.00258; and M. Chatoo et al., "Involvement of Corticotropin-Releasing Factor and Receptors in Immune Cells in Irritable Bowel Syndrome," *Frontiers in Endocrinology* 9 (2018): art. 21, https://doi.org/10.3389/fendo.2018 .00021.

112. See A. N. Epstein, J. T. Fitzsimons, and B. J. Simons, "Drinking Caused by the Intracranial Injection of Angiotensin into the Rat," *Journal of Physiology* 200 (1969): 98P–100P; A. N. Epstein, J. T. Fitzsimons, and B. J. Rolls, "Drinking Induced by Injection of Angiotensin into the Brain of the Rat," *Journal of Physiology* 210 (1970): 457–474; D. A. Denton, *The Hunger of Salt: An Anthropological, Physiological and Medical Analysis* (Berlin: Springer-Verlag, 1982); and Fitzsimons, "Angiotensin, Thirst."

113. I. P. Pavlov, *The Work of the Digestive Glands* (London: Charles Griffin, 1902); and I. P. Pavlov, *Lectures on Conditioned Reflexes* (1927; New York: International Publishing, 1960).

114. See T. L. Powley, "The Ventromedial Hypothalamic Syndrome, Satiety, and a Cephalic Phase Hypothesis," *Psychological Review* 84 (1977): 89–126; and T. L. Powley, "Vagal Input to the Enteric Nervous System," *Gut* 47, supp. IV (2000): iv30–iv32.

115. For example, S. C. Woods, R. A. Hutton, and W. Makous, "Conditioned Insulin Secretion in the Albino Rat," *Proceedings of the Society of Experimental Biology and Medicine* 133 (1970): 965–968; D. W. Pfaff, I. Phillips, and R. T. Rubin, *Principles of Hormone/ Behavior Relations* (Boston: Academic Press, 2004), 57; and M. L. Power and J. Schulkin, *The Evolution of Obesity* (Baltimore, MD: Johns Hopkins University Press, 2009), chap. 9.

116. Gibson, *Ecological Approach*.

117. Norgren, "Gustatory System."

118. Parrott and Schulkin, "Neuropsychology." For a general overview, see J. Schulkin, *Sodium Hunger: The Search for a Salty Taste* (New York: Cambridge University Press, 1991).

119. See J. Schulkin, "Behavior of Sodium-Deficient Rats: The Search for a Salty Taste," *Journal of Comparative and Physiological Psychology* 96 (1982): 628–634.

120. R. J. Contreras, "Changes in Gustatory Nerve Discharges with Sodium Deficiency: A Single Unit Analysis," *Brain Research* 121 (1977): 373–378.

121. K. C. Berridge et al., "Sodium Depletion Enhances Salt Palatability in Rats," *Behavioral Neuroscience* 98 (1984): 652–660.

122. See Crippen, "Aesthetics and Action."

123. Gibson, *Ecological Approach*.

124. J. Gibson, "The Theory of Affordances," in *Perceiving, Acting, and Knowing: Toward an Ecological Psychology*, ed. R. Shaw and J. Bransford (Hillsdale, NJ: Lawrence Erlbaum, 1977).

125. Gibson, *Ecological Approach*, 253.

126. Kaplan and Kaplan, *Experience of Nature*, 32.

127. Kaplan and Kaplan, *Experience of Nature*, 32.

128. For example, Proffitt et al., "Perceiving Geographical Slant"; M. Bhalla and D. R. Proffitt, "Visual-Motor Recalibration in Geographical Slant Perception," *Journal of Experimental Psychology: Human Perception and Performance* 25 (1999): 1076–1096; and D. R. Proffitt, "Embodied Perception and the Economy of Action," *Perspectives on Psychological Science* 1 (2006): 110–122.

129. For example, Brooks, *Cambrian Intelligence*, chap. 8; R. Pfeifer and J. Bongard, *How the Body Shapes the Way We Think: A New View of Intelligence* (Cambridge, MA: MIT Press, 2007), chaps. 6 and 12; M. L. Anderson, *After Phrenology: Neural Reuse and the Interactive Brain* (Cambridge, MA: MIT Press, 2014), chaps. 1 and 2; and M. M. de Wit et al., "Affordances and Neuroscience: Steps Towards a Successful Marriage," *Neuroscience and Biobehavioral Reviews* 80 (2017): 622–629.

130. J. G. White et al., "The Structure of the Nervous System of the Nematode *Caenorhabditis elegans*," *Philosophical Transactions of the Royal Society B: Biological Sciences* 314 (1986): 1–340.

131. Z. F. Altun and D. H. Hall, "Nervous System, General Description," in *WormAtlas*, ed. L. A. Herndon, 2011, last revised June 19, 2013, http://www.wormatlas.org/hermaphrodite /nervous/mainframe.htm. See also Anderson, *After Phrenology*.

132. See C. J. Cela-Conde et al., "Activation of the Prefrontal Cortex in the Human Visual Aesthetic Perception," *Proceedings of the National Academy of Sciences of the United States of America* 101 (2004): 6321–6325; H. Kawabata and S. Zeki, "Neural Correlates of Beauty," *Journal of Neurophysiology* 91 (2004): 1699–1705; and O. Vartanian and V. Goel, "Neuroanatomical Correlates of Aesthetic Preference for Paintings," *NeuroReport* 15 (2004): 893–897.

133. Anderson, *After Phrenology*.

134. De Wit et al., "Affordances and Neuroscience."

135. Gibson, *Ecological Approach*, 129.

136. A. Chemero, "Against Smallism and Localism" (paper presented at *Thinking with Hands, Eyes and Things: Trends in Interdisciplinary Studies; An International Avant-Conference*, Torun, Poland, November 8, 2013).

137. Long, *Darwin's Devices*.

6. BROADENING ECOLOGIES

1. M. Merleau-Ponty, *Phenomenology of Perception*, trans. C. Smith (New York: Routledge and Kegan Paul, 1962; first published 1945 in French).

2. A. Ait-Belgnaoui et al., "Prevention of Gut Leakiness by a Probiotic Treatment Leads to Attenuated HPA Response to an Acute Psychological Stress in Rats," *Psychoneuroendocrinology* 37 (2012): 1885–1895; S. C. Bischoff et al., "Intestinal Permeability—a New Target for Disease Prevention and Therapy," *BMC Gastroenterology* 14 (2014): art. 189, https:// doi.org/10.1186/s12876-014-0189-7; J. A. Foster and K. A. McVey Neufeld, "Gut-Brain Axis: How the Microbiome Influences Anxiety and Depression," *Trends in Neuroscience* 36 (2013): 305–312; M. Carabotti et al., "The Gut-Brain Axis: Interactions Between Enteric Microbiota, Central and Enteric Nervous Systems," *Annals of Gastroenterology* 28 (2015): 203–209; and J. T. K. Ho, G. C. F. Chan, and J. C. B. Li, "Systemic Effects of Gut Microbiota and Its Relationship with Disease and Modulation," *BMC Immunology* 16 (2015): art. 21, https://doi.org/10.1186/s12865-015-0083-2.

3. E. G. Severance, R. H. Yolken, and W. W. Eaton, "Autoimmune Diseases, Gastrointestinal Disorders and the Microbiome in Schizophrenia: More Than a Gut Feeling," *Schizophrenia Research* 176 (2016): 23–35.

4. See M. Clapp et al., "Gut Microbiota's Effect on Mental Health: The Gut-Brain Axis," *Clinics and Practice* 7 (2017): 131–136.

5. See Clapp et al., "Gut Microbiota's Effect."

6. R. Sender, S. Fuchs, and R. Milo, "Revised Estimates for the Number of Human and Bacteria Cells in the Body," *PLoS Biology* 14 (2016): art. e1002533, https://doi.org/10.1371 /journal.pbio.1002533.

7. P. B. Eckburg et al., "Diversity of the Human Intestinal Microbial Flora," *Science* 308 (2005): 1635–1638.

8. See L. V. Hooper, T. Midtvedt, and J. I. Gordon, "How Host-Microbial Interactions Shape the Nutrient Environment of the Mammalian Intestine," *Annual Review of Nutrition* 22 (2002): 283–307.

9. M. Lyte, "Probiotics Function Mechanistically as Delivery Vehicles for Neuroactive Compounds: Microbial Endocrinology in the Design and Use of Probiotics," *Bioessays* 33 (2011): 574–581; and T. G. Dinan, C. Stanton, and J. F. Cryan, "Psychobiotics: A Novel Class of Psychotropic," *Biological Psychiatry* 74 (2013): 720–726.

10. See T. Jenkins et al., "Influence of Tryptophan and Serotonin on Mood and Cognition with a Possible Role of the Gut-Brain Axis," *Nutrients* 8 (2016): 1–15.

11. Clapp et al., "Gut Microbiota's Effect."

12. See J. F. Cryan and T. G. Dinan, "Mind-Altering Microorganisms: The Impact of the Gut Microbiota on Brain and Behaviour," *Nature Reviews Neuroscience* 13 (2012): 701–712; and Clapp et al., "Gut Microbiota's Effect."

13. See P. M. Smith and W. S. Garrett, "The Gut Microbiota and Mucosal T Cells," *Frontiers in Microbiology* 2 (2011): art. 111, https://doi.org/10.3389/fmicb.2011.00111.

14. For example, N. Sudo et al., "Postnatal Microbial Colonization Programs the Hypothalamic-Pituitary-Adrenal System for Stress Response in Mice," *Journal of Physiology* 558 (2004): 263–275.

15. See Ait-Belgnaoui et al., "Prevention of Gut Leakiness"; C. J. Smith et al., "Probiotics Normalize the Gut-Brain-Microbiota Axis in Immunodeficient Mice," *American Journal of Physiology: Gastrointestinal and Liver Physiology* 307 (2014): G793–G802; M. Daulatzai, "Non-celiac Gluten Sensitivity Triggers Gut Dysbiosis, Neuroinflammation, Gut-Brain Axis Dysfunction, and Vulnerability for Dementia," *CNS and Neurological Disorders* 14 (2015): 110–131.

16. For example, D. Benton, C. Williams, and A. Brown, "Impact of Consuming a Milk Drink Containing a Probiotic on Mood and Cognition," *European Journal of Clinical Nutrition* 61 (2006): 355–361; and M. I. Pinto-Sanchez et al., "Probiotic *Bifidobacterium longum* NCC3001 Reduces Depression Scores and Alters Brain Activity: A Pilot Study in Patients with Irritable Bowel Syndrome," *Gastroenterology* 153 (2017): 448–459.

17. R. Huang, K. Wang, and J. Hu, "Effect of Probiotics on Depression: A Systematic Review and Meta-Analysis of Randomized Controlled Trials," *Nutrients* 8 (2016): art. 483, https://doi.org/10.3390/nu8080483.

18. Benton, Williams, and Brown, "Impact of Consuming a Milk Drink." See also Clapp et al., "Gut Microbiota's Effect."

19. Clapp et al., "Gut Microbiota's Effect," 134. See also M. Messaoudi et al., "Assessment of Psychotropic-like Properties of a Probiotic Formulation (*Lactobacillus helveticus* R0052 and *Bifidobacterium longum* R0175) in Rats and Human Subjects," *British Journal of Nutrition* 105 (2010): 755–764; K. Schmidt et al., "Prebiotic Intake Reduces the Waking Cortisol Response and Alters Emotional Bias in Healthy Volunteers," *Psychopharmacology* 232 (2015): 1793–1801; and L. Steenbergen et al., "A Randomized Controlled Trial to Test the Effect of Multispecies Probiotics on Cognitive Reactivity to Sad Mood," *Brain, Behavior, and Immunity* 48 (2015): 258–264.

20. H. M. Savignac, "*Bifidobacteria* Exert Strain-Specific Effects on Stress-Related Behavior and Physiology in BALB/c Mice," *Neurogastroenterology and Motility* 26 (2014): 1615–1627.

21. F. Strati et al., "New Evidences on the Altered Gut Microbiota in Autism Spectrum Disorders," *Microbiome* 5 (2017): art. 24, https://doi.org/10.1186/s40168-017-0242-1.

22. C. J. Burrus, "A Biochemical Rationale for the Interaction Between Gastrointestinal Yeast and Autism," *Medical Hypotheses* 79 (2012): 784–785; I. D. Iliev et al., "Interactions Between Commensal Fungi and the C-type Lectin Receptor Dectin-1 Influence Colitis," *Science*

336 (2012): 1314–1317; and C. Luan et al., "Dysbiosis of Fungal Microbiota in the Intestinal Mucosa of Patients with Colorectal Adenomas," *Scientific Reports* 5 (2015): art. 7980, https://doi.org/10.1038/srep07980.

23. A. S. Kantarcioglu, N. Kiraz, and A. Aydin, "Microbiota-Gut-Brain Axis: Yeast Species Isolated from Stool Samples of Children with Suspected or Diagnosed Autism Spectrum Disorders and In Vitro Susceptibility Against Nystatin and Fluconazole," *Mycopathologia* 181 (2016): 1–7.

24. Clapp et al., "Gut Microbiota's Effect."

25. Burrus, "Biochemical Rationale."

26. J. P. Webster, C. F. Brunton, and D. W. MacDonald, "Effect of *Toxoplasma gondii* Upon Neophobic Behaviour in Wild Brown Rats, *Rattus norvegicus*," *Parasitology* 108 (1994): 407–411.

27. J. P. Webster, "The Effect of *Toxoplasma gondii* on Animal Behavior: Playing Cat and Mouse," *Schizophrenia Bulletin* 33 (2007): 752–756.

28. H. Hodková, P. Kodym, and J. Flegr, "Poorer Results of Mice with Latent Toxoplasmosis in Learning Tests: Impaired Learning Processes or the Novelty Discrimination Mechanism?," *Parasitology* 134 (2007): 1329–1337; and Webster, "Effect of *Toxoplasma gondii* on Animal Behavior."

29. A. Vyas et al., "Behavioral Changes Induced by Toxoplasma Infection of Rodents Are Highly Specific to Aversion of Cat Odors," *Proceedings of the National Academy of Sciences of the United States of America* 104 (2007): 6442–6447; G. Kannan et al., "*Toxoplasma gondii* Strain-Dependent Effects on Mouse Behaviour," *Folia Parasitol* 57 (2010): 151–155; and W. M. Ingram et al., "Mice Infected with Low-Virulence Strains of *Toxoplasma gondii* Lose Their Innate Aversion to Cat Urine, Even After Extensive Parasite Clearance," *PLoS ONE* 8 (2013): art. e75246, https://doi.org/10.1371/journal.pone.0075246.

30. P. A. Witting, "Learning Capacity and Memory of Normal and *Toxoplasma*-Infected Laboratory Rats and Mice," *Zeitschrift für Parasitenkunde* 61 (1979): 29–51.

31. C. P. Richter, "Experimentally Produced Behavioral Reactions to Food Poisoning in Wild and Domesticated Rats," *Annals of the New York Academy of Sciences* 56 (1953): 225–239.

32. M. Berdoy, J. P. Webster, and D. W. Macdonald, "Fatal Attraction in Rats Infected with *Toxoplasma gondii*," *Proceedings of the Royal Society B: Biological Sciences* 267 (2000): 1591–1594; A. Vyas et al., "Behavioral Changes"; A. Vyas, S.-K. Kim, and R. M. Sapolsky, "The Effects of Toxoplasma Infection on Rodent Behaviors Are Dependent on Dose of the Stimulus," *Neuroscience* 148 (2007): 342–348; and P. Lamberton, C. Donnelly, and J. Webster, "Specificity of the *Toxoplasma gondii*–Altered Behaviour to Definitive Versus Nondefinitive Host Predation Risk," *Parasitology* 135 (2008): 1143–1150.

33. J. Flegr et al., "Fatal Attraction Phenomenon in Humans: Cat Odour Attractiveness Increased for *Toxoplasma*-Infected Men While Decreased for Infected Women," *PLoS Neglected Tropical Diseases* 5 (2011): art. e1389, https://doi.org/10.1371%2Fjournal.pntd.0001389.

34. J. Flegr et al., "Increased Risk of Traffic Accidents in Subjects with Latent Toxoplasmosis: A Retrospective Case-Control Study," *BMC Infectious Diseases* 2 (2002): art. 11, http://www.biomedcentral.com/1471-2334/2/11; J. Flegr et al., "Increased Incidence of Traffic Accidents in *Toxoplasma*-Infected Military Drivers and Protective Effect RhD Molecule Revealed by a Large-Scale Prospective Cohort Study," *BMC Infectious Diseases* 9 (2009): art. 72, https://doi.org/10.1186/1471-2334-9-72; and B. Kocazeybek et al., "Higher Prevalence of Toxoplasmosis in Victims of Traffic Accidents Suggest Increased Risk of Traffic Accident in *Toxoplasma*-Infected Inhabitants of Istanbul and Its Suburbs," *Forensic Science International* 187 (2009): 103–108.

35. Flegr et al., "Increased Incidence of Traffic Accidents."

36. J. Havlíček et al., "Decrease of Psychomotor Performance in Subjects with Latent 'Asymptomatic' Toxoplasmosis," *Parasitology* 122 (2001): 515–520.

37. H. Hodokvá et al., "Higher Perceived Dominance in *Toxoplasma* Infected Men—a New Evidence for Role of Increased Level of Testosterone in Toxoplasmosis-Associated Changes in Human Behavior," *Neuroendocrinology Letters* 28 (2007): 110–114; and J. Flegr, J. Lindová, and P. Kodym, "Sex-Dependent Toxoplasmosis-Associated Differences in Testosterone Concentration in Humans," *Parasitology* 135 (2008): 427–431.

38. E. F. Coccaro et al., "*Toxoplasma gondii* Infection: Relationship with Aggression in Psychiatric Subjects," *Journal of Clinical Psychiatry* 77 (2016): 334–341.

39. E. F. Torrey and R. H. Yolken, "*Toxoplasma gondii* and Schizophrenia," *Emerging Infectious Diseases* 9 (2003): 1375–1380; E. F. Torrey et al., "Antibodies to *Toxoplasma gondii* in Patients with Schizophrenia: A Meta-Analysis," *Schizophrenia Bulletin* 33 (2007): 729–736; T. Çelik et al., "Is There a Relationship Between *Taxoplasma gondii* Infection and Idiopathic Parkinson's Disease?," *Scandinavian Journal of Infectious Diseases* 42 (2010): 604–608; A. Fekadu, T. Shibre, and A. J. Cleare, "Toxoplasmosis as a Cause for Behaviour Disorders—Overview of Evidence and Mechanisms," *Folia Parasitol* 57 (2010): 105–113; O. Miman et al., "The Probable Relation Between *Toxoplasma gondii* and Parkinson's Disease," *Neuroscience Letters* 475 (2010): 129–131; J. Prandota, "Autism Spectrum Disorders May be Due to Cerebral Toxoplasmosis Associated with Chronic Neuroinflammation Causing Persistent Hypercytokinemia That Resulted in an Increased Lipid Peroxidation, Oxidative Stress, and Depressed Metabolism of Endogenous and Exogenous Substances," *Research in Autism Spectrum Disorders* 4 (2010): 119–155; O. Y. Kusbeci et al., "Could *Toxoplasma gondii* Have Any Role in Alzheimer Disease?," *Alzheimer Disease and Associated Disorders* 25 (2011): 1–3; and L. Torres et al., "*Toxoplasma gondii* Alters NMDAR Signaling and Induces Signs of Alzheimer's Disease in Wild-Type, C57BL/6 Mice," *Journal of Neuroinflammation* 15 (2018): art. 57, https://doi.org/10.1186/s12974-018-1086-8.

40. B.-K. Jung et al., "*Toxoplasma gondii* Infection in the Brain Inhibits Neuronal Degeneration and Learning and Memory Impairments in a Murine Model of Alzheimer's Disease," *PLoS ONE* 7 (2012): art. e33312, https://doi.org/10.1371%2Fjournal.pone.0033312.

41. T. A. Arling et al., "*Toxoplasma gondii* Antibody Titers and History of Suicide Attempts in Patients with Recurrent Mood Disorders," *Journal of Nervous and Mental Disease* 197 (2009): 905–908; and V. J. Ling et al., "*Toxoplasma gondii* Seropositivity and Suicide Rates in Women," *Journal of Nervous and Mental Disease* 199 (2011): 440–444.

42. G. Enders, *Gut: The Inside Story of Our Body's Most Underrated Organ*, trans. D. Shaw (Vancouver, BC: Greystone Books, 2015).

43. Enders, *Gut*, 125.

44. See Y. K. Cho, C. Li, and D. V. Smith, "Gustatory Projections from the Nucleus of the Solitary Tract to the Parabrachial Nuclei in the Hamster," *Chemical Senses* 27 (2002): 81–90; S. J. McDougall, J. H. Peters, and M. C. Andresen, "Convergence of Cranial Visceral Afferents Within the Solitary Tract Nucleus," *Journal of Neuroscience* 29 (2009): 12886–12895; D. B. Zoccal et al., "The Nucleus of the Solitary Tract and the Coordination of Respiratory and Sympathetic Activities," *Frontiers in Physiology* 5 (2014): art. 238, https://doi.org/10.3389/fphys.2014.00238; and E. Frangos, J. Ellrich, and B. R. Komisaruk, "Non-invasive Access to the Vagus Nerve Central Projections via Electrical Stimulation of the External Ear: fMRI Evidence in Humans," *Brain Stimulation* 8 (2015): 624–636.

45. See P. J. Davern, "A Role for the Lateral Parabrachial Nucleus in Cardiovascular Function and Fluid Homeostasis," *Frontiers in Physiology* 5 (2014): art. 436, https://doi.org/10.3389/fphys.2014.00436; K. C. Berridge and M. Kringelbach, "Pleasure Systems in the Brain," *Neuron* 86 (2015): 646–664; A. C. Spector, "Behavioral Analyses of Taste Function and Ingestion in Rodent Models," *Physiology and Behavior* 152 (2015): 516–526; and T. Yahiro et al., "The Lateral Parabrachial Nucleus, but Not the Thalamus, Mediates

Thermosensory Pathways for Behavioural Thermoregulation," *Scientific Reports* 7 (2017): art. 5031, https://doi.org/10.1038/s41598-017-05327-8.

46. For example, G. Aston-Jones and J. D. Cohen, "An Integrative Theory of the Locus Coeruleus-Norepinephrine Function: Adaptive Gland and Optimal Performance," *Annual Review of Neuroscience* 28 (2005): 403–450; and C. Cirelli et al., "Locus Ceruleus Control of Slow-Wave Homeostasis," *Journal of Neuroscience* 25 (2005): 4503–4511.

47. R. M. Buijs and C. G. van Eden, "The Integration of Stress by the Hypothalamus, Amygdala and Prefrontal Cortex: Balance Between the Autonomic Nervous System and the Neuroendocrine System," *Progress in Brain Research* 126 (2000): 117–132; C. G. van Eden and R. M. Buijs, "Functional Neuroanatomy of the Prefrontal Cortex: Autonomic Interactions," *Progress in Brain Research* 126 (2000): 49–62; V. RajMohan and E. Mohandas, "The Limbic System," *Indian Journal of Psychiatry* 49 (2007): 132–139; and K. Hwang et al., "The Human Thalamus Is an Integrative Hub for Functional Brain Networks," *Journal of Neuroscience* 37 (2017): 5594–5607.

48. See C. Guilleminault, M. Quera-Salva, and M. P. Goldberg, "Pseudo-hypersomnia and Pre-sleep Behaviour with Bilateral Paramedian Thalamic Lesions," *Brain* 116 (1993): 1549–1563; N. D. Schiff, "Central Thalamic Contributions to Arousal Regulation and Neurological Disorders of Consciousness," *Annals of the New York Academy of Sciences* 1129 (2008): 105–118; and Y. B. Saalmann and S. Kastner, "The Cognitive Thalamus," *Frontiers in Systems Neuroscience* 9 (2015): art. 39, https://doi.org/10.3389/fnsys.2015.00039.

49. See RajMohan and Mohandas, "Limbic System."

50. See C. B. Saper, T. C. Chou, and T. E. Scammell, "The Sleep Switch: Hypothalamic Control of Sleep and Wakefulness," *Trends in Neurosciences* 24 (2001): 726–731; and J. Biran et al., "Role of Developmental Factors in Hypothalamic Function," *Frontiers in Neuroanatomy* 9 (2015): art. 47, https://doi.org/10.3389/fnana.2015.00047.

51. F. W. Flynn et al., "Central Gustatory Lesions: II. Effects on Sodium Appetite, Taste Aversion Learning, and Feeding Behaviors," *Behavioral Neuroscience* 105 (1991): 944–954; A. C. Spector, R. Norgren, and H. J. Grill, "Parabrachial Gustatory Lesions Impair Taste Aversion Learning in Rats," *Behavioral Neuroscience* 106 (1992): 147–161; H. D. Critchley, C. J. Mathias, and R. J. Dolan, "Neural Activity in the Human Brain Relating to Uncertainty and Arousal During Anticipation," *Neuron* 29 (2001): 537–545; and H. D. Critchley, C. J. Mathias, and R. J. Dolan, "Neuroanatomical Basis for First- and Second-Order Representations of Bodily States," *Nature Neuroscience* 4 (2001): 207–212.

52. R. Norgren, "Gustatory System," in *The Rat Nervous System*, ed. G. Paxinos, 751–771 (San Diego, CA: Academic Press, 1995); and C. B. Saper and R. L. Stornetta, "Central Autonomic System," in *The Rat Nervous System*, 4th ed., ed. G. Paxinos, 629–673 (New York: Academic Press, 2015).

53. See K. Harris et al., "Is the Gut Microbiota a New Factor Contributing to Obesity and Its Metabolic Disorders?," *Journal of Obesity* 2012: art. 879151, https://doi.org/10.1155/2012/879151; J. Breton et al., "Gut Commensal *E. coli* Proteins Activate Host Satiety Pathways Following Nutrient-Induced Bacterial Growth," *Cell Metabolism* 23 (2016): 324–334; and S. O. Fetissov, "Role of the Gut Microbiota in Host Appetite Control: Bacterial Growth to Animal Feeding Behaviour," *Nature Reviews Endocrinology* 13 (2017): 11–25.

54. G. Gainotti and C. Marra, "Differential Contribution of Right and Left Temporooccipital and Anterior Lesions to Face Recognition Disorders," *Frontiers in Human Neuroscience* 5 (2011): art. 55, https://doi.org/10.3389/fnhum.2011.00055.

55. See M. Riddoch and G. Humphreys, "Visual Agnosia," *Clinical Neurology* 21 (2003): 501–520; and M. J. E. van Zandvoort, T. Nijboer, and E. Dehaan, "Developmental Colour Agnosia," *Cortex* 43 (2007): 750–757.

56. For example, M. L. Anderson, *After Phrenology: Neural Reuse and the Interactive Brain* (Cambridge, MA: MIT Press, 2014); and A. Kind, "Twentieth-Century Philosophy of

Mind: Themes, Problems, and Scientific Context," in *Philosophy of Mind in the Twentieth and Twenty-First Centuries*, ed. A. Kind, 1–20 (New York: Routledge, 2019).

57. See J. Schulkin, *Bodily Sensibility: Intelligent Action* (New York: Oxford University Press, 2004); J. Schulkin, *Reflections on the Musical Mind: An Evolutionary Perspective* (Princeton, NJ: Princeton University Press, 2013); J. Schulkin, *Sport: A Biological, Philosophical, and Cultural Perspective* (New York: Columbia University Press, 2016); and Anderson, *After Phrenology*.

58. Anderson, *After Phrenology*.

59. See V. S. Ramachandran and D. Rogers-Ramachandran, "Phantom Limbs and Neural Plasticity," *Archives of Neurology* 57 (2000): 317–320.

60. For example, Anderson, *After Phrenology*, chaps. 1 and 7.

61. C. Darwin, *The Descent of Man, and Selection in Relation to Sex*, vol. 2 (London: J. Murray, 1871), chap. 14.

62. See D. Timmann et al., "The Human Cerebellum Contributes to Motor, Emotional and Cognitive Associative Learning: A Review," *Cortex* 46 (2010): 845–857.

63. A. Noë, *Out of Our Heads: Why You Are Not Your Brain, and Other Lessons from the Biology of Consciousness* (New York: Hill and Wang, 2009).

64. A. Martin, "The Organization of Semantic Knowledge and the Origin of Words in the Brain," in *The Origins and Diversification of Language*, ed. N. G. Jablonski and L. C. Aiello, 69–88 (San Francisco: California Academy of Sciences, 1998).

65. V. S. Ramachandran and W. Hirstein, "The Perception of Phantom Limbs: The D. O. Hebb Lecture," *Brain* 121 (1998): 1603–1630; and V. S. Ramachandran and E. Altschuler, "The Use of Visual Feedback, in Particular Mirror Visual Feedback, in Restoring Brain Function," *Brain* 132 (2009): 1693–1710.

66. Merleau-Ponty, *Phenomenology of Perception*, chaps. 2–3.

67. W. James, "Does 'Consciousness' Exist?," *Journal of Philosophy, Psychology and Scientific Methods* 1 (1904): 477–491.

68. Aristotle, *De Anima*, trans. J. A. Smith, in *The Basic Works of Aristotle*, ed. R. McKeon (New York: Random House, 1941), 412a.

69. Aristotle, *De Anima*, 412a–414a.

70. W. James, "What Is an Emotion?," *Mind* 9 (1884): 188–205; and W. James, *The Principles of Psychology* (New York: Henry Holt, 1890), vol. 2, chap. 25.

71. M. C. Nussbaum, *The Therapy of Desire* (Princeton, NJ: Princeton University Press, 1994); M. C. Nussbaum, *Upheavals of Thought: The Intelligence of Emotion* (Cambridge: Cambridge University Press, 2001); and R. C. Solomon, *Love* (New York: Prometheus, 1990).

72. See E. Fromm, *The Art of Loving* (1956; New York: Harper and Row, 1975); and C. S. Lewis, *The Four Loves* (London: Geoffrey Bles, 1960).

73. For example, J. Dewey, *Human Nature and Conduct: An Introduction to Social Psychology* (New York: Henry Holt, 1922); and J. Dewey, *Art as Experience* (New York: Minton, Balch, 1934).

74. See P. E. Downing et al., "A Cortical Area Selective for Visual Processing of the Human Body," *Science* 293 (2001): 2470–2473.

75. J. Herbert and J. Schulkin, "Neurochemical Coding of Adaptive Responses in the Limbic Systems," in *Hormones, Brain and Behavior*, ed. D. Pfaff, 659–689 (New York: Elsevier, 2002); L. W. Swanson, "The Cerebral Hemisphere Regulation of Motivated Behavior," *Brain Research* 836 (2000): 113–164; L. W. Swanson, "What Is the Brain?," *Trends in Neural Science* 23 (2000): 519–527; and L. W. Swanson, *Brain Architecture* (New York: Oxford University Press, 2003).

76. For example, Dewey, *Experience and Nature* (Chicago: Open Court, 1925); J. Dewey, *Affective Thought* [1926], in *Philosophy and Civilization* (New York: Minton, Balch, 1931); and Dewey, *Art as Experience*.

77. See, generally, R. J. Davidson and A. Harrington, Visions of Compassion (New York: Oxford University Press, 2002); R. J. Davidson, K. R. Scherer, and H. H. Goldsmith, eds., *Handbook of Affective Neuroscience* (New York: Oxford University Press, 2003); J. E. LeDoux, "Cognitive-Emotional Interactions in the Brain," *Cognition and Emotion* 3 (1989): 267–289; and L. Feldman Barrett, *How Emotions Are Made: The Secret Life of the Brain* (Boston: Houghton Mifflin Harcourt, 2017).

78. J. Panksepp, *Affective Neuroscience: The Foundations of Human and Animal Emotions* (New York: Oxford University Press, 1998); J. B. Rosen and J. Schulkin, "From Normal Fear to Pathological Anxiety," *Psychological Review* 105 (1998): 325–350; and K. C. Berridge, "Evolving Concepts of Emotion and Motivation," *Frontiers in Psychology* 9 (2018): art. 1647, https://doi.org/10.3389/fpsyg.2018.01647.

79. M. Davis, "Are Different Parts of the Extended Amygdala Involved in Fear Versus Anxiety?," *Biological Psychiatry* 44 (1998): 1239–1248; and Rosen and Schulkin, "From Normal Fear."

80. See J. E. LeDoux, *Anxious: Using the Brain to Understand and Treat Fear and Anxiety* (New York: Viking, 2015).

81. See J. Moll et al., "The Neural Basis of Human Moral Cognition," *Nature Reviews Neuroscience* 6 (2005): 799–809; R. Zahn et al., "Social Concepts Are Represented in the Superior Anterior Temporal Cortex," *Proceedings of the National Academy of Sciences of the United States of America* 104 (2007): 6430–6435; and R. Zahn et al., "Social Conceptual Impairments in Frontotemporal Lobar Degeneration with Right Anterior Temporal Hypometabolism," *Brain* 132 (2009): 604–616.

82. See A. R. Damasio, *The Feeling of What Happens: Body and Emotion in the Making of Consciousness* (New York: Harcourt Brace, 1999); and D. A. Denton et al., "Neuroimaging of Genesis and Satiation of Thirst and an Interoceptor-Driven Theory of Origins of Primary Consciousness," *Proceedings of the National Academy of Sciences of the United States of America* 96 (1999): 5304–5309.

83. Norgren, "Gustatory System."

84. G. Scalera, A. C. Spector, and R. Norgren, "Excitotoxic Lesions of the Parabrachial Nuclei Prevent Conditioned Taste Aversions and Sodium Appetite in Rats," *Behavioural Neuroscience* 109 (1995): 997–1008.

85. For example, Flynn et al., "Central Gustatory Lesions"; Spector, Norgren, and Grill, "Parabrachial Gustatory Lesions"; and A. C. Spector, "Gustatory Function in Parabrachial Nuclei: Implications from Lesion Studies," *Revolutions in Neuroscience* 6 (1995): 143–175.

86. A. R. Damasio et al., "Subcortical and Cortical Brain Activity During the Feeling of Self-Generated Emotions," *Nature Neuroscience* 3 (2000): 1049–1105.

87. C. S. Peirce, *Pragmatism as a Principle and Method of Right Thinking: The 1903 Harvard Lectures on Pragmatism*, ed. P. A. Turrisi (Albany, NY: State University of New York Press, 1997), 211.

88. See M. Crippen, "Dewey, Enactivism and Greek Thought," in *Pragmatism and Embodied Cognitive Science: From Bodily Interaction to Symbolic Articulation*, ed. R. Madzia and M. Jung, 229–246 (Boston: De Gruyter, 2016).

89. C. Darwin, *On the Origin of Species* (London: John Murray, 1859), 73.

90. T. Solymosi, "Moral First Aid for a Scientific Age," in *Neuroscience, Neurophilosophy and Pragmatism: Brains at Work with the World*, ed. T. Solymosi and J. Shook (New York: Palgrave Macmillan, 2014), 291.

91. T. Solymosi, "A Reconstruction of Freedom in the Age of Neuroscience: A View from Neuropragmatism," *Contemporary Pragmatism* 8 (2011): 154.

92. J. Schulkin, "Foraging for Coherence in Neuroscience: A Pragmatist Orientation," *Contemporary Pragmamtism* 13 (2016): 1–28.

93. M. Crippen, "The Emotionality of Rationality: James and Neuroscience" (paper presented at *Pragmatism and the Brain*, University of North Carolina, Asheville, NC, June 4, 2016).
94. See especially Darwin, *Descent of Man*.
95. M. Landgrave and A. Nowrasteh, "Criminal Immigrants: Their Numbers, Demographics, and Countries of Origin" (Immigration Research and Policy Brief No. 1, Cato Institute, March 15, 2017, https://www.cato.org/publications/immigration-reform-bulletin /criminal-immigrants-their-numbers-demographics-countries).
96. S. Schachter and J. E. Singer, "Cognitive, Social, and Physiological Determinants of Emotional State," *Psychological Review* 69 (1962): 379–399.
97. See W. W. Blessing, *The Lower Brainstem and Bodily Homeostasis* (New York: Oxford University Press, 1997).
98. See Crippen, "Dewey, Enactivism and Greek Thought"; and M. Crippen, "Pragmatic Evolutions of the Kantian *a priori*: From the Mental to the Bodily," in *Pragmatist Kant: Pragmatism, Kant, and Kantianism in the Twenty-First Century*, ed. K. Skowroński and S. Pihlström, 19–40 (Helsinki: Nordic Pragmatism Network, 2019).
99. R. Kaplan et al., "Human Hippocampal Processing of Environmental Novelty During Spatial Navigation," *Hippocampus* 24 (2014): 740–750; and J. Zheng et al., "Amygdala-Hippocampal Dynamics During Salient Information Processing," *Nature Communications* 8 (2017): art. 14413, https://doi.org/10.1038/ncomms14413.
100. R. Kaplan and S. Kaplan, *The Experience of Nature: A Psychological Perspective* (New York: Cambridge University Press, 1989).
101. Dewey, *Experience and Nature*, 1925 ed., 295.

BIBLIOGRAPHY

Aggleton, J. *The Amygdala*. New York: Oxford University Press, 2000.

Ait-Belgnaoui, A., H. Durand, C. Cartier, G. Chaumaz, H. Eutamene, L. Ferrier, E. Houdeau, J. Fioramonti, L. Bueno, and V. Theodorou. "Prevention of Gut Leakiness by a Probiotic Treatment Leads to Attenuated HPA Response to an Acute Psychological Stress in Rats." *Psychoneuroendocrinology* 37 (2012): 1885–1895.

Aldridge J. W., K. C. Berridge, and A. R. Rosen. "Basal Ganglia Neural Mechanisms of Natural Movement Sequences." *Canadian Journal of Physiology and Pharmacology* 82 (2004): 732–739.

Algoe, S. B., L. E. Kurtz, and K. Grewen. "Oxytocin and Social Bonds: The Role of Oxytocin in Perceptions of Romantic Partners' Bonding Behavior." *Psychological Science* 28 (2017): 1763–1772.

Alibali, M. W., R. Boncoddo, and A. B. Hostetter. "Gesture in Reasoning: An Embodied Perspective." In *The Routledge Handbook of Embodied Cognition*, ed. L. Shapiro, 150–159. New York: Routledge, 2014.

Altun, Z. F., and D. H. Hall. "Nervous System, General Description." In *WormAtlas*, ed. L. A. Herndon. Article published 2011; last revised June 19, 2013. http://www.wormatlas.org /hermaphrodite/nervous/mainframe.htm.

Amedi, A., G. Jacobson, T. Hendler, R. Malach, and E. Zohary. "Convergence of Visual and Tactile Shape Processing in the Human Lateral Occipital Complex." *Cerebral Cortex* 12 (2002): 1202–1212.

Anderson, M. L. *After Phrenology: Neural Reuse and the Interactive Brain*. Cambridge, MA: MIT Press, 2014.

Arimura, G. I., K. Matsui, and J. Takabayashi. "Chemical and Molecular Ecology of Herbivore-Induced Plant Volatiles: Proximate Factors and Their Ultimate Functions." *Plant and Cell Physiology* 50 (2009): 911–923.

Aristotle. *De Anima*. Trans. J. A. Smith. In *The Basic Works of Aristotle*, ed. R. McKeon, 533–603. New York: Random House, 1941.

Aristotle. *Metaphysics*. Trans. W. D. Ross. In *The Basic Works of Aristotle*, ed. R. McKeon, 681–926. New York: Random House, 1941.

Aristotle. *On Dreams.* Trans. J. I. Beare. In *The Basic Works of Aristotle*, ed. R. McKeon, 618–625. New York: Random House, 1941.

Aristotle. *Politics.* Trans. B. Jowett. In *The Basic Works of Aristotle*, ed. R. McKeon, 1113–1316. New York: Random House, 1941.

Aristotle. *Posterior Analytics.* Trans. G. R. G. Mure. In *The Basic Works of Aristotle*, ed. R. McKeon, 108–186. New York: Random House, 1941.

Aristotle. *Rhetorica.* Trans. W. R. Roberts. In *The Basic Works of Aristotle*, ed. R. McKeon, 1317–1451. New York: Random House, 1941.

Arling, T. A., R. H. Yolken, M. Lapidus, P. Langenberg, F. B. Dickerson, S. A. Zimmerman, T. Balis, J. A. Cabassa, D. A. Scrandis, L. H. Tonelli, and T. T. Postolache. "*Toxoplasma gondii* Antibody Titers and History of Suicide Attempts in Patients with Recurrent Mood Disorders." *Journal of Nervous and Mental Disease* 197 (2009): 905–908.

Arno, P., A. G. Volder, A. Vanlierde, M. Wanet-Defalque, E. Streel, A. Robert, S. Sanabria-Bohórquez, and C. Veraart. "Occipital Activation by Pattern Recognition in the Early Blind Using Auditory Substitution for Vision." *NeuroImage* 13 (2001): 632–645.

Aston-Jones, G., and J. D. Cohen. "An Integrative Theory of the Locus Coeruleus-Norepinephrine Function: Adaptive Gland and Optimal Performance." *Annual Review of Neuroscience* 28 (2005): 403–450.

Austin, J. L. "A Plea for Excuses." *Proceedings of the Aristotelian Society* 57 (1956–1957): 1–30.

Azzalini, D., I. Rebollo, and C. Tallon-Baudry. "Visceral Signals Shape Brain Dynamics and Cognition." *Trends in Cognitive Sciences* 23 (2019): 488–509.

Bach-y-Rita, P. "Brain Plasticity as a Basis of Sensory Substitution." *Journal of Neurologic Rehabilitation* 2 (1987): 67–71.

Bach-y-Rita, P. "The Relationship Between Motor Processes and Cognition in Tactile Vision Substitution." In *Cognition and Motor Processes*, ed. A. F. Sanders and W. Prinz, 150–159. Berlin: Springer, 1984.

Bach-y-Rita, P. "Tactile Vision Substitution: Past and Future." *International Journal of Neuroscience* 19 (1983): 29–36.

Bach-y-Rita, P., and S. Kercel. "Sensory Substitution and Augmentation: Incorporating Humans-in-the-Loop." *Intellectica* 2 (2002): 287–297.

Baron-Cohen, S., H. A. Ring, S. Wheelwright, E. T. Bullmore, M. J. Brammer, A. Simmons, and S. C. R. Williams. "Social Intelligence in the Normal and Autistic Brain: An fMRI Study." *European Journal of Neuroscience* 11 (1999): 1891–1898.

Barrett, L. *Beyond the Brain: How Body and Environment Shape Animal and Human Minds.* Princeton, NJ: Princeton University Press, 2011.

Baumann, O., R. Borra, J. Bower, K. Cullen, C. Habas, R. Ivry, M. Leggio, J. Mattingley, M. Molinari, E. Moulton, M. Paulin, M. Pavlova, J. Schmahmann, and A. Sokolov. "Consensus Paper: The Role of the Cerebellum in Perceptual Processes." *Cerebellum* 14 (2015): 197–220.

Bechara, A., H. Damasio, D. Tranel, and A. R. Damasio. "Deciding Advantageously Before Knowing the Advantageous Strategy." *Science* 275 (1997): 1293–1295.

Bedrosian, T. A., and R. J. Nelson. "Timing of Light Exposure Affects Mood and Brain Circuits." *Translational Psychiatry* 7 (2017): art. e1017. https://doi.org/10.1038%2Ftp.2016.262.

Ben-Attia, M., A. Reinberg, M. H. Smolensky, W. Gadacha, A. Khedaier, M. Sani, Y. Touitou, and N. G. Boughamni. "Blooming Rhythms of Cactus *Cereus peruvianus* with Nocturnal Peak at Full Moon During Seasons of Prolonged Daytime Photoperiod." *Chronobiology International* 33 (2016): 419–430.

Benton, D., C. Williams, and A. Brown. "Impact of Consuming a Milk Drink Containing a Probiotic on Mood and Cognition." *European Journal of Clinical Nutrition* 61 (2006): 355–361.

Berdoy, M., J. P. Webster, and D. W. Macdonald. "Fatal Attraction in Rats Infected with *Toxoplasma gondii.*" *Proceedings of the Royal Society B: Biological Sciences* 267 (2000): 1591–1594.

BIBLIOGRAPHY

Berlyne, D. E. *Conflict, Arousal, and Curiosity.* New York: McGraw-Hill, 1960.

Berlyne, D. E. "Ends and Means of Experimental Aesthetics." *Canadian Journal of Psychology* 26 (1972): 303–325.

Berlyne, D. E. "Novelty, Complexity, and Hedonic Values." *Perception and Psychophysics* 8 (1970): 279–285.

Bermúdez-Rattoni, F., C. V. Grijalva, S. W. Kiefer, and J. Garcia. "Flavor-Illness Aversions: The Role of the Amygdala in the Acquisition of Taste-Potentiated Odor Aversions." *Physiology and Behavior* 38 (1986): 503–508.

Bernstein, C., and G. Driessen. "Patch-Marking and Optimal Search Patterns in the Parasitoid *Venturia canescens.*" *Journal of Animal Ecology* 65 (1996): 211–219.

Berridge, K. C. "Evolving Concepts of Emotion and Motivation." *Frontiers in Psychology* 9 (2018): art. 1647. https://doi.org/10.3389/fpsyg.2018.01647.

Berridge, K. C. "Measuring Hedonic Impact in Animals and Infants: Microstructure of Affective Taste Reactivity Patterns." *Neuroscience and Biobehavioral Reviews* 24 (2000): 173–198.

Berridge, K. C., B. F. Flynn, J. Schulkin, and H. G. Grill. "Sodium Depletion Enhances Salt Palatability in Rats." *Behavioral Neuroscience* 98 (1984): 652–660.

Berridge, K. C., and M. Kringelbach. "Pleasure Systems in the Brain." *Neuron* 86 (2015): 646–664.

Bhalla, M., and D. R. Proffitt. "Visual-Motor Recalibration in Geographical Slant Perception." *Journal of Experimental Psychology: Human Perception and Performance* 25 (1999): 1076–1096.

Binkofski, F., K. Amunts, K. M. Stephan, S. Posse, T. Schormann, H. J. Freund, K. Zilles, and R. J. Seitz. "Broca's Region Subserves Imagery of Motion: A Combined Cytoarchitectonic and fMRI Study." *Human Brain Mapping* 11 (2000): 273–285.

Biran, J., M. Tahor, E. Wircer, and G. Levkowitz. "Role of Developmental Factors in Hypothalamic Function." *Frontiers in Neuroanatomy* 9 (2015): art. 47. https://doi.org/10.3389/fnana.2015.00047.

Bischoff, S. C., G. Barbara, W. Buurman, T. Ockhuizen, J. Schulzke, M. Serino, H. Tilg, A. Watson, and J. M. Wells. "Intestinal Permeability—a New Target for Disease Prevention and Therapy." *BMC Gastroenterology* 14 (2014): art. 189. https://doi.org/10.1186/s12876-014-0189-7.

Blessing, W. W. *The Lower Brainstem and Bodily Homeostasis.* New York: Oxford University Press, 1997.

Böhme, G. "Atmosphere as the Fundamental Concept of a New Aesthetic." *Thesis Eleven* 36 (1993): 113–126.

Brazelton, T. B., E. Tronick, L. Adamson, H. Als, and S. Wise. "Early Mother-Infant Reciprocity." In *Parent-Infant Interaction*, ed. M. Hofer, 137–154. New York: Elsevier, 1975.

Brendl, C. M., A. B. Markman, and C. Messner. "The Devaluation Effect: Activating a Need Devalues Unrelated Objects." *Journal of Consumer Research* 29 (2003): 463–473.

Breton, J., N. Tennoune, N. Lucas, M. Francois, R. Legrand, J. Jacquemot, A. Goichon, et al. "Gut Commensal *E. coli* Proteins Activate Host Satiety Pathways Following Nutrient-Induced Bacterial Growth." *Cell Metabolism* 23 (2016): 324–334.

Brooks, R. A. *Cambrian Intelligence: The Early History of the New AI.* Cambridge, MA: MIT Press, 1999.

Brothers, L. "Neurophysiology of the Perception of Intentions in Primates." In *The Cognitive Neurosciences*, ed. M. S. Gazzaniga, 1107–1115. Cambridge, MA: MIT Press, 1994.

Brown, P., and C. D. Marsden. "What Do the Basal Ganglia Do?" *Lancet* 351 (1998): 1801–1804.

Brown, R. G., and C. D. Marsden. "How Common Is Dementia in Parkinson's Disease?" *Lancet* 324 (1984): 1262–1265.

Brown, S., and L. M. Parsons. "The Neuroscience of Dance." *Scientific American* 299 (2008): 78–83.

Büchel, C., C. Price, R. S. Frackowiak, and K. J. Friston. "Different Activation Patterns in the Visual Cortex of Late and Congenitally Blind Subjects." *Brain* 121 (1998): 409–419.

Buijs, R. M., and C. G. van Eden. "The Integration of Stress by the Hypothalamus, Amygdala and Prefrontal Cortex: Balance Between the Autonomic Nervous System and the Neuro-endocrine System." *Progress in Brain Research* 126 (2000): 117–132.

Burrus, C. J. "A Biochemical Rationale for the Interaction Between Gastrointestinal Yeast and Autism." *Medical Hypotheses* 79 (2012): 784–785.

Burton, H. "Visual Cortex Activity in Early and Late Blind People." *Journal of Neuroscience* 23 (2003): 4005–4011.

Bushara, K. O., J. Grafman, and M. Hallett. "Neural Correlates of Auditory-Visual Stimulus Onset Asynchrony Detection." *Journal of Neuroscience* 21 (2001): 300–304.

Calder, A. J., A. D. Lawrence, and A. W. Young. "Neuropsychology of Fear and Loathing." *Nature Neuroscience* 2 (2001): 352–363.

Calderwood, H. *The Relations of Mind and Brain.* London: Macmillan, 1879.

Calvert, G. A., E. T. Bullmore, M. J. Brammer, R. Campbell, S. C. R. Williams, P. K. McQuire, P. W. R. Woodruff, S. D. Iverson, and A. S. David. "Activation of Auditory Cortex During Silent Lipreading." *Science* 276 (1997): 593–596.

Calvert, W. H. "Monarch Butterfly (*Danaus plexippus* L., Nymphalidae) Fall Migration: Flight Behavior and Direction in Relation to Celestial and Physiographic Cues." *Journal of the Lepidopterists' Society* 55 (2001): 162–168.

Cameron, E. "From Plato to Socrates: Wittgenstein's Journey on Collingwood's Map." *AE: Canadian Aesthetics Journal* 10 (2004): 1–29. https://www.uqtr.ca/AE/Vol_10/wittgenstein /cameron.htm.

Canli, T., H. Silvers, S. L. Whitfield, I. H. Gotlib, and J. Gabrieli. "Amygdala Response to Happy Faces as a Function of Extraversion." *Science* 296 (2002): 2191.

Cannon, P. R., A. E. Hayes, and S. P. Tipper. "An Electromyographic Investigation of the Impact of Task Relevance on Facial Mimicry." *Cognition and Emotion* 23 (2009): 918–929.

Cappella, J. N. "The Biological Origins of Automated Patterns of Human Interaction." *Communication Theory* 1 (1991): 4–35.

Carabotti, M., A. Scirocco, M. A. Maselli, and S. Carola. "The Gut-Brain Axis: Interactions Between Enteric Microbiota, Central and Enteric Nervous Systems." *Annals of Gastroenterology* 28 (2015): 203–209.

Carlile, M. J. "Nutrition and Chemotaxis in the Myxomycete *Physarum polycephalum*: The Effect of Carbohydrates on the Plasmodium." *Journal of General Microbiology* 63 (1970): 221–226.

Cela-Conde, C. J., G. Marty, F. Maestu, T. Ortiz, E. Munar, A. Fernandez, M. Roca, J. Rossello, and F. Quesney. "Activation of the Prefrontal Cortex in the Human Visual Aesthetic Perception." *Proceedings of the National Academy of Sciences of the United States of America* 101 (2004): 6321–6325.

Çelik, T., Ö. Kamişli, C. Babür, M. Ö. Çevik, D. Öztuna, and S. Altinayar. "Is There a Relationship Between *Toxoplasma gondii* Infection and Idiopathic Parkinson's Disease?" *Scandinavian Journal of Infectious Diseases* 42 (2010): 604–608.

Charney, E. J. "Postural Configuration in Psychotherapy." *Psychosomatic Medicine* 28 (1966): 305–315.

Chartrand, J. L., and J. A. Bargh. "The Chameleon Effect: The Perception-Behavior Link and Social Interaction." *Journal of Personality and Social Psychology* 76 (1999): 893–910.

Chatoo, M., Y. Li, Z. Ma, J. Coote, J. Du, and X. Chen. "Involvement of Corticotropin-Releasing Factor and Receptors in Immune Cells in Irritable Bowel Syndrome." *Frontiers in Endocrinology* 9 (2018): art. 21. https://doi.org/10.3389/fendo.2018.00021.

Chemero, A. "Against Smallism and Localism." Paper presented at *Thinking with Hands, Eyes and Things: Trends in Interdisciplinary Studies; An International Avant-Conference*, Torun, Poland, November 8, 2013.

Chemero, A. *Radical Embodied Cognitive Science*. Cambridge, MA: MIT Press, 2009.

Chemero, A. "Sensorimotor Empathy." *Journal of Consciousness Studies* 23 (2016): 138–152.

Chemero, A., and S. Käufer. "Pragmatism, Phenomenology, and Extended Cognition." In *Pragmatism and Embodied Cognitive Science: From Bodily Interaction to Symbolic Articulation*, ed. R. Madzia and M. Jung, 55–70. Berlin: De Gruyter, 2016.

Chet, I., A. Naveh, and Y. Henis. "Chemotaxis of *Physarum polycephalum* Towards Carbohydrates, Amino Acids and Nucleotides." *Journal of General Microbiology* 102 (1977): 145–148.

Cho, Y. K., C. Li, and D. V. Smith. "Gustatory Projections from the Nucleus of the Solitary Tract to the Parabrachial Nuclei in the Hamster." *Chemical Senses* 27 (2002): 81–90.

Chomsky, N. A. *Language and Mind*. New York: Cambridge University Press, 2006.

Chomsky, N. A. "A Review of B. F. Skinner's Verbal Behavior." *Language* 35 (1959): 26–58.

Chudoba, E. "What Controls and What Is Controlled? Deweyan Aesthetic Experience and Shusterman's Somatic Experience." *Contemporary Pragmatism* 14 (2017): 112–134.

Cirelli, C., R. Huber, A. Gopalakrishnan, T. L. Southard, and G. Tononi. "Locus Ceruleus Control of Slow-Wave Homeostasis." *Journal of Neuroscience* 25 (2005): 4503–4511.

Clapp, M., N. Aurora, L. Herrera, M. Bhatia, E. Wilen, and S. Wakefield. "Gut Microbiota's Effect on Mental Health: The Gut-Brain Axis." *Clinics and Practice* 7 (2017): 131–136.

Clark, A. *Being There: Putting Brain, Body, and World Together Again*. Cambridge, MA: MIT Press, 1997.

Clark, A. *Supersizing the Mind: Embodiment, Action, and Cognitive Extension*. New York: Oxford University Press, 2008.

Clark, A. *Surfing Uncertainty: Prediction, Action, and the Embodied Mind*. New York: Oxford University Press, 2015.

Clark, A. "Whatever Next? Predictive Brains, Situated Agents, and the Future of Cognitive Science." *Behavioral and Brain Sciences* 36 (2013): 181–204.

Clark, A., and D. J. Chalmers. "The Extended Mind." *Analysis* 58 (1998): 7–19.

Clément, F., and L. Kaufmann. "How Culture Comes to Mind: From Social Affordances to Cultural Analogies." *Intellectica* 46 (2007): 221–250.

Coccaro, E. F., R. Lee, M. W. Groer, A. Can, M. Coussons-Read, and T. T. Postolache. "*Toxoplasma gondii* Infection: Relationship with Aggression in Psychiatric Subjects." *Journal of Clinical Psychiatry* 77 (2016): 334–341.

Coley, P. D. "Costs and Benefits of Defense by Tannins in a Neotropical Tree." *Oecologia* 70 (1986): 238–241.

Collingwood, R. G. *An Autobiography*. New York: Oxford University Press, 1939.

Collingwood, R. G. *An Essay on Metaphysics*. Oxford: Oxford University Press, 1940.

Colombetti, G. *The Feeling Body: Affective Science Meets the Enactive Mind*. Cambridge, MA: MIT Press, 2014.

Contreras, R. J. "Changes in Gustatory Nerve Discharges with Sodium Deficiency: A Single Unit Analysis." *Brain Research* 121 (1977): 373–378.

Cook, N. *Music: A Very Short Introduction*. Oxford: Oxford University Press, 1998.

Corballis, M. C. *From Hand to Mouth*. Princeton, NJ: Princeton University Press, 2002.

Costantini, M., and M. Stapleton. "How the Body Narrows the Interaction with the Environment." In *Foundations of Embodied Cognition: Perceptual and Emotional Embodiment*, ed. Y. Coello and M. H. Fischer, 181–197. New York: Routledge, 2016.

Craig, W. "Appetites and Aversions as Constituents of Instinct." *Biological Bulletin* 34 (1918): 91–107.

Craighero, L. "The Role of the Motor System in Cognitive Functions." In *The Routledge Handbook of Embodied Cognition*, ed. L. Shapiro, 51–58. New York: Routledge, 2014.

Crippen, M. "Aesthetics and Action: Situations, Emotional Perception and the Kuleshov Effect." *Synthese* [issue unassigned] (2019). https://doi.org/10.1007/s11229-019-02110-2.

Crippen, M. "Art and Pragmatism: James and Dewey on the Reconstructive Presuppositions of Experience." PhD diss., York University, 2010.

Crippen, M. "Asleep at the Press: Thoreau, Egyptian Revolt and Nuances of Democracy." *Arab Media and Society*, no. 20 (2015). https://www.arabmediasociety.com/wp-content/uploads /2017/12/20150302085925_Crippen_AsleepAtThePress.pdf.

Crippen, M. "Body Phenomenology, Somaesthetics and Nietzschean Themes in Medieval Art." *Pragmatism Today* 5 (2014): 40–45.

Crippen, M. "Body Politics: Revolt and City Celebration." In *Bodies in the Streets: Somaesthetics of City Life*, ed. R. Shusterman, 89–110. Boston: Brill, 2019.

Crippen, M. "Contours of Cairo Revolt: Semiology and Political Affordances in Street Discourses." *Topoi* [issue unassigned] (2019). https://doi.org/10.1007/s11245-019-09650-9.

Crippen, M. "Dewey, Enactivism and Greek Thought." In *Pragmatism and Embodied Cognitive Science: From Bodily Interaction to Symbolic Articulation*, ed. R. Madzia and M. Jung, 229–246. Boston: De Gruyter, 2016.

Crippen, M. "Dewey on Arts, Sciences and Greek Philosophy." In *In the Beginning Was the Image: The Omnipresence of Pictures; Time, Truth, Tradition*, ed. A. Benedek and A. Veszelszki, 153–159. Visual Learning 6. Frankfurt am Main: Peter Lang, 2016.

Crippen, M. "Digital Fabrication and Its Meaning for Film." In *Conceiving Virtuality: From Art to Technology*, ed. J. Braga, 119–131. New York: Springer, 2019.

Crippen, M. "Egypt and the Middle East: Democracy, Anti-Democracy and Pragmatic Faith." *Saint Louis University Public Law Review* 35 (2016): 281–302.

Crippen, M. "Embodied Cognition and Perception: Dewey, Science and Skepticism." *Contemporary Pragmatism* 14 (2017): 121–134.

Crippen, M. "The Emotionality of Rationality: James and Neuroscience." Paper presented at *Pragmatism and the Brain*, University of North Carolina, Asheville, NC, June 4, 2016.

Crippen, M. "Group Cognition in Pragmatism, Developmental Psychology and Aesthetics." *Pragmatism Today* 8 (2017): 185–197.

Crippen, M. "Intuitive Cities: Pre-reflective, Aesthetic and Political Aspects of Urban Design." *Journal of Aesthetics and Phenomenology* 3 (2016): 125–145.

Crippen, M. "Pictures, Experiential Learning and Phenomenology." In *Beyond Words: Pictures, Parables, Paradoxes*, ed. A. Benedek and K. Nyíri, 83–90. Visual Learning 5. Frankfurt am Main: Peter Lang, 2015.

Crippen, M. "Pragmatic Evolutions of the Kantian *a priori*: From the Mental to the Bodily." In *Pragmatist Kant: Pragmatism, Kant, and Kantianism in the Twenty-First Century*, ed. K. Skowroński and S. Pihlström, 19–40. Helsinki: Nordic Pragmatism Network, 2019.

Crippen, M. "Pragmatism and the Valuative Mind." *Transactions of the Charles S. Peirce Society* 54 (2018): 341–360.

Crippen, M. "Screen Performers Playing Themselves." *British Journal of Aesthetics* 56 (2016): 163–177.

Crippen, M. "The Soma in City Life: Cultural, Political and Bodily Aesthetics of Mandalay's Water Festival." *Pragmatism Today* 9 (2018): 29–40.

Crippen, M. "The Totalitarianism of Therapeutic Philosophy: Reading Wittgenstein Through Critical Theory." *Essays in Philosophy* 8 (2007): 29–55.

Crippen, M. "William James and His Darwinian Defense of Freewill." In *150 Years of Evolution: Darwin's Impact on Contemporary Thought and Culture*, ed. M. Wheeler, 68–89. San Diego, CA: San Diego State University Press, 2011.

Crippen, M. "William James on Belief: Turning Darwinism Against Empiricistic Skepticism." *Transactions of the Charles S. Peirce Society* 46 (2010): 477–502.

Crippen, M., and M. Dixon. "Echoes of Past and Present." In *Tom Petty and Philosophy: We Need to Know*, ed. R. Auxier and M. Volpert, 16–25. Chicago: Open Court, 2019.

Critchley, H. D., and N. Harrison. "Visceral Influences on Brain and Behavior." *Neuron* 77 (2013): 624–638.

Critchley, H. D., C. J. Mathias, and R. J. Dolan. "Neural Activity in the Human Brain Relating to Uncertainty and Arousal During Anticipation." *Neuron* 29 (2001): 537–545.

Critchley, H. D., C. J. Mathias, and R. J. Dolan. "Neuroanatomical Basis for First- and Second-Order Representations of Bodily States." *Nature Neuroscience* 4 (2001): 207–212.

Cryan, J. F., and T. G. Dinan. "Mind-Altering Microorganisms: The Impact of the Gut Microbiota on Brain and Behaviour." *Nature Reviews Neuroscience* 13 (2012): 701–712.

Csikszentmihalyi, M. *Flow and the Foundations of Positive Psychology: The Collected Works of Mihaly Csikszentmihalyi*. New York: Springer, 2014.

Damasio, A. R. *Descartes' Error: Emotion, Reason, and the Human Brain*. New York: G. P. Putnam, 1994.

Damasio, A. R. *The Feeling of What Happens: Body and Emotion in the Making of Consciousness*. New York: Harcourt Brace, 1999.

Damasio, A. R. *Self Comes to Mind: Constructing the Conscious Brain*. New York: Pantheon Books, 2010.

Damasio, A. R. *The Strange Order of Things: Life, Feeling, and the Making of Cultures*. New York: Pantheon Books, 2018.

Damasio, A. R., and G. B. Carvalho. "The Nature of Feelings: Evolutionary and Neurobiological Origins." *Nature Reviews Neuroscience* 14 (2013): 143–152.

Damasio, A. R., T. J. Grabowski, A. Bechara, H. Damasio, L. L. B. Ponto, J. Parvizi, and R. D. Hichwa. "Subcortical and Cortical Brain Activity During the Feeling of Self-Generated Emotions." *Nature Neuroscience* 3 (2000): 1049–1105.

Darwin, C. *The Descent of Man, and Selection in Relation to Sex*. 2 vols. London: J. Murray, 1871.

Darwin, C. *On the Origin of Species*. London: John Murray, 1859.

Daugman, J. "Brain Metaphor and Brain Theory." In *Philosophy and the Neurosciences: A Reader*, ed. W. Bechtel, P. Mandik, J. Mundale, and R. Stufflebeam, 23–36. Oxford: Blackwell, 2001.

Daulatzai, M. "Non-celiac Gluten Sensitivity Triggers Gut Dysbiosis, Neuroinflammation, Gut-Brain Axis Dysfunction, and Vulnerability for Dementia." *CNS and Neurological Disorders* 14 (2015): 110–131.

Davern, P. J. "A Role for the Lateral Parabrachial Nucleus in Cardiovascular Function and Fluid Homeostasis." *Frontiers in Physiology* 5 (2014): art. 436. https://doi.org/10.3389/fphys.2014.00436.

Davidson, R. J., and A. Harrington. *Visions of Compassion*. New York: Oxford University Press, 2002.

Davidson, R. J., K. R. Scherer, and H. H. Goldsmith, eds. *Handbook of Affective Neuroscience*. New York: Oxford University Press, 2003.

Davis, J. I., A. Senghas, F. Brandt, and K. N. Ochsner. "The Effects of Botox Injections on Emotional Experience." *Emotion* 10 (2010): 433–440.

Davis, J. I., A. Senghas, and K. N. Ochsner. "How Does Facial Feedback Modulate Emotional Experience?" *Journal of Research in Personality* 43 (2009): 822–829.

Davis, M. "Are Different Parts of the Extended Amygdala Involved in Fear Versus Anxiety?" *Biological Psychiatry* 44 (1998): 1239–1248.

Dawkins, R. *The God Delusion*. Boston: Houghton Mifflin, 2006.

Deady, D. K., N. T. North, D. Allan, M. J. Law Smith, and R. E. O'Carroll. "Examining the Effect of Spinal Cord Injury on Emotional Awareness, Expressivity and Memory for Emotional Material." *Psychology, Health and Medicine* 15 (2010): 406–419.

De Bruin, L., and S. de Haan. "Enactivism and Social Cognition: In Search of the Whole Story." *Cognitive Semiotics* 4 (2009): 225–250.

Decety, J. "Do Imagined and Executed Actions Share the Same Neural Substrate?" *Cognitive Brain Research* 3 (1996): 87–93.

Decety, J., J. Grèzes, N. Costes, D. Perani, M. Jeannerod, E. Procyk, F. Grassi, and F. Fazio. "Brain Activity During Observation of Actions: Influence of Action Content and Subject's Strategy." *Brain* 120 (1997): 1763–1777.

Delafield-Butt, J. T., and C. Trevarthen. "The Ontogenesis of Narrative: From Moving to Meaning." *Frontiers in Psychology* 6 (2015): art. 1157. https://doi.org/10.3389/fpsyg.2015.01157.

Deng, W., J. B. Aimone, and F. H. Gage. "New Neurons and New Memories: How Does Adult Hippocampal Neurogenesis Affect Learning and Memory?" *Nature Reviews Neuroscience* 11 (2010): 339–350.

Dennett, D. C. *Consciousness Explained*. New York: Little, Brown, 1991.

Denton, D. A. *The Hunger for Salt: An Anthropological, Physiological and Medical Analysis*. Berlin: Springer-Verlag, 1982.

Denton, D. A., M. J. McKinley, and R. S. Wessinger. "Hypothalamic Integration of Body Fluid Regulation." *Proceedings of the National Academy of Sciences of the United States of America* 93 (1996): 7397–7404.

Denton, D. A., R. Shade, F. Zamarippa, G. Egan, J. Blair-West, M. McKinley, J. Lancaster, and P. Fox. "Neuroimaging of Genesis and Satiation of Thirst and an Interoceptor-Driven Theory of Origins of Primary Consciousness," *Proceedings of the National Academy of Sciences of the United States of America* 96 (1999): 5304–5309.

Depue, B. E., J. H. Olson-Madden, H. R. Smolker, M. Rajamani, L. A. Brenner, and M. T. Banich. "Reduced Amygdala Volume Is Associated with Deficits in Inhibitory Control: A Voxel- and Surface-Based Morphometric Analysis of Comorbid PTSD/Mild TBI." *BioMed Research International* (2014): art. 691505. https://doi.org/10.1155/2014/691505.

Descartes, R. Descartes to Mersenne, April 1, 1640. In *The Philosophical Writings of Descartes*, trans. J. Cottingham, R. Stoothoff, D. Murdoch, and A. Kenny, vol. 3, *Correspondence*, 145–146. New York: Cambridge University Press, 1991.

Desimone, R., and J. Duncan. "Neural Mechanisms of Selective Visual Attention." *Annual Review of Neuroscience* 18 (1995): 193–222.

De Sousa, R. *The Rationality of Emotion*. Cambridge, MA: MIT Press, 1987.

Devenport, L., J. Devenport, and C. Kokesh. "The Role of Urine Marking in the Foraging Behaviour of Least Chipmunks." *Animal Behaviour* 57 (1999): 557–563.

De Volder, A. G., M. Catalan-Ahumada, A. Robert, A. Bol, D. Labar, A. Coppens, C. Michel, and C. Veraart. "Changes in Occipital Cortex Activity in Early Blind Humans Using a Sensory Substitution Device." *Brain Research* 826 (1999): 128–134.

Dewey, J. *Affective Thought*, 1926. In *Philosophy and Civilization*, 117–125. New York: Minton, Balch, 1931.

Dewey, J. *Art as Experience*. New York: Minton, Balch, 1934.

Dewey, J. *Democracy and Education: An Introduction to the Philosophy of Education*. New York: Macmillan, 1916.

Dewey, J. *Essays in Experimental Logic*. Chicago: University of Chicago Press, 1916.

Dewey, J. *Experience and Nature*. Chicago: Open Court, 1925.

Dewey, J. *Experience and Nature*. 2nd ed. 1929. Reprint, New York: Dover, 1958.

Dewey, J. "Experience and Objective Idealism." *Philosophical Review* 15 (1906): 465–481.

Dewey, J. *Human Nature and Conduct: An Introduction to Social Psychology*. New York: Henry Holt, 1922.

Dewey, J. "The Need for a Recovery of Philosophy." In *Creative Intelligence: Essays in the Pragmatic Attitude*, by J. Dewey, A. W. Moore, H. C. Brown, G. H. Mead, B. H. Bode, H. W. Stuart, J. H, Tufts, and H. M. Kallen, 3–69. New York: Henry Holt, 1917.

BIBLIOGRAPHY

Dewey, J. *Philosophy and Education in Their Historic Relation*. Ed. J. J. Chambliss from 1910–1911 lectures transcribed by E. R. Clapp. Boulder, CO: Westview Press, 1993.

Dewey, J. "The Practical Character of Reality," 1908. In *Philosophy and Civilization*, 35–56. New York: Minton, Balch, 1931.

Dewey, J. *The Quest for Certainty*. New York: Minton, Balch, 1929.

Dewey, J. *Reconstruction in Philosophy*. New York: Henry Holt, 1920.

Dewey, J. "The Reflex Arc Concept in Psychology." *Psychological Review* 3 (1896): 357–370.

Dewey, J. "Syllabus: Types of Philosophical Thought," 1922–1923. In *The Middle Works, 1899–1922*, vol. 13, ed. J. A. Boydston, 349–396. Carbondale, IL: Southern Illinois University Press, 1983.

Dewey, J. "The Theory of Emotions II: The Significance of Emotions." *Psychological Review* 2 (1895): 13–32.

Dewey, J. *Theory of Valuation*. Chicago: University of Chicago Press, 1939.

Dewey, J. "Unfinished Introduction" [excerpt], 1951. In *The Later Works, 1925–1953*, vol. 1, ed. J. A. Boydston, 361–364. Carbondale, IL: Southern Illinois University Press, 1981.

De Wit, M. M., S. de Vries, J. van der Kamp, and R. Withagen. "Affordances and Neuroscience: Steps Towards a Successful Marriage." *Neuroscience and Biobehavioral Reviews* 80 (2017): 622–629.

Dietrich, A. "Neurocognitive Mechanism Underlying the Experience of Flow." *Consciousness and Cognition* 13 (2004): 746–761.

Dimberg, U. "Facial Reactions to Facial Expressions." *Psychophysiology* 19 (1982): 643–647.

Dinan, T. G., C. Stanton, and J. F. Cryan. "Psychobiotics: A Novel Class of Psychotropic." *Biological Psychiatry* 74 (2013): 720–726.

Doucet, S., R. Soussignan, P. Sagot, and B. Schaal. "The 'Smellscape' of Mother's Breast: Effects of Odor Masking and Selective Unmasking on Neonatal Arousal, Oral, and Visual Responses." *Developmental Psychobiology* 49 (2007): 129–138.

Downing, P. E., Y. Jiang, M. Shuman, and N. Kanwisher. "A Cortical Area Selective for Visual Processing of the Human Body." *Science* 293 (2001): 2470–2473.

Dreyfus, H. *Being-in-the-World: A Commentary on Heidegger's* Being and Time, *Division I*. Cambridge, MA: MIT Press, 1991.

Dreyfus, H. "Existential Phenomenology and the Brave New World of *The Matrix*." *Harvard Review of Philosophy* 11 (2003): 18–31.

Driver, J., and T. Noesselt. "Multisensory Interplay Reveals Crossmodal Influences on 'Sensory-Specific' Brain Regions, Neural Responses, and Judgments." *Neuron* 57 (2008): 11–23.

Dussutour, A., T. Latty, M. Beekman, and S. J. Simpson. "Amoeboid Organism Solves Complex Nutritional Challenges." *Proceedings of the National Academy of Sciences of the United States of America* 107 (2010): 4607–4611.

Eckburg, P. B., E. M. Bik, C. N. Bernstein, E. Purdom, L. Dethlefsen, M. Sargent, S. R. Gill, K. E. Nelson, and D. A. Relman. "Diversity of the Human Intestinal Microbial Flora." *Science* 308 (2005): 1635–1638.

Eisenberg, N. I. "Identifying the Neural Correlates Underlying Social Pain: Implications for Developmental Processes." *Human Development* 49 (2006): 273–293.

Emery, N. J. "The Eyes Have It: The Neuroethology, Function, and Evolution of Social Gaze." *Neuroscience and Biobehavioral Reviews* 24 (2000): 581–604.

Enders, G. *Gut: The Inside Story of Our Body's Most Underrated Organ*. Trans. D. Shaw. Vancouver, BC: Greystone Books, 2015.

Engel, A. K., K. J. Friston, and D. Kragic. *The Pragmatic Turn: Toward Action-Oriented Views in Cognitive Science*. Cambridge, MA: MIT Press, 2015.

Epstein, A. N., J. T. Fitzsimons, and B. J. Rolls. "Drinking Induced by Injection of Angiotensin into the Brain of the Rat." *Journal of Physiology* 210 (1970): 457–474.

Epstein, A. N., J. T. Fitzsimons, and B. J. Simons. "Drinking Caused by the Intracranial Injection of Angiotensin into the Rat." *Journal of Physiology* 200 (1969): 98P–100P.

Farah, M. J. "The Neurobiological Basis of Visual Imagery: A Componential Analysis." *Cognition* 18 (1984): 245–272.

Fekadu, A., T. Shibre, and A. J. Cleare. "Toxoplasmosis as a Cause for Behaviour Disorders— Overview of Evidence and Mechanisms." *Folia Parasitol* 57 (2010): 105–113.

Feldman Barrett, L. *How Emotions Are Made: The Secret Life of the Brain*. Boston: Houghton Miffli Harcourt, 2017.

Ferrari, P. F., V. Gallese, G. Rizzolatti, and L. Fogassi. "Mirror Neurons Responding to the Observation of Ingestive and Communicative Mouth Actions in the Monkey Ventral Premotor Cortex." *European Journal of Neuroscience* 17 (2003): 1703–1714.

Festinger, L. *A Theory of Cognitive Dissonance*. Evanston, IL: Row, Peterson, 1957.

Fetissov, S. O. "Role of the Gut Microbiota in Host Appetite Control: Bacterial Growth to Animal Feeding Behaviour." *Nature Reviews Endocrinology* 13 (2017): 11–25.

Fingerhut, J. "Sensorimotor Signature, Skill, and Synaesthesia: Two Challenges to Enactive Theories of Perception." In *Habitus in Habitat III: Synaesthesia and Kinaesthetics*, ed. J. Fingerhut, S. Flach, and J. Söffner, 101–120. New York: Peter Lang, 2011.

Finkelstein, D. R. "Physical Process and Physical Law." In *Physics and Whitehead: Quantum, Process, and Experience*, ed. T. E. Eastman and H. Keeton, 180–186. Albany, NY: State University of New York Press, 2004.

Fisk, A. S., S. K. Tam, L. A. Brown, V. V. Vyazovskiy, D. M. Bannerman, and S. N. Peirson. "Light and Cognition: Roles for Circadian Rhythms, Sleep, and Arousal." *Frontiers in Neurology* 9 (2018): art. 56. https://doi.org/10.3389/fneur.2018.00056.

Fitzsimons, J. T. "Angiotensin, Thirst, and Sodium Appetite." *Physiological Review* 78 (1999): 583–686.

Flegr, J. "Effects of Toxoplasma on Human Behavior." *Schizophrenia Bulletin* 33 (2007): 757–760.

Flegr, J., J. Havlícek, P. Kodym, M. Malý, and Z. Smahel. "Increased Risk of Traffic Accidents in Subjects with Latent Toxoplasmosis: A Retrospective Case-Control Study." *BMC Infectious Diseases* 2 (2002): art. 11. http://www.biomedcentral.com/1471-2334/2/11.

Flegr, J., J. Klose, M. Novotná, M. Berenreitterová, and J. Havlíček. "Increased Incidence of Traffic Accidents in *Toxoplasma*-Infected Military Drivers and Protective Effect RhD Molecule Revealed by a Large-Scale Prospective Cohort Study." *BMC Infectious Diseases* 9 (2009): art. 72. https://doi.org/10.1186/1471-2334-9-72.

Flegr, J., P. Lenochová, Z. Hodný, and M. Vondrová. "Fatal Attraction Phenomenon in Humans: Cat Odour Attractiveness Increased for *Toxoplasma*-Infected Men While Decreased for Infected Women." *PLoS Neglected Tropical Diseases* 5 (2011): art. e1389. https://doi.org/10.1371%2Fjournal.pntd.0001389.

Flegr, J., J. Lindová, and P. Kodym. "Sex-Dependent Toxoplasmosis-Associated Differences in Testosterone Concentration in Humans." *Parasitology* 135 (2008): 427–431.

Flynn, F. W., H. G. Grill, J. Schulkin, and R. Norgren. "Central Gustatory Lesions: II. Effects on Sodium Appetite, Taste Aversion Learning, and Feeding Behaviors." *Behavioral Neuroscience* 105 (1991): 944–954.

Fong, T., I. Nourbakhsh, and K. Dautenhahn. "A Survey of Socially Interactive Robots." *Robotics and Autonomous Systems* 42 (2003): 143–166.

Forestell, C. A., and J. A. Mennella. "Children's Hedonic Judgments of Cigarette Smoke Odor: Effect of Parental Smoking and Maternal Mood." *Psychology of Addiction and Behavior* 19 (2005): 423–432.

Foster, J. A., and K. A. McVey Neufeld. "Gut-Brain Axis: How the Microbiome Influences Anxiety and Depression." *Trends in Neuroscience* 36 (2013): 305–312.

Frangos, E., J. Ellrich, and B. R. Komisaruk. "Non-invasive Access to the Vagus Nerve Central Projections via Electrical Stimulation of the External Ear: fMRI Evidence in Humans." *Brain Stimulation* 8 (2015): 624–636.

Franks, N. R., and C. R. Fletcher. "Spatial Patterns in Army Ant Foraging and Migration: *Eciton burchelli* on Barro Colorado Island, Panama." *Behavioral Ecology and Sociobiology* 12 (1983): 261–270.

Frega, R. "Evolutionary Prolegomena to a Pragmatist Epistemology of Belief." In *Pragmatist Epistemologies*, ed. R. Frega, 127–152. New York: Lexington Books, 2011.

Fried, I., C. L. Wilson, J. W. Morrow, K. A. Camerone, E. D. Behnke, L. C. Ackerson, and N. T. Maidment. "Increased Dopamine Release in the Human Amygdala During Performance of Cognitive Tasks." *Nature Neuroscience* 4 (2001): 201–206.

Friederici, A. D. "Towards a Neural Basis of Auditory Sentence Processing." *Trends in Cognitive Sciences* 6 (2002): 78–84.

Frijda, N. "Emotion Experiences and Its Varieties." *Emotion Review* 1 (2009): 264–271.

Frijda, N. *The Emotions*. New York: Cambridge University Press, 1986.

Frijda, N. "Impulsive Action and Motivation." *Biological Psychology* 84 (2010): 570–579.

Frijda, N. "The Laws of Emotion." *American Psychologist* 43 (1988): 349–358.

Friston, K. J. "The Free-Energy Principle: A Rough Guide to the Brain?" *Trends in Cognitive Sciences* 13 (2009): 293–301.

Fromm, E. *The Art of Loving*. New York: Harper and Row, 1975. First published 1956.

Frost, C. J., M. C. Mescher, J. E. Carlson, and C. M. De Moraes. "Plant Defense Priming Against Herbivores: Getting Ready for a Different Battle." *Plant Physiology* 146 (2008): 818–824.

Fuchs, T. *Ecology of the Brain: The Phenomenology and Biology of the Embodied Mind*. New York: Oxford University Press, 2018.

Fuster, J. M. *The Prefrontal Cortex*. New York: Elsevier, 2008.

Gainotti, G., and C. Marra. "Differential Contribution of Right and Left Temporooccipital and Anterior Temporal Lesions to Face Recognition Disorders." *Frontiers in Human Neuroscience* 5 (2011): art. 55. https://doi.org/10.3389/fnhum.2011.00055.

Gallagher, M., and P. C. Holland. "The Amygdala Complex: Multiple Roles in Associative Learning and Emotion." *Proceedings of the National Academy of Sciences of the United States of America* 91 (1994): 11771–11776.

Gallagher, S. *Enactivist Interventions: Rethinking the Mind*. New York: Oxford University Press, 2017.

Gallagher, S. "Philosophical Antecedents of Situated Cognition." In *The Cambridge Handbook of Situated Cognition*, ed. P. Robbins and M. Aydede, 35–52. New York: Cambridge University Press, 2009.

Gallese, V., and A. Goldman. "Mirror Neurons and the Simulation Theory of Mind-Reading." *Trends in Cognitive Science* 2 (1998): 493–501.

Gallistel, C. R. *The Organization of Learning*. Cambridge, MA: MIT Press, 1992.

Galton, F. *Inquiries into Human Faculty and Its Development*. New York: Macmillan, 1883.

Gamble, K. L., R. Berry, S. J. Frank, and M. E. Young. "Circadian Clock Control of Endocrine Factors." *Nature Reviews Endocrinology* 10 (2014): 466–475.

Garcia, J., W. G. Hankins, and K. W. Rusiniak. "Behavioral Regulation of the Milieu Interne in Man and Rat." *Science* 185 (1974): 824–831.

Giard, M. H., and F. Peronnet. "Auditory-Visual Integration During Multimodal Object Recognition in Humans: A Behavioral and Electrophysiological Study." *Journal of Cognitive Neuroscience* 11 (1999): 473–490.

Gibbs, R. "Embodied Experience and Linguistic Meaning." *Brain and Language* 84 (2003): 1–15.

Gibson, J. J. *The Ecological Approach to Visual Perception*. Boston: Houghton Mifflin, 1979.

Gigerenzer, G. *Adaptive Thinking: Rationality in the Real World*. New York: Oxford University Press, 2000.

Gigerenzer, G. *Gut Feelings: The Intelligence of the Unconscious*. New York: Viking, 2007.

Goehler, L. E., R. P. Gaykema, M. K. Hansen, K. Anderson, S. F. Maier, and L. R. Watkins. "Vagal Immune-to-Brain Communication: A Visceral Chemosensory Pathway." *Autonomic Neuroscience* 85 (2000): 49–59.

Goldin-Meadow, S. "The Role of Gesture in Communication and Thinking." *Trends in Cognitive Sciences* 3 (1999): 419–429.

Gordon, I., O. Zagoory-Sharon, J. F. Leckman, and R. Feldman. "Oxytocin and the Development of Parenting in Humans." *Biological Psychiatry* 68 (2010): 377–382.

Gould, S. J. "Shades of Lamarck." In *The Panda's Thumb*, ed. S. J. Gould, 76–84. New York: Norton, 1979.

Grahn, J. A., and J. B. Rowe. "Feeling the Beat: Premotor and Striatal Interactions in Musicians and Nonmusicians During Beat Perception." *Journal of Neuroscience* 29 (2009): 7540–7548.

Graybiel, A. M. "Habits, Rituals, and the Evaluative Brain." *Annual Review of Neuroscience* 31 (2008): 359–387.

Graybiel, A. M., T. Aosaki, A. W. Flaherty, and M. Kimura. "The Basal Ganglia and Adaptive Motor Control." *Science* 265 (1994): 1826–1831.

Graybiel, A. M., and S. T. Grafton. "The Striatum: Where Skills and Habits Meet." *Cold Spring Harbor Perspectives in Biology* 7 (2015): 1–13.

Green, C. D. "Darwinian Theory, Functionalism, and the First American Psychological Revolution." *American Psychologist* 64 (2009): 75–83.

Gregory, L., L. Yaguez, S. C. Williams, C. Altmann, S. J. Coen, V. Ng, J. M. Brammer, D. G. Thompson, and Q. Aziz. "Cognitive Modulation of the Cerebral Processing of Human Oesophageal Sensation Using Functional Magnetic Resonance Imaging." *Gut* 52 (2003): 1671–1677.

Guilleminault, C., M. Quera-Salva, and M. P. Goldberg. "Pseudo-hypersomnia and Pre-sleep Behaviour with Bilateral Paramedian Thalamic Lesions." *Brain* 116 (1993): 1549–1563.

Guttes, E., and S. Guttes. "Mitotic Synchrony in the Plasmodia of *Physarum polycephalum* and Mitotic Synchronization by Coalescence of Microplasmodia." *Methods in Cell Biology* 1 (1964): 43–54.

Hamann, S. B., T. D. Ely, S. T. Grafton, and C. D. Kilts. "Amygdala Activity Related to Enhanced Memory for Pleasant and Aversive Stimuli." *Nature Neuroscience* 2 (1999): 289–294.

Hamzei, F., M. Rijntjes, C. Dettmers, V. Glauche, C. Weiller, and C. Büchel. "The Human Action Recognition System and Its Relationship to Broca's Area: An fMRI Study." *NeuroImage* 19 (2003): 637–644.

Harmon-Jones, E. "Cognitive Dissonance and Experienced Negative Affect: Evidence That Dissonance Increases Experienced Negative Affect Even in the Absence of Aversive Consequences." *Personality and Social Psychology Bulletin* 26 (2000): 1490–1501.

Harrington, F. H. "Urine-Marking and Caching Behavior in the Wolf." *Behaviour* 76 (1981): 280–288.

Harris, K., A. Kassis, G. Major, and C. J. Chou. "Is the Gut Microbiota a New Factor Contributing to Obesity and Its Metabolic Disorders?" *Journal of Obesity* 2012: art. 879151. https://doi.org/10.1155/2012/879151.

Harrison, N. A., T. Singer, P. Rotshtein, R. J. Dolan, and H. D. Critchley. "Pupillary Contagion: Central Mechanisms Engaged in Sadness Processing." *Social Cognitive and Affective Neuroscience* 1 (2006): 5–17.

Hasson, U., A. A. Ghazanfar, B. Galantucci, S. Garrod, and C. Keysers. "Brain-to-Brain Coupling: A Mechanism for Creating and Sharing a Social World." *Trends in Cognitive Sciences* 16 (2012): 114–121.

Hauk, O., I. Johnsrude, and F. Pulvermüller. "Somatotopic Representation of Action Words in Human Motor and Premotor Cortex." *Neuron* 41 (2004): 301–307.

Havlíček, J., Z. Gašová, A. P. Smith, K. Zvára, and J. Flegr. "Decrease of Psychomotor Performance in Subjects with Latent 'Asymptomatic' Toxoplasmosis." *Parasitology* 122 (2001): 515–520.

Hawking, S., ed. *On the Shoulders of Giants: The Great Works of Physics and Astronomy*. With commentary by S. Hawking. Philadelphia: Running Press, 2004.

Heft, H. *Ecological Psychology in Context: James Gibson, Roger Barker, and the Legacy of William James's Radical Empiricism*. Mahwah, NJ: Lawrence Erlbaum Associates, 2001.

Heidegger, M. *Being and Time*. Trans. J. Macquarrie and E. Robinson. New York: Harper and Row, 1962. First published 1927 in German.

Heil, M. "Indirect Defence via Tritrophic Interactions." *New Phytologist* 178 (2008): 41–61.

Held, R., and A. Hein. "Movement-Produced Stimulation in the Development of Visually Guided Behavior." *Journal of Comparative and Physiological Psychology* 56 (1963): 872–876.

Heller, J. *Catch-22*. New York: Simon and Schuster, 1961.

Helmholtz, H. *Handbook of Physiological Optics*. 1867. Reprint, New York: Dover, 1963.

Henry, J. D. "The Use of Urine Marking in Scavenging Behavior of Red Fox (*Vulpes vulpes*)." *Behaviour* 61 (1977): 82–106.

Herbert, J. "Peptides in the Limbic System: Neurochemical Codes for Co-ordinated Adaptive Responses to Behavioral and Physiological Demand." *Progress in Neurobiology* 41 (1993): 723–791.

Herbert, J., and J. Schulkin. "Neurochemical Coding of Adaptive Responses in the Limbic System." In *Hormones, Brain and Behavior*, ed. D. Pfaff, 659–689. New York: Elsevier, 2002.

Ho, J. T. K., G. C. F. Chan, and J. C. B. Li. "Systemic Effects of Gut Microbiota and Its Relationship with Disease and Modulation." *BMC Immunology* 16 (2015): art. 21. https://doi.org /10.1186/s12865-015-0083-2.

Hodges, B. H., and R. M. Baron. "Values as Constraints on Affordances: Perceiving and Acting Properly." *Journal for the Theory of Social Behaviour* 22 (1992): 263–294.

Hodková, H., P. Kodym, and J. Flegr. "Poorer Results of Mice with Latent Toxoplasmosis in Learning Tests: Impaired Learning Processes or the Novelty Discrimination Mechanism?" *Parasitology* 134 (2007): 1329–1337.

Hodková, H., P. Kolbeková, A. Skallová, J. Lindová, and J. Flegr. "Higher Perceived Dominance in *Toxoplasma* Infected Men—a New Evidence for Role of Increased Level of Testosterone in Toxoplasmosis-Associated Changes in Human Behavior." *Neuroendocrinology Letters* 28 (2007): 110–114.

Hoffman, D. "The Construction of Visual Reality." In *Hallucination: Research and Practice*, ed. J. D. Blom and I. Sommer, 7–15. New York: Springer, 2011.

Hoffman, D. "The Interface Theory of Perception: Natural Selection Drives True Perception to Swift Extinction." In *Object Categorization: Computer and Human Vision Perspectives*, ed. S. Dickinson, A. Leonardis, B. Schiele, and M. J. Tarr, 148–165. New York: Cambridge University Press, 2009.

Hoffman, D., and C. Prakash. "Objects of Consciousness." *Frontiers in Psychology* 5 (2014): art. 577. https://doi.org/10.3389/fpsyg.2014.00577.

Hoffmeister, T. S., and B. D. Roitberg. "To Mark the Host or the Patch: Decisions of a Parasitoid Searching for Concealed Host Larvae." *Evolutionary Ecology* 11 (1997): 145–168.

Hooper, L. V., T. Midtvedt, and J. I. Gordon. "How Host-Microbial Interactions Shape the Nutrient Environment of the Mammalian Intestine." *Annual Review of Nutrition* 22 (2002): 283–307.

Huang, R., K. Wang, and J. Hu. "Effect of Probiotics on Depression: A Systematic Review and Meta-Analysis of Randomized Controlled Trials." *Nutrients* 8 (2016): art. 483. https://doi .org/10.3390/nu8080483.

Hume, D. *A Treatise of Human Nature*. Ed. D. Norton and M. Norton. New York: Oxford University Press, 2000. First published 1740.

Hurley, S. *Consciousness in Action*. Cambridge, MA: Harvard University Press, 1998.

Hutto, D., and E. Myin. *Evolving Enactivism: Basic Minds Meet Content*. Cambridge, MA: MIT Press, 2017.

Hutto, D., and E. Myin. *Radicalizing Enactivism: Basic Minds Without Content*. Cambridge, MA: MIT Press, 2013.

Huxley, T. H. "On the Hypothesis That Animals Are Automata, and Its History." *Nature* 10 (1874): 362–366.

Hwang, K., M. A. Bertolero, W. B. Liu, and M. Desposito. "The Human Thalamus Is an Integrative Hub for Functional Brain Networks." *Journal of Neuroscience* 37 (2017): 5594–5607.

Iacoboni, M., R. P. Woods, M. Brass, H. Bekkering, J. C. Mazziotta, and G. Rizzolatti. "Cortical Mechanisms of Human Imitation." *Science* 286 (1999): 2526–2528.

Iliev, I. D., V. A. Funari, K. D. Taylor, Q. Nguyen, C. N. Reyes, S. P. Strom, J. Brown, et al. "Interactions Between Commensal Fungi and the C-type Lectin Receptor Dectin-1 Influence Colitis." *Science* 336 (2012): 1314–1317.

Ingram, W. M., L. M. Goodrich, E. A. Robey, and M. B. Eisen. "Mice Infected with Low-Virulence Strains of *Toxoplasma gondii* Lose Their Innate Aversion to Cat Urine, Even After Extensive Parasite Clearance." *PLoS ONE* 8 (2013): art. e75246. https://doi.org/10.1371/journal.pone.0075246.

Jackman, H. "William James's Naturalistic Account of Concepts and His 'Rejection of Logic.'" In *Philosophy of Mind in the Nineteenth Century*, ed. S. Lapointe, 133–146. The History of the Philosophy of Mind 5. New York: Routledge, 2018.

Jackson, J. H. "Evolution and Dissolution of the Nervous System," 1884. In *Selected Writings of John Hughlings Jackson*, vol. 1, ed. J. Taylor, 45–91. London: Staples Press, 1958.

James, W. "Are We Automata?" *Mind* 4 (1879): 1–22.

James, W. "Brute and Human Intellect," 1878. In *William James: Writings, 1878–1899*, ed. G. E. Myers, 910–949. New York: Library of America, 1992.

James, W. "Does 'Consciousness' Exist?" *Journal of Philosophy, Psychology and Scientific Methods* 1 (1904): 477–491.

James, W. "Great Men and Their Environment," 1880. In *The Will to Believe and Other Essays in Popular Philosophy*. In *William James: Writings, 1878–1899*, ed. G. E. Myers, 618–646. New York: Library of America, 1992.

James, W. *The Letters of William James*. Ed. H. James. Vol. 1. Boston: Atlantic Monthly Press, 1920.

James, W. "On Some Omissions of Introspective Psychology," 1884. In *William James: Writings, 1878–1899*, ed. G. E. Myers, 986–1014. New York: Library of America, 1992.

James, W. "Philosophical Conceptions and Practical Results," 1898. In *William James: Writings, 1878–1899*, ed. G. E. Myers, 1077–1097. New York: Library of America, 1992.

James, W. *A Pluralistic Universe*, 1909. In *William James: Writings, 1902–1910*, ed. B. Kuklick, 625–819. New York: Library of America, 1987.

James, W. *Pragmatism: A New Name for Some Old Ways of Thinking*, 1907. In *William James: Writings, 1902–1910*, ed. B. Kuklick, 479–624. New York: Library of America, 1987.

James, W. *The Principles of Psychology*. 2 vols. New York: Henry Holt, 1890.

James, W. "The Psychology of Belief," 1889. In *William James: Writings, 1878–1899*, ed. G. E. Myers, 1021–1056. New York: Library of America, 1992.

James, W. "Rationality, Activity and Faith." *Princeton Review* 2 (1882): 58–86.

James, W. "Reflex Action and Theism," 1881. Reprinted in *The Will to Believe and Other Essays in Popular Philosophy*. In *William James: Writings, 1878–1899*, ed. G. E. Myers, 540–565. New York: Library of America, 1992.

James, W. "Remarks on Spencer's Definition of Mind as Correspondence," 1878. In *William James: Writings, 1878–1899*, ed. G. E. Myers, 893–909. New York: Library of America, 1992.

[James, W.]. Review of *Lectures on the Elements of Comparative Anatomy*, by T. H. Huxley. *North American Review* 100 (1865): 290–298.

[James, W.]. Review of "The Origin of Human Races and the Antiquity of Man Deduced from the Theory of 'Natural Selection,'" by A. R. Wallace. *North American Review* 101 (1865): 261–263.

James, W. "The Sentiment of Rationality," 1879. In *William James: Writings, 1878–1899*, ed. G. E. Myers, 950–985. New York: Library of America, 1992.

James, W. *Some Problems of Philosophy*, 1911. In *William James: Writings, 1902–1910*, ed. B. Kuklick, 979–1106. New York: Library of America, 1987.

James, W. "What Is an Emotion?" *Mind* 9 (1884): 188–205.

James, W. "The Will to Believe," 1896. In *William James: Writings, 1878–1899*, ed. G. E. Myers, 457–479. New York: Library of America, 1992.

Jeannerod, M. *The Cognitive Neuroscience of Action*. Oxford: Blackwell, 1997.

Jeannerod, M. *The Neural and Behavioural Organization of Goal-Directed Movements*. New York: Oxford University Press, 1988.

Jeannerod, M. "The Representing Brain: Neural Correlates of Motor Intention and Imagery." *Behavioral and Brain Sciences* 17 (1994): 187–201.

Jeannerod, M. "To Act or Not to Act: Perspectives on the Representation of Action." *Quarterly Journal of Experimental Psychology* 52 (1999): 1–29.

Jenkins, T., J. Nguyen, K. Polglaze, and P. Bertrand. "Influence of Tryptophan and Serotonin on Mood and Cognition with a Possible Role of the Gut-Brain Axis." *Nutrients* 8 (2016): 1–15.

Johnson, M. *The Aesthetics of Meaning and Thought: The Bodily Roots of Philosophy, Science, Morality, and Art*. Chicago: University of Chicago Press, 2018.

Johnson, M. *The Body in the Mind: The Bodily Basis of Meaning, Imagination, and Reason*. Chicago: University of Chicago Press, 1987.

Johnson, M. *Embodied Mind, Meaning, and Reason: How Our Bodies Give Rise to Understanding*. Chicago: University of Chicago Press, 2017.

Johnson, M. "Keeping the Pragmatism in Neuropragmatism." In *Neuroscience, Neurophilosophy and Pragmatism: Brains at Work with the World*, ed. T. Solymosi and J. Shook, 37–56. New York: Palgrave Macmillan, 2014.

Johnson, M. *The Meaning of the Body: Aesthetics of Human Understanding*. Chicago: University of Chicago Press, 2007.

Jovanovic, T., and S. D. Norrholm. "Neural Mechanisms of Impaired Fear Inhibition in Post-traumatic Stress Disorder." *Frontiers in Behavioral Neuroscience* 5 (2011): art. 44. https://doi.org/10.3389/fnbeh.2011.00044.

Jung, B.-K., K.-H. Pyo, K. Y. Shin, Y. S. Hwang, H. Lim, S. J. Lee, J.-H. Moon, et al. "*Toxoplasma gondii* Infection in the Brain Inhibits Neuronal Degeneration and Learning and Memory Impairments in a Murine Model of Alzheimer's Disease." *PLoS ONE* 7 (2012): art. e33312. https://doi.org/10.1371%2Fjournal.pone.0033312.

Kannan, G., K. Moldovan, J. C. Xiao, R. H. Yolken, L. Jones-Brando, and M. V. Pletnikov. "*Toxoplasma gondii* Strain-Dependent Effects on Mouse Behaviour." *Folia Parasitologica* 57 (2010): 151–155.

Kant, I. *Critique of Pure Reason* [synthesis of 1781 and 1787 eds.]. Trans. P. Guyer and A. W. Wood. New York: Cambridge University Press, 1998.

Kantarcioglu, A. S., N. Kiraz, and A. Aydin. "Microbiota-Gut-Brain Axis: Yeast Species Isolated from Stool Samples of Children with Suspected or Diagnosed Autism Spectrum Disorders and In Vitro Susceptibility Against Nystatin and Fluconazole." *Mycopathologia* 181 (2016): 1–7.

Kaplan, R., A. J. Horner, P. A. Bandettini, C. F. Doeller, and N. Burgess. "Human Hippocampal Processing of Environmental Novelty During Spatial Navigation." *Hippocampus* 24 (2014): 740–750.

Kaplan, R., and S. Kaplan. *The Experience of Nature: A Psychological Perspective.* New York: Cambridge University Press, 1989.

Kaplan, S. "Aesthetics, Affect, and Cognition: Environmental Preference from an Evolutionary Perspective." *Environment and Behavior* 1 (1987): 4–32.

Kaplan, S. "Environmental Preference in a Knowledge-Seeking, Knowledge-Using Organism." In *The Adapted Mind*, ed. J. Barkow, L. Cosmides, and J. Tooby, 581–598. New York: Oxford University Press, 1992.

Kaplan, S. "Perception and Landscape: Conceptions and Misconceptions," 1979. In *Environmental Aesthetics*, ed. J. Nasar, 45–55. New York: Cambridge University Press, 1988.

Kaplan, S. "Where Cognition and Affect Meet: A Theoretical Analysis of Preference," 1982. In *Environmental Aesthetics*, ed. J. Nasar, 56–63. New York: Cambridge University Press, 1988.

Kaplan, S., R. Kaplan, and J. S. Wendt. "Rated Preference and Complexity for Natural and Urban Visual Material." *Perception and Psychophysics* 12 (1972): 354–356.

Kauffmann, O. "Brain Plasticity and Phenomenal Consciousness." *Journal of Consciousness Studies* 18 (2011): 46–70.

Kawabata, H., and S. Zeki. "Neural Correlates of Beauty." *Journal of Neurophysiology* 91 (2004): 1699–1705.

Keeler, J. R., E. A. Roth, B. L. Neuser, J. M. Spitsbergen, D. J. Waters, and J. Vianney. "The Neurochemistry and Social Flow of Singing: Bonding and Oxytocin." *Frontiers in Human Neuroscience* 9 (2015): art. 518. https://doi.org/10.3389/fnhum.2015.00518.

Kestenbaum, V. *The Phenomenological Sense of John Dewey: Habit and Meaning.* Atlantic Highlands, NJ: Humanities Press, 1977.

Keil, J., N. Muller, N. Ihssen, and N. Weisz. "On the Variability of the McGurk Effect: Audiovisual Integration Depends on Prestimulus Brain States." *Cerebral Cortex* 22 (2011): 221–231.

Kendon, A. "Movement Coordination in Social Interactions: Some Examples Described." *Acta Psychologica* 32 (1970): 101–125.

Kidd, C., S. T. Piantadosi, and R. N. Aslin. "The Goldilocks Effect: Human Infants Allocate Attention to Visual Sequences That Are Neither Too Simple nor Too Complex." *PLoS ONE* 7 (2012): art. e36399. https://doi.org/10.1371%2Fjournal.pone.0036399.

Kincaid, R. L., and T. E. Mansour. "Chemotaxis Toward Carbohydrates and Amino Acids in *Physarum polycephalum*." *Experimental Cell Research* 116 (1978): 377–385.

Kind, A. "Twentieth-Century Philosophy of Mind: Themes, Problems, and Scientific Context." In *Philosophy of Mind in the Twentieth and Twenty-First Centuries*, ed. A. Kind, 1–20. The History of the Philosophy of Mind 6. New York: Routledge, 2018.

King, D. W., G. A. Leskin, L. A. King, and F. W. Weathers. "Confirmatory Factor Analysis of the Clinician-Administered PTSD Scale: Evidence for the Dimensionality of Posttraumatic Stress Disorder." *Psychological Assessment* 10 (1998): 90–96.

Knowlton, B., J. Mangels, and L. Squire. "A Neostriatal Habit Learning System in Humans." *Science* 273 (1996): 1399–1402.

Kocazeybek, B., Y. A. Oner, R. Turksoy, C. Babur, H. Cakan, N. Sahip, A. Unal, et al. "Higher Prevalence of Toxoplasmosis in Victims of Traffic Accidents Suggest Increased Risk of Traffic Accident in *Toxoplasma*-Infected Inhabitants of Istanbul and Its Suburbs." *Forensic Science International* 187 (2009): 103–108.

Koelsch, S. "Significance of Broca's Area and Ventral Premotor Cortex for Music-Syntactic Processing." *Cortex* 42 (2006): 518–520.

Koelsch, S., T. Fritz, and G. Schlaug. "Amygdala Activity Can Be Modulated by Unexpected Chord Functions During Music Listening." *NeuroReport* 19 (2008): 1815–1819.

Koelsch, S., T. Fritz, D. Y. von Cramon, K. Muller, and A. D. Friederici. "Investigating Emotion with Music: An fMRI Study." *Human Brain Mapping* 27 (2006): 239–250.

Koelsch, S., T. C. Gunter, D. Y. von Cramon, S. Zysset, G. Lohmann, and A. D. Friederici. "Bach Speaks: A Cortical 'Language-Network' Serves the Processing of Music." *NeuroImage* 17 (2002): 956–966.

Koelsch, S., and N. Steinbeis. "Shared Neural Resources Between Music and Language Indicate Semantic Processing of Musical Tension-Resolution Patterns." *Cerebral Cortex* 18 (2008): 1169–1178.

Koffka, K. *Principles of Gestalt Psychology*. New York: Harcourt, Brace, 1935.

Kohler, E., C. Keysers, M. A. Umiltá, L. Fogassi, V. Gallese, and G. Rizzolatti. "Hearing Sounds, Understanding Actions: Action Representation in Mirror Neurons." *Science* 297 (2002): 846–848.

Kokal, I., A. Engel, S. Kirschner, and C. Keysers. "Synchronized Drumming Enhances Activity in the Caudate and Facilitates Prosocial Commitment—if the Rhythm Comes Easily." *PLoS ONE* 6 (2011): art. e27272. https://doi.org/10.1371/journal.pone.0027272.

Kosfeld, M., M. Heinrichs, P. J. Zak, U. Fischbacher, and E. Fehr. "Oxytocin Increases Trust in Humans." *Science* 435 (2005): 673–676.

Kosslyn, S. M. *Image and Brain: The Resolution of the Imagery Debate*. Cambridge, MA: MIT Press, 1994.

Kosslyn, S. M. *Image and Mind*. Cambridge, MA: Harvard University Press, 1986.

Kotz, S. A., and M. Schmidt-Kassow. "Basal Ganglia Contribution to Rule Expectancy and Temporal Predictability in Speech." *Cortex* 68 (2015): 48–60.

Kotz, S. A., M. Schwartze, and M. Schmidt-Kassow. "Non-motor Basal Ganglia Functions: A Review and Proposal for a Model of Sensory Predictability in Auditory Language Perception." *Cortex* 45 (2009): 982–990.

Krueger, J. "Extended Cognition and the Space of Social Interaction." *Consciousness and Cognition* 20 (2011): 643–655.

Kugiumutzakis, G. "Genesis and Development of Early Infant Mimesis to Facial and Vocal Models." In *Imitation in Infancy*, ed. J. Nadel and G. Butterworth, 127–185. New York: Cambridge University Press, 1999.

Kühle, L. "William James and the Embodied Mind." *Contemporary Pragmatism* 14 (2017): 51–75.

Kunert, R., R. M. Willems, D. Casasanto, A. D. Patel, and P. Hagoort. "Music and Language Syntax Interact in Broca's Area: An fMRI Study." *PLoS ONE* 10 (2015): art. e0141069. https://doi.org/10.1371%2Fjournal.pone.0141069.

Kupers, R., and M. Ptito. "Insights from Darkness: What the Study of Blindness Has Taught Us About Brain Structure and Function." *Progress in Brain Research* 192 (2011): 17–31.

Kusbeci, O. Y., O. Miman, M. Yaman, O. C. Aktepe, and S. Yazar. "Could *Toxoplasma gondii* Have Any Role in Alzheimer Disease?" *Alzheimer Disease and Associated Disorders* 25 (2011): 1–3.

Laeng, B., and U. Sulutvedt. "The Eye Pupil Adjusts to Imaginary Light." *Psychological Science* 25 (2013): 188–197.

Lakoff, G., and M. Johnson. *Metaphors We Live By*. Chicago: University of Chicago Press, 1980.

Lakoff, G., and M. Johnson. *Philosophy in the Flesh*. New York: Basic Books, 1999.

Lamberton, P., C. Donnelly, and J. Webster. "Specificity of the *Toxoplasma gondii*–Altered Behaviour to Definitive Versus Non-definitive Host Predation Risk." *Parasitology* 135 (2008): 1143–1150.

Landgrave, M., and A. Nowrasteh. "Criminal Immigrants: Their Numbers, Demographics, and Countries of Origin." Immigration Research and Policy Brief No. 1, Cato Institute, March 15, 2017. https://www.cato.org/publications/immigration-reform-bulletin/criminal-immigrants-their-numbers-demographics-countries.

Langfeld, H. S. "Fifty Volumes of the *Psychological Review*." *Psychological Review* 50 (1943): 141–155.

Latty, T., and M. Beekman. "Food Quality Affects Search Strategy in the Acellular Slime Mould, *Physarum polycephalum*." *Behavioral Ecology* 20 (2009): 1160–1167.

LeDoux, J. E. *Anxious: Using the Brain to Understand and Treat Fear and Anxiety*. New York: Viking, 2015.

LeDoux, J. E. "Cognitive-Emotional Interactions in the Brain." *Cognition and Emotion* 3 (1989): 267–289.

Lee, D. N., and J. R. Lishman. "Visual Proprioceptive Control of Stance." *Journal of Human Movement Studies* 1 (1975): 87–95.

Lee, P., M. M. Swarbrick, and K. K. Ho. "Brown Adipose Tissue in Adult Humans: A Metabolic Renaissance." *Endocrine Reviews* 34 (2013): 413–438.

Leisman, G., A. Moustafa, and T. Shafir. "Thinking, Walking, Talking: Integratory Motor and Cognitive Brain Function." *Frontiers in Public Health* 4 (2016): art. 94. https://doi.org/10.3389/fpubh.2016.00094.

Lennon, J. "The Notion of Experience." *Thomist: A Speculative Quarterly Review* 23 (1960): 315–343.

Leproult, R., E. F. Colecchia, M. L'Hermite-Balériaux, and E. V. Cauter. "Transition from Dim to Bright Light in the Morning Induces an Immediate Elevation of Cortisol Levels." *Journal of Clinical Endocrinology and Metabolism* 86 (2001): 151–157.

Lessard, D. A., S. A. Linkenauger, and D. R. Proffitt. "Look Before You Leap: Jumping Ability Affects Distance Perception." *Perception* 38 (2009): 1863–1866.

Lévêque, Y., and D. Schön. "Modulation of the Motor Cortex During Singing-Voice Perception." *Neuropsychologia* 70 (2015): 58–63.

Lewicki, P., T. Hill, and M. Czyzewska. "Nonconscious Acquisition of Information." *American Psychologist* 47 (1992): 796–801.

Lewis, C. I. *Mind and the World-Order*. New York: Charles Scribner's Sons, 1929.

Lewis, C. S. *The Four Loves*. London: Geoffrey Bles, 1960.

Lickliter, R. "The Integrated Development of Sensory Organization." *Clinics in Perinatology* 38 (2011): 591–603.

Ling, V. J., D. Lester, P. B. Mortensen, P. W. Langenberg, and T. T. Postolache. "*Toxoplasma gondii* Seropositivity and Suicide Rates in Women." *Journal of Nervous and Mental Disease* 199 (2011): 440–444.

Linson, A., A. Clark, S. Ramamoorthy, and K. J. Friston. "The Active Inference Approach to Ecological Perception: General Information Dynamics for Natural and Artificial Embodied Cognition." *Frontiers in Robotics and AI* 5 (2018): art. 21. https://doi.org/10.3389/frobt.2018.00021.

Loewenstein, G. "The Psychology of Curiosity: A Review and Reinterpretation." *Psychological Bulletin* 116 (1994): 75–98.

Long, J. *Darwin's Devices: What Evolving Robots Can Teach Us About the History of Life and the Future of Technology*. New York: Basic Books, 2011.

Love, D. "Who Was Johannes Kepler?" *Astronomy and Geophysics* 50 (2009): 6.15–6.17.

Luan, C., L. Xie, X. Yang, H. Miao, N. Lv, R. Zhang, X. Xiao, Y. Hu, Y. Liu, N. Wu, Y. Zhu, and B. Zhu. "Dysbiosis of Fungal Microbiota in the Intestinal Mucosa of Patients with Colorectal Adenomas." *Scientific Reports* 5 (2015): art. 7980. https://doi.org/10.1038/srep07980.

Luria, A. R. *The Mind of a Mnemonist: A Little Book About a Vast Memory*. Trans. L. Solotaroff. New York: Basic Books, 1968.

Lyte, M. "Probiotics Function Mechanistically as Delivery Vehicles for Neuroactive Compounds: Microbial Endocrinology in the Design and Use of Probiotics." *Bioessays* 33 (2011): 574–581.

Macaluso, E., C. D. Frith, and J. Driver. "Modulation of Human Visual Cortex by Crossmodal Spatial Attention." *Science* 289 (2000): 1206–1208.

MacKenzie M. "Dewey, Enactivism, and the Qualitative Dimension." *Humana Mente: Journal of Philosophical Studies* 31 (2016): 21–36.

Madzia, R. "Chicago Pragmatism and the Extended Mind Theory: Mead and Dewey on the Nature of Cognition." *European Journal of Pragmatism and American Philosophy* 5 (2013): 193–211.

Madzia, R. "Constructive Realism: In Defense of the Objective Reality of Perspectives." *Human Affairs* 23 (2013): 645–657.

Madzia, R. "Root-Brains: The Frontiers of Cognition in the Light of John Dewey's Philosophy of Nature." *Contemporary Pragmatism* 14 (2017): 93–111.

Madzia, R., and M. Jung, eds. *Pragmatism and Embodied Cognitive Science: From Bodily Intersubjectivity to Symbolic Articulation*. Boston: De Gruyter, 2016.

Maess, B., S. Koelsch, T. C. Gunter, and A. D. Friederici. "Musical Syntax Is Processed in Broca's Area: An MEG study." *Nature Neuroscience* 4 (2001): 540–545.

Magnusson, A., and T. Partonen. "The Diagnosis, Symptomatology, and Epidemiology of Seasonal Affective Disorder." *CNS Spectrums* 10 (2005): 625–634.

Malloch, S. "Mothers and Infants and Communicative Musicality." In "Rhythm, Musical Narrative, and the Origins of Human Communication," ed. C. Trevarthen. Supplement, *Musicae Scientiae* 3, no. S1 (1999–2000): 29–57.

Malloch, S., and C. Trevarthen. "The Neuroscience of Emotion in Music." In *Communicative Musicality: Exploring the Basis of Human Companionship*, ed. S. Malloch and C. Trevarthen, 105–146. New York: Oxford University Press, 2009.

Maratos, O. "Trends in Development of Imitation in Early Infancy." In *Regressions in Mental Development: Basis Phenomena and Theories*, ed. T. G. Bever, 81–101. Hillsdale, NJ: Erlbaum, 1982.

Maravita, A., and A. Iriki. "Tools for the Body (Schema)." *Trends in Cognitive Sciences* 8 (2004): 79–86.

Martin, A. "The Organization of Semantic Knowledge and the Origin of Words in the Brain." In *The Origins and Diversification of Language*, ed. N. G. Jablonski and L. C. Aiello, 69–88. San Francisco: California Academy of Sciences, 1998.

Mather, M., and J. F. Thayer. "How Heart Rate Variability Affects Emotion Regulation Brain Networks." *Current Opinion in Behavioral Sciences* 19 (2018): 98–104.

Matthias, E., R. Schandry, S. Duschek, and O. Pollatos. "On the Relationship Between Interoceptive Awareness and the Attentional Processing of Visual Stimuli." *International Journal of Psychophysiology* 72 (2009): 154–159.

McDermott, R. "The Feeling of Emotionality: The Meaning of Neuroscientific Advances for Political Science." *Perspectives on Politics* 2 (2004): 691–706.

McDougall, S. J., and M. C. Andresen. "Independent Transmission of Convergent Visceral Primary Afferents in the Solitary Tract Nucleus." *Journal of Neurophysiology* 109 (2013): 507–517.

McDougall, S. J., J. H. Peters, and M. C. Andresen. "Convergence of Cranial Visceral Afferents Within the Solitary Tract Nucleus." *Journal of Neuroscience* 29 (2009): 12886–12895.

McGranahan, L. *Darwinism and Pragmatism: William James on Evolution and Self-Transformation*. New York: Routledge, 2017.

McGregor, I., I. R. Newby-Clark, and M. P. Zanna. "'Remembering' Dissonance: Simultaneous Accessibility of Inconsistent Cognitive Elements Moderates Epistemic Discomfort." In *Cognitive Dissonance: Progress on a Pivotal Theory in Social Psychology*, ed. E. Harmon-Jones and J. Mills, 325–353. Washington, DC: American Psychological Association, 1999.

McGregor, I., M. P. Zanna, J. G. Holmes, and S. J. Spencer. "Compensatory Conviction in the Face of Personal Uncertainty: Going to Extremes and Being Oneself." *Journal of Personality and Social Psychology* 80 (2001): 472–488.

McGurk, H., and J. MacDonald. "Hearing Lips and Seeing Voices." *Nature* 264 (1976): 746–748.

McReynolds, P. "Autopoiesis and Transaction." *Transactions of the Charles S. Peirce Society* 53 (2017): 312–334.

Mead, G. H. *Mind, Self, and Society: From the Standpoint of a Social Behaviorist.* Ed. C. W. Morris. Chicago: University of Chicago Press, 1934.

Mead, G. H. *The Philosophy of the Act.* Ed. C. W. Morris with J. M. Brewster, A. M. Dunham, and D. Miller. Chicago: University of Chicago Press, 1938.

Mead, G. H. *The Philosophy of the Present.* With introductory remarks by J. Dewey and A. Murphy. LaSalle, IL: Open Court, 1932.

Mead, G. H. "The Relations of Psychology and Philology." *Psychological Bulletin* 1 (1904): 375–391.

Mead, G. H. "Social Psychology as Counterpart to Physiological Psychology." *Psychological Bulletin* 6 (1909): 401–408.

Mealey, L., and P. Theis. "The Relationship Between Mood and Preferences Among Natural Landscapes: An Evolutionary Perspective." *Ethology and Sociobiology* 16 (1995): 247–256.

Meltzoff, A. N., and M. K. Moore. "Explaining Facial Imitation: A Theoretical Model." *Early Development and Parenting* 6 (1997): 179–192.

Meltzoff, A. N., and M. K. Moore. "Imitation in Newborn Infants: Exploring the Range of Gestures Imitated and the Underlying Mechanisms." *Developmental Psychology* 25 (1989): 954–962.

Meltzoff, A. N., and M. K. Moore. "Imitation, Memory, and the Representation of Persons." *Infant Behavior and Development* 17 (1994): 83–99.

Meltzoff, A. N., and M. K. Moore. "Imitation of Facial and Manual Gestures by Human Neonates." *Science* 198 (1977): 75–78.

Menary, R. *Cognitive Integration: Mind and Cognition Unbounded.* New York: Palgrave Macmillan, 2007.

Menary, R. "Pragmatism and the Pragmatic Turn in Cognitive Science." In *The Pragmatic Turn: Toward Action-Oriented Views in Cognitive Science,* ed. A. K. Engel, K. J. Friston, and D. Kragic, 215–223. Cambridge, MA: MIT Press, 2015.

Mennella, J. A., and C. A. Forestell. "Children's Hedonic Responses to the Odors of Alcoholic Beverages: A Window to Emotions." *Alcohol* 42 (2008): 249–260.

Mennella, J. A., and P. L. Garcia. "Children's Hedonic Response to the Smell of Alcohol: Effects of Parental Drinking Habits." *Alcohol: Clinical and Experimental Research* 24 (2000): 1167–1171.

Merabet, L. B., R. Hamilton, G. Schlaug, J. D. Swisher, E. T. Kiriakapoulos, N. B. Pitskel, T. Kauffman, and A. Pascual-Leone. "Rapid and Reversible Recruitment of Early Visual Cortex for Touch." *PLoS ONE* 3 (2008): art. e3046. https://doi.org/10.1371%2Fjournal.pone.0003046.

Merleau-Ponty, M. *Consciousness and the Acquisition of Language.* Trans. H. J. Silverman. Evanston, IL: Northwestern University Press, 1991. First published 1964 in French.

Merleau-Ponty, M. "Film and the New Psychology," 1947. In *Sense and Non-Sense,* trans. H. Dreyfus and P. Dreyfus, 43–59. Evanston, IL: Northwestern University Press, 1964.

Merleau-Ponty, M. *Phenomenology of Perception.* Trans. C. Smith. New York: Routledge and Kegan Paul, 1962. First published 1945 in French.

Messaoudi, M., R. Lalonde, N. Violle, H. Javelot, D. Desor, A. Nejdi, J. F. Bisson, et al. "Assessment of Psychotropic-like Properties of a Probiotic Formulation (*Lactobacillus helveticus* R0052 and *Bifidobacterium longum* R0175) in Rats and Human Subjects." *British Journal of Nutrition* 105 (2010): 755–764.

BIBLIOGRAPHY

Miceli, M., and C. Castelfranchi. *Expectancy and Emotion*. New York: Oxford University Press, 2015.

Miman, O., O. Y. Kusbeci, O. C. Aktepe, and Z. Cetinkaya. "The Probable Relation Between *Toxoplasma gondii* and Parkinson's Disease." *Neuroscience Letters* 475 (2010): 129–131.

Moffat, D. "On the Positive Value of Affect." Paper presented at the AISB 2001 Symposium on Emotion, Cognition, and Affective Computing; York, UK, March 23–24, 2001.

Moll, J., R. Zahn, R. de Oliveira-Souza, F. Krueger, and J. Grafman. "The Neural Basis of Human Moral Cognition." *Nature Reviews Neuroscience* 6 (2005): 799–809.

Morrot, G., F. Brochet, and D. Dubourdieu. "The Color of Odors." *Brain and Language* 79 (2001): 309–320.

Murphy, M. C., A. C. Nau, C. Fisher, S. Kim, J. S. Schuman, and K. C. Chan. "Top-Down Influence on the Visual Cortex of the Blind During Sensory Substitution." *NeuroImage* 125 (2016): 932–940.

Myin, E., and J. Degenaar. "Enactive Vision." In *The Routledge Handbook of Embodied Cognition*, ed. L. Shapiro, 90–98. New York: Routledge, 2014.

Nagy, E. "Innate Intersubjectivity: Newborns' Sensitivity to Communication Disturbance." *Developmental Psychology* 44 (2008): 1779–1784.

Nagy, E., and P. Molnár. "Homo Imitans or Homo Provocans? The Phenomenon of Neonatal Initiation." *Infant Behavior and Development* 27 (2004): 57–63.

Nakagaki, T., M. Iima, T. Ueda, Y. Nishiura, T. Saigusa, A. Tero, R. Kobayashi, and K. Showalter. "Minimum-Risk Path Finding by an Adaptive Amoebal Network." *Physical Review Letters* 99 (2007): art. 068104. https://doi.org/10.1103/PhysRevLett.99.068104.

Nakagaki, T., R. Kobayashi, Y. Nishiura, and T. Ueda. "Obtaining Multiple Separate Food Sources: Behavioural Intelligence in the *Physarum* Plasmodium." *Proceedings of the Royal Society B: Biological Sciences* 27 (2004): 2305–2310.

Nakagaki, T., H. Yamada, and M. Hara. "Smart Network Solutions in an Amoeboid Organism." *Biophysical Chemistry* 107 (2004): 1–5.

Nakagaki, T., H. Yamada, and Á. Tóth. "Maze-Solving by an Amoeboid Organism." *Nature* 407 (2000): 470.

Nakagaki, T., H. Yamada, and Á. Tóth. "Path Finding by Tube Morphogenesis in an Amoeboid Organism." *Biophysical Chemistry* 92 (2001): 47–52.

Nakashima, Y., M. Teshiba, and Y. Hirose. "Flexible Use of Patch Marks in an Insect Predator: Effect of Sex, Hunger State, and Patch Quality." *Ecological Entomology* 27 (2002): 581–587.

Nakayama, K. 2010. "Introduction: Vision Going Social." In *The Science of Social Vision*, ed. R. N. Adams Jr., R. Ambady, K. Nakayama, and S. Shimojo, 3–17. New York: Oxford University Press, 2010.

Nath, A. R., and M. S. Beauchamp. "A Neural Basis for Interindividual Differences in the McGurk Effect, a Multisensory Speech Illusion." *NeuroImage* 59 (2012): 781–787.

Neal, D. T., and T. Chartrand. "Embodied Emotion Perception: Amplifying and Dampening Facial Feedback Modulates Emotion Perception Accuracy." *Social Psychology and Personality Science* 2 (2011): 673–678.

Neville, R. C. *The Cosmology of Freedom*. New Haven, CT: Yale University Press, 1974.

Niedenthal, P. M., M. Brauer, J. B. Halberstadt, and A. H. Inner-Ker. "When Did Her Smile Drop? Facial Mimicry and the Influences of Emotional State on the Detection of Change in Emotional Expression." *Cognition and Emotion* 15 (2001): 853–864.

Niederhoffer, K. G., and J. W. Pennebaker. "Linguistic Style Matching in Social Interaction." *Journal of Language and Social Psychology* 10 (2002): 59–65.

Nietzsche, F. *Twilight of the Idols*, 1888. Trans. W. Kaufman. In *The Portable Nietzsche*, ed. W. Kaufman, 463–563. New York: Penguin, 1954.

Nietzsche, F. *The Will to Power*. Ed. W. Kaufmann. Trans. W. Kaufman and R. J. Hollingdale. New York: Vintage Books, 1968.

Nishitani, N., M. Schürmann, K. Amunts, and R. Hari. "Broca's Region: From Action to Language." *Physiology* 20 (2005): 60–69.

Noë, A. *Action in Perception*. Cambridge, MA: MIT Press, 2004.

Noë, A. *Out of Our Heads: Why You Are Not Your Brain, and Other Lessons from the Biology of Consciousness*. New York: Hill and Wang, 2009.

Norgren, R. "Gustatory System." In *The Rat Nervous System*, ed. G. Paxinos, 751–771. San Diego, CA: Academic Press, 1995.

Norman, S. B., M. B. Stein, and J. R. Davidson. "Profiling Posttraumatic Functional Impairment." *Journal of Nervous and Mental Disease* 195 (2007): 48–53.

Nussbaum, C. *The Musical Representation: Meaning, Ontology, and Emotion*. Cambridge, MA: MIT Press, 2007.

Nussbaum, M. C. *The Therapy of Desire*. Princeton, NJ: Princeton University Press, 1994.

Nussbaum, M. C. *Upheavals of Thought: The Intelligence of Emotion*. Cambridge: Cambridge University Press, 2001.

Nyíri, K. *Meaning and Motoricity: Essays on Image and Time*. New York: Peter Lang, 2014.

Olianas, M. C., V. Loi, M. Lai, E. Mosca, and P. Onali. "Corticotropin-Releasing Hormone Stimulates Adenylyl Cyclase Activity in the Retinas of Different Animal Species." *Regulatory Peptides* 47 (1993): 127–132.

Olianas, M. C., and P. Onali. "G Protein-Coupled Corticotropin-Releasing Hormone Receptors in Rat Retina." *Regulatory Peptides* 56 (1995): 61–70.

Ooishi, Y., H. Mukai, K. Watanabe, S. Kawato, and M. Kashino. "Increase in Salivary Oxytocin and Decrease in Salivary Cortisol After Listening to Relaxing Slow-Tempo and Exciting Fast-Tempo Music." *PLoS ONE* 12 (2017): art. e0189075. https://doi.org/10.1371%2Fjournal.pone.0189075.

O'Regan, K., and A. Noë. "A Sensorimotor Account of Vision and Visual Consciousness." *Behavior and Brain Sciences* 24 (2001): 939–973.

Packard, M. G., L. Cahill, and J. L. McGaugh. "Amygdala Modulation of Hippocampal-Dependent and Caudate Nucleus-Dependent Memory Processes." *Proceedings of the National Academy of Sciences of the United States of America* 91 (1994): 8477–8481.

Pangborn, R. M., H. W. Berg, and B. Hansen. "The Influence of Color on Discrimination of Sweetness in Dry Table-Wine." *American Journal of Psychology* 76 (1963): 492–495.

Panksepp, J. *Affective Neuroscience: The Foundations of Human and Animal Emotions*. New York: Oxford University Press, 1998.

Papoušek, H., and M. Papoušek. "Intuitive Parenting: A Dialectic Counterpart to the Infant's Integrative Competence." In *Handbook of Infant Development*, ed. J. D. Osofsky, 669–720. New York: Wiley, 1987.

Papoušek, H., and M. Papoušek. "Mothering and Cognitive Head Start: Psychobiological Considerations." In *Studies in Mother-Infant Interaction: The Loch Lomond Symposium*, ed. H. R. Schaffer, 63–85. New York: Academic Press, 1977.

Parr, W. V., K. G. White, and D. Heatherbell. "The Nose Knows: Influence of Colour on Perception of Wine Aroma." *Journal of Wine Research* 14 (2003): 79–101.

Parrott, G., and J. Schulkin. "Neuropsychology and the Cognitive Nature of the Emotions." *Cognition and Emotion* 7 (1993): 43–59.

Patel, A. D. "Language, Music, Syntax and the Brain." *Nature Neuroscience* 6 (2003): 674–681.

Patel, A. D. "The Relationship of Music to the Melody of Speech and to Syntactic Processing Disorders in Aphasia." *Annals of the New York Academy of Sciences* 1060 (2005): 59–70.

Patel, A. D. "Syntactic Processing in Language and Music: Different Cognitive Operations, Similar Neural Resources." *Music Perception* 16 (1998): 27–42.

Patel, A. D. "Talk of the Tone." *Nature* 453 (2008): 726–727.

Patel, A. D., E. Gibson, J. Ratner, M. Besson, and P. J. Holcomb. "Processing Syntactic Relations in Language and Music: An Event-Related Potential Study." *Journal of Cognitive Neuroscience* 10 (1998): 717–733.

Patel, A. D., J. R. Iversen, M. Wassenaar, and P. Hagoort. "Musical Syntactic Processing in Agrammatic Broca's Aphasia." *Aphasiology* 22 (2008): 776–789.

Pavlov, I. P. *Lectures on Conditioned Reflexes.* New York: International Publishing, 1960. First published 1927.

Pavlov, I. P. *The Work of the Digestive Glands.* London: Charles Griffin, 1902.

Peciña, S., J. Schulkin, and K. C. Berridge. "Nucleus Accumbens Corticotropin-Releasing Factor Increases Cue-Triggered Motivation for Sucrose Reward: Paradoxical Positive Incentive Effects in Stress?" *BMC Biology* 4 (2006): art. 8. https://doi.org/10.1186/1741 -7007-4-8.

Peirce, C. S. "The Fixation of Belief." *Popular Science Monthly* 12 (1877): 1–15.

Peirce, C. S. "How to Make Our Ideas Clear." *Popular Science Monthly* 12 (1878): 286–302.

Peirce, C. S. *Pragmatism as a Principle and Method of Right Thinking: The 1903 Harvard Lectures on Pragmatism.* Ed. P. A. Turrisi. Albany: State University of New York Press, 1997.

Peirce, C. S. "Some Consequences of Four Incapacities." *Journal of Speculative Philosophy* 2 (1868): 140–157.

Pelagio-Flores, R., R. Ortíz-Castro, A. Méndez-Bravo, L. Macías-Rodríguez, and J. López-Bucio. "Serotonin, a Tryptophan-Derived Signal Conserved in Plants and Animals, Regulates Root System Architecture Probably Acting as a Natural Auxin Inhibitor in *Arabidopsis thaliana.*" *Plant and Cell Physiology* 52 (2011): 490–508.

Perret, D. I., and N. J. Emery. "Understanding the Intentions of Others from Visual Signals: Neurophysiological Evidence." *Cahiers de psychologie cognitive* 13 (1994): 683–694.

Perry, R. B. *The Thought and Character of William James, as Revealed in Unpublished Correspondence and Notes, Together with His Published Writings.* Vol. 1. Boston: Little, Brown, 1935.

Pessoa, L. *The Cognitive-Emotional Brain: From Interactions to Integration.* Cambridge, MA: MIT Press, 2013.

Petersen, S. E., and M. I. Posner. "The Attention System of the Human Brain: 20 Years After." *Annual Review of Neuroscience* 35 (2012): 73–89.

Petro, L. S., A. T. Paton, and L. Muckli. "Contextual Modulation of Primary Visual Cortex by Auditory Signals." *Philosophical Transactions of the Royal Society B: Biological Sciences* 372 (2017): art. 20160104. https://doi.org/10.1098/rstb.2016.0104.

Petrovich, G. D., N. W. Canteras, and L. W. Swanson. "Combinatorial Amygdalar Inputs to Hippocampal Domains and the Hypothalamic Behavior Systems." *Brain Research Reviews* 38 (2001): 247–289.

Pfaff, D. W., I. Phillips, and R. T. Rubin. *Principles of Hormone/Behavior Relations.* Boston: Academic Press, an imprint of Elsevier, 2004.

Pfeifer, R. and J. Bongard. *How the Body Shapes the Way We Think: A New View of Intelligence.* Cambridge, MA: MIT Press, 2007.

Pfeifer, R., M. Lungarella, O. Sporns, and Y. Kuniyoshi. "On the Information Theoretic Implications of Embodiment—Principles and Methods." In *50 Years of Artificial Intelligence: Essays Dedicated to the 50th Anniversary of Artificial Intelligence,* ed. M. Lungarella, F. Iida, J. Bongard, and R. Pfeifer, 6–86. New York: Springer, 2007.

Phelps, E. A. "Human Emotion and Memory: Interactions of the Amygdala and Hippocampal Complex." *Current Opinion in Neurobiology* 14 (2004): 198–202.

Pinto-Sanchez, M. I., G. B. Hall, K. Ghajar, A. Nardelli, C. Bolino, J. T. Lau, F. P. Martin, et al. "Probiotic *Bifidobacterium longum* NCC3001 Reduces Depression Scores and Alters Brain Activity: A Pilot Study in Patients with Irritable Bowel Syndrome." *Gastroenterology* 153 (2017): 448–459.

Plato. *Gorgias*. Trans. W. D. Woodhead. In *Plato: The Collected Dialogues*, ed. E. Hamilton and H. Cairns, 229–307. Princeton, NJ: Princeton University Press, 1963.

Plato. *Laws*. Trans. A. E. Taylor. In *Plato: The Collected Dialogues*, ed. E. Hamilton and H. Cairns, 1125–1518. Princeton, NJ: Princeton University Press, 1963.

Plato. *Republic*. Trans. P. Shorey. In *Plato: The Collected Dialogues*, ed. E. Hamilton and H. Cairns, 575–844. Princeton, NJ: Princeton University Press, 1963.

Portugal, S. J., T. Y. Hubel, J. Fritz, S. Heese, D. Trobe, B. Voelkl, S. Hailes, A. M. Wilson, and J. R. Usherwood. "Upwash Exploitation and Downwash Avoidance by Flap Phasing in Ibis Formation Flight." *Nature* 505 (2014): 399–402.

Posner, M. I., and S. E. Petersen. "The Attention System of the Human Brain." *Annual Review of Neuroscience* 13 (1990): 25–42.

Power, M. L. "Viability as Opposed to Stability: An Evolutionary Perspective on Physiological Regulation." In *Allostasis, Homeostasis, and the Costs of Physiological Adaptation*, ed. J. Schulkin, 343–364. New York: Cambridge University Press, 2004.

Power, M. L., and J. Schulkin. *The Evolution of Obesity*. Baltimore, MD: Johns Hopkins University Press, 2009.

Power, M. L., and J. Schulkin. *The Evolution of the Human Placenta*. Baltimore, MD: Johns Hopkins University Press, 2012.

Powley, T. L. "Vagal Input to the Enteric Nervous System." *Gut* 47, supp. IV (2000): iv30–iv32.

Powley, T. L. "The Ventromedial Hypothalamic Syndrome, Satiety, and a Cephalic Phase Hypothesis." *Psychological Review* 84 (1977): 89–126.

Powley, T. L., F. A. Martinson, R. J. Phillips, S. Jones, E. A. Baronowsky, and S. E. Swithers. "Gastrointestinal Projection Maps of the Vagus Nerve Are Specified Permanently in the Perinatal Period." *Developmental Brain Research* 129 (2001): 570–572.

Prandota, J. "Autism Spectrum Disorders May Be Due to Cerebral Toxoplasmosis Associated with Chronic Neuroinflammation Causing Persistent Hypercytokinemia That Resulted in an Increased Lipid Peroxidation, Oxidative Stress, and Depressed Metabolism of Endogenous and Exogenous Substances." *Research in Autism Spectrum Disorders* 4 (2010): 119–155.

Prinz, J. J. *The Conscious Brain: How Attention Engenders Experience*. New York: Oxford University Press, 2012.

Prinz, J. J. *Gut Reactions: A Perceptual Theory of Emotion*. New York: Oxford University Press, 2004.

Prinz. J. J. "Is Emotion a Form of Perception?" *Canadian Journal of Philosophy* 36 (2006): 137–160.

Pritchard, R. M. "Stabilized Images on the Retina." *Scientific American* 204 (1961): 72–78.

Proffitt, D. R. "Embodied Perception and the Economy of Action." *Perspectives on Psychological Science* 1 (2006): 110–122.

Proffitt, D. R., M. Bhalla, R. Grossweiller, and J. Midgett. "Perceiving Geographical Slant." *Psychonomic Bulletin and Review* 16 (1995): 970–972.

Proffitt, D. R., and S. A. Linkenauger. "Perception Viewed as a Phenotypic Expression." In *Action Science: Foundations of an Emerging Discipline*, ed. W. Prinz, M. Beisert, and A. Herwig, 171–197. Cambridge, MA: MIT Press, 2013.

Ptito, M., and R. Kupers. "Cross-Modal Plasticity in Early Blindness." *Journal of Integrative Neuroscience* 4 (2005): 479–88.

Ptito, M., S. M. Moesgaard, A. Gjedde, and R. Kupers. "Cross-Modal Plasticity Revealed by Electrotactile Stimulation of the Tongue in the Congenitally Blind." *Brain* 128 (2005): 606–614.

Putnam, H. *The Collapse of the Fact/Value Dichotomy and Other Essays*. Cambridge, MA; Harvard University Press, 2002.

Raglan, G. B., L. A. Schmidt, and J. Schulkin. "The Role of Glucocorticoids and Corticotropin-Releasing Hormone Regulation on Anxiety Symptoms and Response to Treatment. *Endocrine Connections* 6 (2017): R1–R7.

RajMohan, V., and E. Mohandas. "The Limbic System." *Indian Journal of Psychiatry* 49 (2007): 132–139.

Ramachandran, V. S., and E. Altschuler. "The Use of Visual Feedback, in Particular Mirror Visual Feedback, in Restoring Brain Function." *Brain* 132 (2009): 1693–1710.

Ramachandran, V. S., and W. Hirstein. "The Perception of Phantom Limbs: The D. O. Hebb Lecture." *Brain* 121 (1998): 1603–1630.

Ramachandran, V. S., and D. Rogers-Ramachandran. "Phantom Limbs and Neural Plasticity." *Archives of Neurology* 57 (2000): 317–320.

Ratcliffe, M. *Feelings of Being: Phenomenology, Psychiatry and the Sense of Reality.* New York: Oxford University Press, 2008.

Reed, E. *James J. Gibson and the Psychology of Perception.* New Haven, CT: Yale University Press, 1988.

Reid, C. R., M. Beekman, T. Latty, and A. Dussutour. "Amoeboid Organism Uses Extracellular Secretions to Make Smart Foraging Decisions." *Behavioral Ecology* 24 (2013): 812–818.

Reid, C. R., T. Latty, A. Dussutourc, and M. Beekman. "Slime Mold Uses an Externalized Spatial 'Memory' to Navigate in Complex Environments." *Proceedings of the National Academy of Sciences of the United States of America* 109 (2012): 17490–17494.

Reisenzein, R. "What Is an Emotion in the Belief-Desire Theory of Emotion?" In *The Goals of Cognition: Essays in Honour of Cristiano Castelfranchi*, ed. F. Paglieri, L. Tummolini, R. Falcone, and M. Miceli, 193–223. London: College Publications, 2012.

Reppert, S. M., R. J. Gegear, and C. Merlin. "Navigational Mechanisms of Migrating Monarch Butterflies." *Trends in Neuroscience* 33 (2010): 399–406.

Rescorla, R. A., and A. R. Wagner. "A Theory of Pavlovian Conditioning: Variations in the Effectiveness of Reinforcement and Nonreinforcement." In *Classical Conditioning II: Current Research and Theory*, ed. A. H. Black and W. Prokasy, 64–99. New York: Appleton-Century-Crofts, 1972.

Reynolds, C. W. "Flocks, Herds, and Schools: A Distributed Behavioral Model." *Computer Graphics* 21 (1987): 25–34.

Richards, R. *Darwin and the Emergence of Evolutionary Theories of Mind and Behavior.* Chicago: University of Chicago Press, 1987.

Richardson, M. J., and A. Chemero. "Complex Dynamical Systems and Embodiment." In *The Routledge Handbook of Embodied Cognition*, ed. L. Shapiro, 39–50. New York: Routledge, 2014.

Richter, C. P. *Biological Clocks in Medicine and Psychiatry.* Springfield, IL: Charles C. Thomas, 1965.

Richter, C. P. "Experimentally Produced Behavioral Reactions to Food Poisoning in Wild and Domesticated Rats." *Annals of the New York Academy of Sciences* 56 (1953): 225–239.

Riddoch, M., and G. Humphreys. "Visual Agnosia." *Clinical Neurology* 21 (2003): 501–520.

Riener C. R., J. K. Stefanucci, D. R. Proffitt, and G. L. Clore. "An Effect of Mood on the Perception of Geographical Slant." *Cognition and Emotion* 25 (2011): 174–182.

Rizzolatti, G., L. Fogassi, and V. Gallese. "Neurophysiological Mechanisms Underlying the Understanding and Imitation of Action." *Nature Reviews Neuroscience* 2 (2001): 661–670.

Rizzolatti, G., and G. Luppino. "The Cortical Motor System." *Neuron* 31 (2001): 889–901.

Robbins, C. T., S. Mole, A. E. Hagerman, and T. A. Hanley. "Role of Tannins in Defending Plants Against Ruminants: Reduction in Dry Matter Digestion?" *Ecology* 68 (1987): 1606–1615.

Rockwell, T. *Neither Brain nor Ghost: A Nondualist Alternative to the Mind-Brain Identity Theory.* Cambridge, MA: MIT Press, 2005.

Rodriguez-Saona, C. R., L. E. Rodriguez-Saona, and C. J. Frost. "Herbivore-Induced Volatiles in the Perennial Shrub, *Vaccinium corymbosum*, and Their Role in Inter-branch Signaling. *Journal of Chemical Ecology* 35 (2009): 163–175.

Rolls, E. T. *The Brain and Emotion*. New York: Oxford University Press, 1999.

Rorty, R. *Philosophy and the Mirror of Nature*. Princeton, NJ: Princeton University Press, 1979.

Rosen, J. B., and J. Schulkin. "From Normal Fear to Pathological Anxiety." *Psychological Review* 105 (1998): 325–350.

Rosenthal, N. E., D. A. Sack, J. C. Gillin, A. J. Lewy, F. K. Goodwin, Y. Davenport, P. S. Mueller, D. A. Newsome, and T. A. Wehr. "Seasonal Affective Disorder: A Description of the Syndrome and Preliminary Findings with Light Therapy." *Archives of General Psychiatry* 41 (1984): 72–80.

Rowlands, M. *The New Science of the Mind: From Extended Mind to Embodied Phenomenology*. Cambridge, MA: MIT Press, 2010.

Rozin, P. "The Selection of Foods by Rats, Humans, and Other Animals." In *Advances in the Study of Behavior*, vol. 6, ed. J. S. Rosenblatt, R. A. Hinde, E. Shaw, and C. Beer, 21–76. New York: Academic Press, 1976.

Rzoska, J. "Bait Shyness, a Study in Rat Behaviour." *British Journal of Animal Behaviour* 1 (1953): 128–135.

Saalmann, Y. B., and S. Kastner. "The Cognitive Thalamus." *Frontiers in Systems Neuroscience* 9 (2015): art. 39. https://doi.org/10.3389/fnsys.2015.00039.

Sadato, N., A. Pascual-Leone, J. Grafman, M. P. Deiber, V. Ibañez, and M. Hallett. "Neural Networks for Braille Reading by the Blind." *Brain* 121 (1998): 1213–1229.

Sander, L. W. *Living Systems, Evolving Consciousness, and the Emerging Person: A Selection of Papers from the Life Work of Louis Sander*, ed. G. Amadei and I. Bianchi. New York: Analytic Press, 2008.

Saper, C. B. "The Central Autonomic Nervous System: Conscious Visceral Perception and Autonomic Pattern Generation." *Annual Review of Neuroscience* 25 (2002): 433–469.

Saper, C. B., T. C. Chou, and T. E. Scammell. "The Sleep Switch: Hypothalamic Control of Sleep and Wakefulness." *Trends in Neurosciences* 24 (2001): 726–731.

Saper, C. B., and R. L. Stornetta. "Central Autonomic System." In *The Rat Nervous System*, 4th ed., ed. G. Paxinos, 629–673. New York: Academic Press, 2015.

Sartre, J.-P. *The Emotions: Outline of a Theory*. Trans. B. Frechtman. New York: Carol Publishing Group, 1993. First published 1948 in French.

Savaskan, E., R. Ehrhardt, A. Schulz, M. Walter, and H. Schächinger. "Post-Learning Intranasal Oxytocin Modulates Human Memory for Facial Identity." *Psychoneuroendocrinology* 33 (2008): 368–374.

Savignac, H. M., B. Kiely, T. G. Dinan, and J. F. Cryan. "*Bifidobacteria* Exert Strain-Specific Effects on Stress-Related Behavior and Physiology in BALB/c Mice." *Neurogastroenterology and Motility* 26 (2014): 1615–1627.

Scalera, G., A. C. Spector, and R. Norgren. "Excitotoxic Lesions of the Parabrachial Nuclei Prevent Conditioned Taste Aversions and Sodium Appetite in Rats." *Behavioral Neuroscience* 109 (1995): 997–1008.

Schaal, B., and K. Durand. "The Role of Olfaction in Human Multisensory Development." In *Multisensory Development*, ed. A. J. Bremner, D. J. Lewkowicz, and C. Spence, 29–62. New York: Oxford University Press, 2012.

Schachter, S., and J. E. Singer. "Cognitive, Social, and Physiological Determinants of Emotional State." *Psychological Review* 69 (1962): 379–399.

Schiff, N. D. "Central Thalamic Contributions to Arousal Regulation and Neurological Disorders of Consciousness." *Annals of the New York Academy of Sciences* 1129 (2008): 105–118.

Schiltz, C., B. Sorger, R. Caldara, F. Ahmed, E. Mayer, R. Goebel, and B. Rossion. "Impaired Face Discrimination in Acquired Prosopagnosia Is Associated with Abnormal Response to Individual Faces in the Right Middle Fusiform Gyrus." *Cerebral Cortex* 16 (2006): 574–586.

Schmidt, K., P. J. Cowen, C. J. Harmer, G. Tzortzis, S. Errington, and P. W. Burnet. "Prebiotic Intake Reduces the Waking Cortisol Response and Alters Emotional Bias in Healthy Volunteers." *Psychopharmacology* 232 (2015): 1793–1801.

Schmitz, H., R. O. Müllan, and J. Slaby. "Emotions Outside the Box—the New Phenomenology of Feeling and Corporeality." *Phenomenology and the Cognitive Sciences* 10 (2011): 241–259.

Schnall, S., J. Zadra, and D. R. Proffitt. "Direct Evidence for the Economy of Actions: Glucose and the Perception of Geographical Slant." *Perception* 39 (2010): 464–482.

Schulkin, J. "Behavior of Sodium-Deficient Rats: The Search for a Salty Taste." *Journal of Comparative and Physiological Psychology* 96 (1982): 628–634.

Schulkin, J. *Bodily Sensibility: Intelligent Action*. New York: Oxford University Press, 2004.

Schulkin, J. "Cognitive Functions, Bodily Sensibility and the Brain." *Phenomenology and the Cognitive Sciences* 5 (2006): 341–349.

Schulkin, J. "Foraging for Coherence in Neuroscience: A Pragmatist Orientation." *Contemporary Pragmatism* 13 (2016): 1–28.

Schulkin, J. *Pragmatism and the Search for Coherence in Neuroscience*. New York: Palgrave Macmillan, 2015.

Schulkin, J. *Reflections on the Musical Mind: An Evolutionary Perspective*. Princeton, NJ: Princeton University Press, 2013.

Schulkin, J. *Roots of Social Sensibility and Neural Function*. Cambridge, MA: MIT Press, 2000.

Schulkin, J. *Sodium Hunger: The Search for a Salty Taste*. New York: Cambridge University Press, 1991.

Schulkin, J. *Sport: A Biological, Philosophical, and Cultural Perspective*. New York: Columbia University Press, 2016.

Schulkin, J., B. L. Thompson, and J. B. Rosen. "Demythologizing the Emotions: Adaptation, Cognition, and Visceral Representations of Emotion in the Nervous System." *Brain and Cognition* 52 (2003): 15–23.

Schultz, W. "Getting Formal with Dopamine and Reward." *Neuron* 36 (2002): 241–263.

Schulz, L. "Infants Explore the Unexpected." *Science* 348 (2015): 42–43.

Sender, R., S. Fuchs, and R. Milo. "Revised Estimates for the Number of Human and Bacteria Cells in the Body." *PLoS Biology* 14 (2016): art. e1002533. https://doi.org/10.1371/journal.pbio .1002533.

Senkowski, D., D. Saint-Amour, T. Gruber, and J. J. Foxe. "Look Who's Talking: The Deployment of Visuo-Spatial Attention During Multisensory Speech Processing Under Noisy Environmental Conditions." *NeuroImage* 43 (2008): 379–387.

Senkowski, D., T. R. Schneider, J. J. Foxe, and A. K. Engel. "Crossmodal Binding Through Neural Coherence: Implications for Multisensory Processing." *Trends in Neurosciences* 31 (2008): 401–409.

Severance, E. G., R. H. Yolken, and W. W. Eaton. "Autoimmune Diseases, Gastrointestinal Disorders and the Microbiome in Schizophrenia: More Than a Gut Feeling." *Schizophrenia Research* 176 (2016): 23–35.

Shannon, C. E. "The Mathematical Theory of Communication." In *The Mathematical Theory of Communication*, by C. Shannon and W. Weaver, 29–115. Chicago: University of Illinois Press, 1949.

Shaw, R., and J. Bransford, eds. *Perceiving, Acting, and Knowing: Toward an Ecological Psychology*. Hillsdale, NJ: Lawrence Erlbaum, 1977.

Simon, H. A. "Motivational and Emotional Controls of Cognition." *Psychological Review* 74 (1967): 29–39.

Simon, H. A. *The Sciences of the Artificial*. Cambridge, MA: MIT Press, 1969.

Simon, L., J. Greenberg, and J. Brehm. "Trivialization: The Forgotten Mode of Dissonance Reduction." *Journal of Personality and Social Psychology* 68 (1995): 247–260.

Singer, T., B. Seymour, J. O'Doherty, H. Kaube, R. J. Dolan, and C. D. Frith. "Empathy for Pain Involves the Affective but Not Sensory Components of Pain." *Science* 303 (2004): 1157–1162.

Skinner, B. F. *Verbal Behavior.* New York: Appleton-Century-Crofts, 1957.

Slaby, J. "Emotions and the Extended Mind." In *Collective Emotions: Perspectives from Psychology, Philosophy, and Sociology,* ed. C. von Scheve and M. Salmela, 32–46. New York: Oxford University Press, 2014.

Smith, C. J., J. R. Emge, K. Berzins, L. Lung, R. Khamishon, P. Shah, D. M. Rodrigues, et al. "Probiotics Normalize the Gut-Brain-Microbiota Axis in Immunodeficient Mice." *American Journal of Physiology: Gastrointestinal and Liver Physiology* 307 (2014): G793–G802.

Smith, P. M., and W. S. Garrett. "The Gut Microbiota and Mucosal T Cells." *Frontiers in Microbiology* 2 (2011): art. 111. https://doi.org/10.3389/fmicb.2011.00111.

Smuts, B. "Encounters with Animal Minds." *Journal of Consciousness Studies* 8 (2001): 293–309.

Solomon, R. C. "Emotions and Choice." *Review of Metaphysics* 27 (1973): 20–41.

Solomon, R. C. *Love.* New York: Prometheus, 1990.

Solomon, R. C. *True to Our Feelings: What Our Emotions Are Really Telling Us.* New York: Oxford University Press, 2007.

Solymosi, T. "Dewey on the Brain: Dopamine, Digital Devices, and Democracy." *Contemporary Pragmatism* 14 (2017): 5–34.

Solymosi, T. "Moral First Aid for a Scientific Age." In *Neuroscience, Neurophilosophy and Pragmatism: Brains at Work with the World,* ed. T. Solymosi and J. Shook, 291–317. New York: Palgrave Macmillan, 2014.

Solymosi, T. "A Reconstruction of Freedom in the Age of Neuroscience: A View from Neuropragmatism." *Contemporary Pragmatism* 8 (2011): 153–171.

Solymosi, T., and J. Shook, eds. *Neuroscience, Neurophilosophy and Pragmatism: Brains at Work with the World.* New York: Palgrave Macmillan, 2014.

Sonea, S., and M. Panisset. *A New Bacteriology.* Boston: Jones and Bartlett, 1983.

Spector, A. C. "Behavioral Analyses of Taste Function and Ingestion in Rodent Models." *Physiology and Behavior* 152 (2015): 516–526.

Spector, A. C. "Gustatory Function in Parabrachial Nuclei: Implications from Lesion Studies." *Revolutions in Neuroscience* 6 (1995): 143–175.

Spector, A. C., R. Norgren, and H. J. Grill. "Parabrachial Gustatory Lesions Impair Taste Aversion Learning in Rats." *Behavioral Neuroscience* 106 (1992): 147–161.

Spencer, H. *The Principles of Psychology.* Westmead, UK: Gregg International, 1970. First published 1855.

Stahl, A. E., and L. Feigenson. "Observing the Unexpected Enhances Infants' Learning and Exploration." *Science* 348 (2015): 91–94.

Steenbergen, L., R. Sellaro, S. V. Hemert, J. A. Bosch, and L. S. Colzato. "A Randomized Controlled Trial to Test the Effect of Multispecies Probiotics on Cognitive Reactivity to Sad Mood." *Brain, Behavior, and Immunity* 48 (2015): 258–264.

Stefanucci, J. K., and M. N. Geuss. "Big People, Little World: The Body Influences Size Perception." *Perception* 38 (2009): 1782–1795.

Stefanucci, J. K., and M. N. Geuss. "Duck! Scaling the Height of a Horizontal Barrier to Body Height." *Attention, Perception, and Psychophysics* 72 (2010): 1338–1349.

Stein, B. E., and T. R. Stanford. "Multisensory Integration: Current Issues from the Perspective of the Individual Neuron." *Nature Reviews Neuroscience* 9 (2008): 255–266.

Stein, B. E., T. R. Stanford, and B. A. Rowland. "Development of Multisensory Integration from the Perspective of the Individual Neuron." *Nature Reviews Neuroscience* 15 (2014): 520–535.

Steinbeck, J. *The Grapes of Wrath*. New York: Viking, 1939.

Steiner, P. "Philosophie, technologie et cognition: État des lieux et perspectives." *Intellectica* 53 (2010): 7–40.

Steiner, P. "Sciences cognitives, tournant pragmatique et horizons pragmatistes." *Tracés: Revue de Sciences humaines* 15 (2008): 85–105.

Stel, M., and A. van Knippenberg. "The Role of Facial Mimicry in the Recognition of Affect." *Psychological Science* 19 (2008): 984–985.

Sterling, P. "Principles of Allostasis: Optimal Design, Predictive Regulation, Pathophysiology, and Rational Therapeutics." In *Allostasis, Homeostasis, and the Costs of Physiological Adaptation*, ed. J. Schulkin, 17–64. New York: Cambridge University Press, 2004.

Sterling, P., and J. Eyer. "Allostasis: A New Paradigm to Explain Arousal Pathology." In *Handbook of Life Stress, Cognition, and Health*, ed. S. Fisher and J. Reason, 629–649. New York: John Wiley and Sons, 1988.

Still, A., and J. Good. "The Ontology of Mutualism." *Ecological Psychology* 10 (1998): 39–63.

Strack, F., L. L. Martin, and S. Stepper. "Inhibiting and Facilitating Conditions of the Human Smile: A Nonobtrusive Test of the Facial Feedback Hypothesis." *Journal of Personality and Social Psychology* 54 (1988): 768–777.

Strati, F., D. Cavalieri, D. Albanese, C. D. Felice, C. Donati, J. Hayek, O. Jousson, et al. "New Evidences on the Altered Gut Microbiota in Autism Spectrum Disorders." *Microbiome* 5 (2017): art. 24. https://doi.org/10.1186/s40168-017-0242-1.

Sudo, N., Y. Chida, Y. Aiba, J. Sonoda, N. Oyama, X. N. Yu, C. Kubo, and Y. Koga. "Postnatal Microbial Colonization Programs the Hypothalamic-Pituitary-Adrenal System for Stress Response in Mice." *Journal of Physiology* 558 (2004): 263–275.

Swanson, L. W. *Brain Architecture*. New York: Oxford University Press, 2003.

Swanson, L. W. "The Cerebral Hemisphere Regulation of Motivated Behavior." *Brain Research* 836 (2000): 113–164.

Swanson, L. W. "What Is the Brain?" *Trends in Neural Science* 23 (2000): 519–527.

Swanson, L. W., and G. D. Petrovich. "What Is the Amygdala?" *Trends in Neural Science* 21 (1998): 323–331.

Takakusaki, K. "Functional Neuroanatomy for Posture and Gait Control." *Journal of Movement Disorders* 10 (2017): 1–17.

Tanaka, S., and T. Inui. "Cortical Involvement for Action Imitation of Hand/Arm Postures Versus Finger Configurations: An fMRI Study." *NeuroReport* 13 (2002): 1599–1602.

Taylor, A. G., L. E. Goehler, D. I. Galper, K. E. Innes, and C. Bourguignon. "Top-Down and Bottom-Up Mechanisms in Mind-Body Medicine: Development of an Integrative Framework for Psychophysiological Research." *Explore* 6 (2010): 29–41.

Tessitore, A., A. R. Hariri, F. Fera, W. G. Smith, T. N. Chase, T. M. Hyde, D. R. Weinberger, and V. S. Mattay. "Dopamine Modulates the Response of the Human Amygdala: A Study in Parkinson's Disease." *Journal of Neuroscience* 22 (2002): 9099–9103.

Tettamanti, M., and D. Weniger. "Broca's Area: A Supramodal Hierarchical Processor?" *Cortex* 42 (2006): 491–494.

Thayer, J. F., and R. D. Lane. "A Model of Neurovisceral Integration in Emotion Regulation and Dysregulation." *Journal of Affective Disorders* 61 (2000): 201–216.

Thoenissen, D., K. Zilles, and I. Toni. "Differential Involvement of Parietal and Precentral Regions in Movement Preparation and Motor Intention." *Journal of Neuroscience* 22 (2002): 9024–9034.

Thompson, E. *Mind in Life: Biology, Phenomenology, and the Sciences of Mind*. Cambridge, MA: Harvard University Press, 2007.

Thompson, E., A. Palacios, and F. J. Varela. "Ways of Coloring: Comparative Color Vision as a Case Study for Cognitive Science." *Behavioral and Brain Sciences* 15 (1992): 1–26.

Timmann, D., J. Drepper, M. Frings, M. Maschke, S. Richter, M. Gerwig, and F. Kolb. "The Human Cerebellum Contributes to Motor, Emotional and Cognitive Associative Learning: A Review." *Cortex* 46 (2010): 845–857.

Titchener, E. *Lectures on the Experimental Psychology of the Thought-Processes*. New York: Macmillan, 1909.

Tomasello, M. *The Cultural Origins of Human Cognition*. Cambridge, MA: Harvard University Press, 1999.

Tomasello, M. *Origins of Human Communication*. Cambridge, MA: MIT Press, 2008.

Tomasello, M., A. C. Kruger, and H. H. Ratner. "Cultural Learning." *Behavioral and Brain Sciences* 16 (1993): 495–511.

Torres, L., S.-A. Robinson, D.-G. Kim, A. Yan, T. A. Cleland, and M. S. Bynoe. "*Toxoplasma gondii* Alters NMDAR Signaling and Induces Signs of Alzheimer's Disease in Wild-Type, C57BL/6 Mice." *Journal of Neuroinflammation* 15 (2018): art. 57. https://doi.org/10.1186/s12974-018-1086-8.

Torrey E. F., J. J. Bartko, Z. R. Lun, and R. H. Yolken. "Antibodies to *Toxoplasma gondii* in Patients with Schizophrenia: A Meta-Analysis." *Schizophrenia Bulletin* 33 (2007): 729–736.

Torrey, E. F., and R. H. Yolken. "*Toxoplasma gondii* and Schizophrenia." *Emerging Infectious Diseases* 9 (2003): 1375–1380.

Touzani, K., and A. Sclafani. "Critical Role of Amygdala in Flavor but Not Taste Preference Learning in Rats." *European Journal of Neuroscience* 22 (2005): 1767–1774.

Tramontin, A. D., and E. A. Brenowitz. "Seasonal Plasticity in the Adult Brain." *Trends in Neurosciences* 23 (2000): 251–258.

Trevarthen, C. "Awareness of Infants: What Do They, and We, Seek?" *Psychoanalytic Inquiry* 35 (2015): 395–416.

Trevarthen, C. "Communication and Cooperation in Early Infancy: A Description of Primary Intersubjectivity." In *Before Speech: The Beginning of Human Communication*, ed. M. Bullowa, 321–347. New York: Cambridge University Press, 1979.

Trevarthen, C. "Descriptive Analyses of Infant Communication Behavior." In *Studies in Mother-Infant Interaction: The Loch Lomond Symposium*, ed. H. R. Schaffer, 227–270. New York: Academic Press, 1977.

Trevarthen, C. "Early Attempts at Speech." In *Child Alive: New Insights into the Development of Young Children*, ed. R. Lewin, 62–80. London: Temple Smith, 1975.

Trevarthen, C. "Embodied Human Intersubjectivity: Imaginative Agency, to Share Meaning." *Journal of Cognitive Semiotics* 4 (2009): 6–56.

Trevarthen, C. "Infant Semiosis: The Psycho-biology of Action and Shared Experience from Birth." *Cognitive Development* 36 (2015): 130–141.

Trevarthen, C. "In Praise of a Doctor Who Welcomes the Newborn Infant Person." *Journal of Child and Adolescent Psychiatric Nursing* 26 (2013): 204–213.

Trevarthen, C. "The Intersubjective Psychobiology of Human Meaning: Learning of Culture Depends on Interest for Co-operative Practical Work—and Affection for the Joyful Art of Good Company." *Psychoanalytic Dialogues* 19 (2009): 507–518.

Trevarthen, C. "What Is It Like to Be a Person Who Knows Nothing? Defining the Active Intersubjective Mind of a Newborn Human Being." *Infant and Child Development* 20 (2011): 119–135.

Trevarthen, C., K. J. Aitken, M. Vandekerckhove, J. Delafield-Butt, and E. Nagy. "Collaborative Regulations of Vitality in Early Childhood: Stress in Intimate Relationships and Postnatal Psychopathology." In *Developmental Psychopathology*, ed. D. Cicchetti and D. J. Cohen, 65–126. New York: Wiley, 2015.

Tronick, E. Z. "Emotions and Emotional Communication in Infants." *American Psychologist* 44 (1989): 112–126.

Tronick, E. Z. "Why Is Connection with Others So Critical? The Formation of Dyadic States of Consciousness: Coherence-Governed Selection and the Co-creation of Meaning Out of Messy Meaning Making." In *Emotional Development: Recent Research Advances*, ed. J. Nadel and D. Muir, 293–315. New York: Oxford University Press, 2005.

Tronick, E. Z., H. Als, L. Adamson, S. Wise, and T. B. Brazelton. "The Infant's Response to Entrapment Between Contradictory Messages in Face-to-Face Interaction." *Journal of the American Academy of Child Psychiatry* 17 (1978): 1–13.

Tronick, E. Z., N. Bruschweiler-Stern, A. M. Harrison, K. Lyons-Ruth, A. C. Morgan, J. P. Nahum, L. Sander, and D. N. Stern. "Dyadically Expanded States of Consciousness and the Process of Therapeutic Change." *Infant Mental Health Journal* 19 (1999): 290–299.

Ueda, T., T. Hirose, and Y. Kobatake. "Membrane Biophysics of Chemoreception and Taxis in the Plasmodium of *Physarum polycephalum*." *Biophysical Chemistry* 11 (1980): 461–473.

Uexküll, J. von. "A Stroll Through the Worlds of Animals and Men: A Picture Book of Invisible Worlds," 1934. In *Instinctive Behavior: The Development of a Modern Concept*, ed. and trans. C. H. Schiller, 5–80. New York: International Universities Press, 1957.

Uexküll, J. von. *Umwelt und Innenwelt der Tiere*. Berlin: Julius Springer, 1909.

Ullman, M. T. "Is Broca's Area Part of a Basal Ganglia Thalamocortical Circuit?" *Cortex* 42 (2006): 480–485.

Ullman, M. T., S. Corkin, M. Coppola, G. Hickok, H. Growdon, W. J. Horoshe, and S. Pinker. "A Neural Dissociation Within Language: Evidence That the Mental Dictionary Is Part of Declarative Memory, and That Grammatical Rules Are Processed by the Procedural System." *Journal of Cognitive Neuroscience* 9 (1997): 266–286.

Ulrich, R. S. "Aesthetic and Affective Response to Natural Environment." In *Behavior and the Natural Environment*, ed. I. Altman and J. F. Wohlwill. New York: Plenum, 1983.

Umiltà, M. A., E. Kohler, V. Gallese, L. Fogassi, L. Fadiga, C. Keysers, and G. Rizzolatti. "I Know What You Are Doing: A Neurophysiological Study." *Neuron* 31 (2001): 155–165.

Van der Lans, A. A., J. Hoeks, B. Brans, G. H. Vijgen, M. G. Visser, M. J. Vosselman, J. Hansen, et al. "Cold Acclimation Recruits Human Brown Fat and Increases Nonshivering Thermogenesis." *Journal of Clinical Investigation* 123 (2013): 3395–3340.

Van Eden, C. G., and R. M. Buijs. "Functional Neuroanatomy of the Prefrontal Cortex: Autonomic Interactions." *Progress in Brain Research* 126 (2000): 49–62.

Van Zandvoort, M. J. E., T. Nijboer, and E. Dehaan. "Developmental Colour Agnosia." *Cortex* 43 (2007): 750–757.

Varela, F., E. Thompson, and E. Rosch. *The Embodied Mind: Cognitive Science and Human Experience*. Cambridge, MA: MIT Press, 1991.

Vartanian, O., and V. Goel. "Neuroanatomical Correlates of Aesthetic Preference for Paintings." *NeuroReport* 15 (2004): 893–897.

Veltkamp, M., H. Aarts, and R. Custers. "Perception in the Service of Goal Pursuit: Motivation to Attain Goals Enhances the Perceived Size of GoalInstrumental Objects." *Social Cognition* 26 (2008): 720–736.

Venkatraman, A., B. L. Edlow, and M. H. Immordino-Yang. "The Brainstem in Emotion: A Review." *Frontiers in Neuroanatomy* 11 (2017): art. 15. https://doi.org/10.3389/fnana.2017 .00015.

Vera, S., and D. Schoeller. "Cognition as a Transformative Process: Re-affirming a Classical Pragmatist Understanding." *European Journal of Pragmatism and American Philosophy* 10 (2018): 1–21. https://doi.org/10.4000/ejpap.1211.

Voss, P., M. Lassonde, F. Gougoux, M. Fortin, J. Guillemot, and F. Lepore. "Early- and Late-Onset Blind Individuals Show Supra-Normal Auditory Abilities in Far-Space." *Current Biology* 14 (2004): 1734–1738.

Vyas, A., S.-K. Kim, N. Giacomini, J. C. Boothroyd, and R. M. Sapolsky. "Behavioral Changes Induced by *Toxoplasma* Infection of Rodents Are Highly Specific to Aversion of Cat Odors."

Proceedings of the National Academy of Sciences of the United States of America 104 (2007): 6442–6447.

Vyas, A., S.-K. Kim, and R. M. Sapolsky. "The Effects of Toxoplasma Infection on Rodent Behavior Are Dependent on Dose of the Stimulus." *Neuroscience* 148 (2007): 342–348.

Vyas, A., R. Mitra, B. S. Rao, and S. Chattarji. "Chronic Stress Induces Contrasting Patterns of Dendritic Remodeling in Hippocampal and Amygdaloid Neurons." *Journal of Neuroscience* 22 (2002): 6810–6818.

Vyas, A., A. Pillai, and S. Chattarji. "Recovery After Chronic Stress Fails to Reverse Amygdaloid Neuronal Hypertrophy and Enhanced Anxiety-like Behavior." *Neuroscience* 128 (2004): 667–673.

Wagenmakers, E.-J., T. Beek, L. Dijkhoff, Q. F. Gronau, A. Acosta, R. B. Adams, D. N. Albohn, et al. "Registered Replication Report." *Perspectives on Psychological Science* 11 (2016): 917–928.

Wallace, A. R. "The Origin of Human Races and the Antiquity of Man Deduced from the Theory of 'Natural Selection.'" *Journal of the Anthropological Society of London* 2 (1864): clviii–clxxxvii.

War, A. R., M. G. Paulraj, T. Ahmad, A. A. Buhroo, B. Hussain, S. Ignacimuthu, and H. C. Sharma. "Mechanisms of Plant Defense Against Insect Herbivores." *Plant Signaling and Behavior* 7 (2012): 1306–1320.

Ward, L. G., and R. Throop. "The Dewey-Mead Analysis of Emotions." *Social Science Journal* 26 (1989): 465–479.

Washburn, M. *Movement and Mental Imagery: Outlines of a Motor Theory of the Complexer Mental Processes*. Boston: Houghton Mifflin, 1916.

Weaver, W. "Recent Contributions to the Mathematical Theory of Communication." In *The Mathematical Theory of Communication*, ed. C. Shannon and W. Weaver, 1–28. Chicago: University of Illinois Press, 1949.

Webster, J. P. "The Effect of *Toxoplasma gondii* on Animal Behavior: Playing Cat and Mouse." *Schizophrenia Bulletin* 33 (2007): 752–756.

Webster, J. P., C. F. Brunton, and D. W. MacDonald. "Effect of *Toxoplasma gondii* Upon Neophobic Behaviour in Wild Brown Rats, *Rattus norvegicus*." *Parasitology* 108 (1994): 407–411.

Weinberg, M. K., and E. Z. Tronick. "Beyond the Face: An Empirical Study of Infant Affective Configurations of Facial, Vocal, Gestural, and Regulatory Behaviors." *Child Development* 65 (1994): 1495–1507.

Weissman, D. *Truth's Debt to Value*. New Haven, CT: Yale University Press, 1993.

Wellman, L. L., L. Yang, and L. D. Sanford. "Effects of Corticotropin Releasing Factor (CRF) on Sleep and Temperature Following Predictable Controllable and Uncontrollable Stress in Mice." *Frontiers in Neuroscience* 9 (2015): art. 258. https://doi.org/10.3389/fnins.2015.00258.

Werner, H. "On Physiognomic Modes of Perception and Their Experimental Investigation," 1927. In *Developmental Processes: Heinz Werner's Selected Writings*, ed. S. S. Barten and M. B. Franklin, vol. 1, *General Theory and Perceptual Experience*, 149–152. New York: International Universities Press, 1978.

Wheaton, K. J., J. C. Thompson, A. Syngeniotis, D. F. Abbott, and A. Puce. "Viewing the Motion of Human Body Parts Activates Different Regions of Premotor, Temporal, and Parietal Cortex." *NeuroImage* 22 (2004): 277–288.

White, J. G., E. Southgate, J. N. Thomson, and S. Brenner. "The Structure of the Nervous System of the Nematode *Caenorhabditis elegans*." *Philosophical Transactions of the Royal Society B: Biological Sciences* 314 (1986): 1–340.

Whitehead, A. N. *Science and the Modern World*. New York: Cambridge University Press, 1953. First published 1926.

Witting, P. A. "Learning Capacity and Memory of Normal and *Toxoplasma*-Infected Laboratory Rats and Mice." *Zeitschrift für Parasitenkunde* 61 (1979): 29–51.

Wohlleben, P. *The Hidden Life of Trees—What They Feel, How They Communicate: Discoveries from a Secret World*. Trans. J. Billinghurst. Vancouver, BC: Greystone Books, 2016.

Whyte, W. *The Social Life of Small Urban Spaces*. Washington, DC: Conservation Foundation, 1980.

Wiener, P. P. *Evolution and the Founders of Pragmatism*. Cambridge, MA: Harvard University Press, 1949.

Willson, S. K., R. Sharp, I. P. Ramler, and A. Sen. "Spatial Movement Optimization in Amazonian *Eciton burchellii* Army Ants." *Insectes Sociaux: International Journal for the Study of Social Arthropods* 58 (2011): 325–334.

Witt, J. K., and D. R. Proffitt. "Action-Specific Influences on Distance Perception: A Role for Motor Simulation." *Journal of Experimental Psychology: Human Perception and Performance* 34 (2008): 1479–1492.

Witt, J. K., D. R. Proffitt, and W. Epstein. "Tool Use Affects Perceived Distance, but Only When You Intend to Use It." *Journal of Experimental Psychology: Human Perception and Performance* 31 (2005): 80–88.

Wittgenstein, L. *Philosophical Investigations* [written 1929–1949]. Trans. G. E. M. Anscombe. Oxford: Basil Blackwell, 1953.

Woods, S. C., R. A. Hutton, and W. Makous. "Conditioned Insulin Secretion in the Albino Rat." *Proceedings of the Society of Experimental Biology and Medicine* 133 (1970): 965–968.

Woodworth, R. *Psychology: A Study of Mental Life*. New York: Henry Holt, 1921.

Wright, C. "Evolution of Self-Consciousness." *North America Review* 116 (1873): 245–310.

Wu, T., A. J. Dufford, M. Mackie, L. J. Egan, and J. Fan. "The Capacity of Cognitive Control Estimated from a Perceptual Decision Making Task." *Scientific Reports* 6 (2016): art. 34025. https://doi.org/10.1038/srep34025.

Xenakis, I., and A. Arnellos. "The Relation Between Interaction Aesthetics and Affordances." *Design Studies* 34 (2013): 57–73.

Yahiro, T., N. Kataoka, Y. Nakamura, and K. Nakamura. "The Lateral Parabrachial Nucleus, but Not the Thalamus, Mediates Thermosensory Pathways for Behavioural Thermoregulation." *Scientific Reports* 7 (2017): art. 5031. https://doi.org/10.1038/s41598-017-05327-8.

Yam, P. "Acacia Trees Kill Antelope in the Transvaal." *Scientific American* 263 (1990): 28.

Yu, L., B. E. Stein, and B. A. Rowland. "Adult Plasticity in Multisensory Neurons: Short-Term Experience-Dependent Changes in the Superior Colliculus." *Journal of Neuroscience* 29 (2009): 15910–15922.

Zadra, J., S. Schnall, A. Weltman, and D. R. Proffitt. "Direct Physiological Evidence for the Economy of Action: Bioenergetics and the Perception of Spatial Layout." *Journal of Vision* 10 (2010): 54.

Zahn, R., J. Moll, V. Iyengar, E. D. Huey, M. Tierney, F. Krueger, and J. Grafman. "Social Conceptual Impairments in Frontotemporal Lobar Degeneration with Right Anterior Temporal Hypometabolism." *Brain* 132 (2009): 604–616.

Zahn, R., J. Moll, F. Krueger, E. D. Huey, G. Garrido, and J. Grafman. "Social Concepts Are Represented in the Superior Anterior Temporal Cortex." *Proceedings of the National Academy of Sciences of the United States of America* 104 (2007): 6430–6435.

Zajac, F. E. "Muscle Coordination of Movement: A Perspective." *Journal of Biomechanics* 26 (1993): 109–124.

Zak, P. J., R. Kurzban, and W. T. Matzner. "Oxytocin Is Associated with Human Trustworthiness." *Hormones and Behavior* 48 (2005): 522–527.

Zald, D. H., and J. V. Pardo. "Emotion, Olfaction, and the Human Amygdala: Amygdala Activation During Aversive Olfactory Stimulation." *Proceedings of the National Academy of Sciences of the United States of America* 94 (1997): 4119–4124.

Zampini, M., and C. Spence. "The Role of Auditory Cues in Modulating the Perceived Crispness and Staleness of Potato Chips." *Journal of Sensory Studies* 19 (2004): 347–363.

Zazzo, R. "Le problème de l'imitation chez le nouveau-né." *Enfance* 10 (1957): 135–142.

Zhao, T. C., and P. K. Kuhl. "Musical Intervention Enhances Infants' Neural Processing of Temporal Structure in Music and Speech." *Proceedings of the National Academy of Sciences of the United States of America* 113 (2016): 5212–5217.

Zheng, J., K. L. Anderson, S. L. Leal, A. Shestyuk, G. Gulsen, L. Mnatsakanyan, S. Vadera, et al. "Amygdala-Hippocampal Dynamics During Salient Information Processing." *Nature Communications* 8 (2017): art. 14413. https://doi.org/10.1038/ncomms14413.

Zink, C. F., L. Kempf, S. Hakimi, C. A. Rainey, J. L. Stein, and A. Meyer-Lindenberg. "Vasopressin Modulates Social Recognition-Related Activity in the Left Temporoparietal Junction in Humans." *Translational Psychiatry* 1 (2011): art. e3. https://doi.org/10.1038/tp .2011.2.

Zink, C. F., and A. Meyer-Lindenberg. "Human Neuroimaging of Oxytocin and Vasopressin in Social Cognition." *Hormones and Behavior* 61 (2012): 400–409.

Zoccal, D. B., W. I. Furuya, M. Bassi, D. S. Colombari, and E. Colombari. "The Nucleus of the Solitary Tract and the Coordination of Respiratory and Sympathetic Activities." *Frontiers in Physiology* 5 (2014): art. 238. https://doi.org/10.3389/fphys.2014.00238.

INDEX

Page numbers in *italics* refer to figures.

action: and affect as cognitive, 49; and anticipating future situations, 138, 140; and appraisal systems, 175; and artificial intelligence, 165–166 (*see also* artificial intelligence); and attention, 165; and beliefs, 27, 35, 146; and the brain, 18, 62, 71, 183; Calderwood and, 61; and cognition, 2, 6, 17, 26–27, 49, 52, 58–59, 64–65, 88–92; constraints on, 4, 24, 30, 34, 36–37, 69–70, 86–87, 97, 121–123; and constraints on belief, 27; Dewey and (*see* Dewey, John); and emotions, 45, 133, 138, 145–146, 151, 165–166, 195; emotions and "action tendencies," 165–166; evolution and the actions of organisms, 23; examples, 19, 28, 52, 60, 97–98; and 4E cognitive science, 4–5; Galton and, 61; and Gibson's affordances (*see* affordance theory); and gut biome, 185; and habits and skills, 52, 69, 73 (*see also* habits; skill); and hearing, 74; and inquiry, 34; integration of action, affect, cognition, and perception, 2, 7,

10–15, 120, 122, 157–167, 169, 178–186, 192, 194–196, 200, 202, 203, 206, 208; integration of action, perception, and cognition, 6, 178–179; integration of emotion, cognition, and action, 125, 152; and intermodal perception, 75–83; Jackson and, 61; James and, 26–27; knowing as a capacity to do something, 92–93; knowing/ influencing reality through interacting with it, 18, 26, 57, 58; and language, 92–93; and meaning, 61; Merleau-Ponty and, 26–27; motor activity as more than brain function, 17; motor and sensory operations combined in roundworm neurons, 181, 194; neurological connections between motor function, perception, and cognition, 61–62, 91–92; and perception, 2, 4, 6, 17, 19, 28, 53, 57–60, 63, 68–75, 80, 85, 145, 173, 178 (*see also* affordance theory); perception constituted through sensorimotor organization, 53, 57–60, 63–64, 69,

CPSIA information can be obtained
at www.ICGtesting.com
Printed in the USA
LVHW091424061220
673467LV00003B/3